Leadership, Management, and Adoption Techniques for Digital Service Innovation

Kamaljeet Sandhu
University of New England, Australia

A volume in the Advances in Logistics, Operations, and Management Science (ALOMS) Book Series

IGI Global
DISSEMINATOR OF KNOWLEDGE

Published in the United States of America by
IGI Global
Business Science Reference (an imprint of IGI Global)
701 E. Chocolate Avenue
Hershey PA, USA 17033
Tel: 717-533-8845
Fax: 717-533-8661
E-mail: cust@igi-global.com
Web site: http://www.igi-global.com

Library of Congress Cataloging-in-Publication Data

Names: Sandhu, Kamaljeet, 1970- editor.
Title: Leadership, management, and adoption techniques for digital service innovation / Kamaljeet Sandhu, editor.
Description: Hershey, PA : Business Science Reference, [2020] | Includes bibliographical references and index. | Summary: “”This book examines digital service innovation processes, newer techniques, problem solving critical areas, leadership, applied solutions, and recommendations”--Provided by publisher”-- Provided by publisher.
Identifiers: LCCN 2019043607 (print) | LCCN 2019043608 (ebook) | ISBN 9781799827993 (hardcover) | ISBN 9781799828006 (paperback) | ISBN 9781799828013 (ebook)
Subjects: LCSH: Technological innovations--Economic aspects. | Technological innovations--Management. | Information technology--Economic aspects. | Information technology--Management.
Classification: LCC HC79.T4 .L4225 2020 (print) | LCC HC79.T4 (ebook) | DDC 658/.05--dc23
LC record available at https://lccn.loc.gov/2019043607
LC ebook record available at https://lccn.loc.gov/2019043608

This book is published in the IGI Global book series Advances in Logistics, Operations, and Management Science (ALOMS) (ISSN: 2327-350X; eISSN: 2327-3518)

British Cataloguing in Publication Data
A Cataloguing in Publication record for this book is available from the British Library.

All work contributed to this book is new, previously-unpublished material. The views expressed in this book are those of the authors, but not necessarily of the publisher.

For electronic access to this publication, please contact: eresources@igi-global.com.

Advances in Logistics, Operations, and Management Science (ALOMS) Book Series

John Wang
Montclair State University, USA

ISSN:2327-350X
EISSN:2327-3518

Mission

Operations research and management science continue to influence business processes, administration, and management information systems, particularly in covering the application methods for decision-making processes. New case studies and applications on management science, operations management, social sciences, and other behavioral sciences have been incorporated into business and organizations real-world objectives.

The **Advances in Logistics, Operations, and Management Science** (ALOMS) Book Series provides a collection of reference publications on the current trends, applications, theories, and practices in the management science field. Providing relevant and current research, this series and its individual publications would be useful for academics, researchers, scholars, and practitioners interested in improving decision making models and business functions.

Coverage

- Information Management
- Decision analysis and decision support
- Production Management
- Services management
- Networks
- Finance
- Risk Management
- Operations Management
- Marketing engineering
- Organizational Behavior

IGI Global is currently accepting manuscripts for publication within this series. To submit a proposal for a volume in this series, please contact our Acquisition Editors at Acquisitions@igi-global.com or visit: http://www.igi-global.com/publish/.

The Advances in Logistics, Operations, and Management Science (ALOMS) Book Series (ISSN 2327-350X) is published by IGI Global, 701 E. Chocolate Avenue, Hershey, PA 17033-1240, USA, www.igi-global.com. This series is composed of titles available for purchase individually; each title is edited to be contextually exclusive from any other title within the series. For pricing and ordering information please visit http://www.igi-global.com/book-series/advances-logistics-operations-management-science/37170. Postmaster: Send all address changes to above address.

Titles in this Series

For a list of additional titles in this series, please visit: www.igi-global.com/book-series

Advancing Skill Development for Business Managers in Industry 4.0 Emerging Research and Opportunities
Sara Fazzin (H-Farm College, UK)
Business Science Reference • © 2020 • 168pp • H/C (ISBN: 9781799820369) • US $185.00

Management and Inter/Intra Organizational Relationships in the Textile and Apparel Industry
Vasilica-Maria Margalina (Technical University of Ambato, Ecuador) and José M. Lavín (CESINE University Centre, Spain)
Business Science Reference • © 2020 • 427pp • H/C (ISBN: 9781799818595) • US $215.00

Neutrosophic Sets in Decision Analysis and Operations Research
Mohamed Abdel-Basset (Zagazig University, Egypt) and Florentin Smarandache (University of New Mexico, USA)
Engineering Science Reference • © 2020 • 300pp • H/C (ISBN: 9781799825555) • US $245.00

Global Perspectives on Green Business Administration and Sustainable Supply Chain Management
Syed Abdul Rehman Khan (Tsinghua University, China)
Business Science Reference • © 2020 • 362pp • H/C (ISBN: 9781799821731) • US $225.00

Handbook of Research on the Applications of International Transportation and Logistics for World Trade
Gökçe Çiçek Ceyhun (Bursa Technical University, Turkey)
Business Science Reference • © 2020 • 578pp • H/C (ISBN: 9781799813972) • US $345.00

Informing View of Organization Strategic Perspective
Bob Travica (University of Manitoba, Canada)
Business Science Reference • © 2020 • 385pp • H/C (ISBN: 9781799827603) • US $245.00

Handbook of Research on Interdisciplinary Approaches to Decision Making for Sustainable Supply Chains
Anjali Awasthi (Concordia University, Canada) and Katarzyna Grzybowska (Poznan University of Technology, Poland)
Business Science Reference • © 2020 • 674pp • H/C (ISBN: 9781522595700) • US $365.00

Global Applications of Multigenerational Management and Leadership in the Transcultural Era
Bryan Christiansen (Global Research Society, LLC, USA)
Business Science Reference • © 2020 • 328pp • H/C (ISBN: 9781522599067) • US $205.00

IGI Global
DISSEMINATOR OF KNOWLEDGE

701 East Chocolate Avenue, Hershey, PA 17033, USA
Tel: 717-533-8845 x100 • Fax: 717-533-8661
E-Mail: cust@igi-global.com • www.igi-global.com

I dedicate this book to the memory of my mother and father.

Editorial Advisory Board

Hamed Taherdoost, *Independent researcher, Malaysia*
Mohan Tanniru, *University of Arizona, USA*
Kanitsorn Terdpaopong, *Rangsit University, Thailand*
Tomislav Videc, *Oman Arab Bank, Oman*
Vesna Bosilj Vukšić, *University of Zagreb, Croatia*
Youmin Xi, *Xian Jiatong Liverpool University, China*
Dina Ziadlou, *Colorado Technical University, USA*
Jovana Zoroja, *University of Zagreb, Croatia*

Table of Contents

Detailed Table of Contents

companies both in the process of innovation generation (value creation), as well as in the implemented digital business model (value capture).

This chapter explores the demands that digital service innovation places on those stakeholders who use the digital service through three lenses: workers and workplace, digital literacy, and, finally, digital education. The chapter includes an overview of digital service innovation. The impact of digital services generally is explored from the perspective of the workforce and the future of work developing an argument for the need for ongoing upskilling in the digital space. The skills required in the age of digital disruption are explored, and a definition and general understanding of what digital literacy entails is presented. Finally, the opportunities for education in digital environments are explored through three vignettes which illustrate different opportunities for upskilling and retraining. The chapter highlights adaptations required in socio-political environments, education and training, and curricula to allow digital service innovations to achieve the expected benefits.

This chapter addresses a conceptual review of the methodology to define strategies and supports for carrying out effective, efficient, and innovative decisions and strengthen management and leadership, given the dynamic changes that are taking place in the current turbulent environment. The main catalyst is the digital revolution and its implications in a new management style. The functions of anticipation and early assessment of threats, risks, and opportunities, constitute the central functionality of the new strategic approach. Its impact is especially notable in the field of strategic decision making, which must be fully timely, rigorous, transparent, and credible in the face of the growing demands of customers and the public. The use of strategic intelligence has become critical, and the skills for its management and application are of great relevance in terms of defining management strategies and decisions, and complementary to the use of digital technology tools.

The emergence of digital technologies provides organizations with affordable, easy-to-use, and easy-to-learn technologies that opens line-of-business (LOB) managers to participate in information technology-led innovations. Such participatory actions by the LOB-managers in information technology (IT) innovations at the department levels will lead the creation of new roles like the 'chief digital marketing officer.' While

digitally-enabled consumer. Much effort has been made to systematic education and training relating to digital health, an important aspect in developing the digital health workforce.

Emerging technologies have an undeniably significant impact on the hospitality industry. Hotel owners and general managers face many challenges to compete in an ever-increasingly competitive market. Many hotels are quick to adopt some of the emerging technologies to maximize brand visibility, gain operational efficiencies, and enhance customer experiences. The digital transformation is well underway. Hotels need to keep up with technological changes; otherwise, they are taking risks being left behind. This chapter encompasses the digital transformation of the hospitality industry to examine the impact of it on operational competence and organizational performance in an emerging country like Taiwan. Five case studies are used in this chapter. In-depth interviews with the executives of the leading hotels in Taiwan are a source of rich information and reflect the gradual transformation of technology in the hotel industry of Taiwan.

The development of digital government leads to the digital transformation of public administration. This requires a mix of competences to design, implement, and manage digital government endeavors and the resulting transformations and to use new digital tools. However, researchers agree that digital government competences are under-studied. Moreover, the rare existing studies focus on singular aspects of digital government competences such as technical skills. Thus, this chapter proposes a synthesis of the extant literature about the digital government competences and how they are approached by researchers. The supply-side perspective and the demand-side perspective are proposed to organize the extant literature. Competences are categorized according to each perspective. An empirical investigation is needed to determine the most important competences and to propose the necessary training programs for the lacking ones.

Cloud computing represents a popular and frequent notion in media and scientific research. However, in post-transition countries, there remains a low offer of cloud computing solutions and users have not still accepted this concept besides its broad spectrum of benefits. This chapter examines the possible reasons why, despite high media attention, there are not many examples of using cloud-computing solutions by companies in post-transition countries, using Croatia as a case study. For that purpose, the survey among Croatian enterprises has been conducted regarding cloud-computing usage in business. Collected data is analyzed using knowledge discovery from databases aiming to define firms' readiness when using cloud-

Preface

The term "Digital and Digital Transformation" in recent times have a profound impact on many businesses, customers, employees, and suppliers. Organisations are categorised as Digitally-ready or Digitally-not ready. Leaders of organisations are fast steering companies towards where the business is Digitally-ready. For an organisation that is 'not' Digitally-ready and to become Digitally-ready is not an easy and straight forward task. Leadership & Management around the organisation have to deal with several issues simultaneously such as technical that is related to IT and non-technical that is related to people e.g. customers, employees, suppliers. Businesses that have adopted both technical and non-technical approach to Digital Service Innovation are successful, and many others are not successful. Main drivers of "Digital Business and Digital Transformation" are people.

The Digital Transformation empowers an organisation and its people, revolves around leadership and management and adoption techniques for digital service innovation. Digital transformation provides an organisation – private or public - to better leverage technology for the customer's, employees, and suppliers benefit and far reaching digital social impact on society through social media. Content development and use is driven by data for digital systems and are created by people. Digital Enterprises seeks to integrate complex processes and functions such as management, marketing, production, manufacturing, accounting & finance, customer relationship management, human resources, logistics, information technology, and many other activities as a core aspect of digital systems. Large investment into digital systems are guided which is based on value for money, real time digital services delivery online, cost effectiveness, and competition among firms to be digitally superior than the others and becoming digitally compliant.

Many governments are also digitalising their core functions and services delivery. Smart phones are increasingly popular and have become the standard platform for digital services delivery for banking, utilities, hospitality, travel, education, healthcare, government, social media, and many more. Organisations that are leading the drive to digitalisation are highly empowering people, as the drivers of digital systems are people creating content and its use, when the digital transformation succeeds people are empowered. This book is a good example of digital research in which every author has used digital tools provided by the publisher to make this book a success. At universities teaching, learning, and research are transformed into essential digital tools that are successfully adopted.

Digital system is not restricted to a particular academic domain, or a geographic region, or a particular industry, rather it is globally widespread phenomenon and merged with many other domains creating a self-evolving digital ecosystem. From traditional mortar brick to new start-up business, are harnessing the power of digital business as a cost-effective model to deliver goods, & services online 24/7. Digital business strategy is adopted for transforming business, for streamline processes, and making best use of technologies to enhance interaction with customers and employees and deliver excellent customer experience in real time. This book covers a broad area of research for Digital business in which the core emphasis is to bind on important topics such as Leadership, Management, Adoption Techniques, and Service Innovation. These topics are considered important as it provides a clear and a forward-thinking direction for the readership on digital projects. Interesting case studies and empirical research are presented by the authors to demonstrate the importance of Digital Service Innovation.

THE DIGITAL RESEARCH, GLOBAL CHALLENGES, & NEW DISCOVERIES

The key areas of digitalisation and global challenges are new and unique and requires new knowledge that is gained from a deep understanding of complex issues (technical e.g. technology related, & non-technical e.g. people related) that have been reported in this book. Digital business is not clearly understood at different levels of management and their leaderships (as many top leaders are not IT experts, and not many IT experts are not top leaders, the competencies required are very different), and there is incomplete information around it. The challenges become more complex when data that is required to power digital systems is either non-existent or digitally-not ready to provide useful solutions on a digital platform. The issues related to digital skills, software, hardware, data, internet connectivity, website interaction & access from different devices such as smartphones & tablets, and cybersecurity, also presents complex challenges for the leadership and management of digital service innovation.

The chapters in this book unravel important information, piece by piece, so that the readers can clearly understand, that is driving the requirements for digital transformation for healthcare, universities, banks, hospitality, computing, government and private sector institutions, new start-up, and many others. Prominent researchers have deeply examined areas of digital business models & strategy and have provided the readers with a clear understanding of important issues & challenges being studied, such as problems, literature and theory, research methodology, new research data findings, solutions and recommendations, and future research direction for Digital Service Innovation that adds to the discovery of new knowledge and direction for further research.

The journey for developing and completing this book is a long one in which all authors have made a valuable contribution to the body of new research knowledge, and adopted new research techniques to study the problems, and found innovative solutions to address those complicated problems. This study will be very useful for many researchers to build on new findings that have been reported in this book. This book will also serve as a valuable tool for many students, practitioners, government, managerial staff and other professionals to learn about new developments in digital transformation of business systems from a global perspective. International studies have been presented with important new findings from America, Australia, Asia, Europe, Africa & the Middle east, which demonstrates the far-reaching impact for Digital Service Innovation, and suggests that digital businesses are global and are continuously evolving online into new ecosystems in different parts of the world via internet.

ORGANIZATION OF THE BOOK

A brief description of each of the chapters follows:

Chapter 1 presents case analysis for complexity theory which argues for bounded instability to allow organizations to run operations at a regular speed while also allowing them to explore innovations at a faster speed in support of digital transformation. HeXie management theory uses a mix of systems engineering and holism to argue for a theme around which empowered employees can explore and couple the dividends from such explorations to the organizational vision and mission. These principles are used to discuss how digital leadership has guided healthcare transformation both inside and outside a hospital in multiple use cases, thus providing insight for thought leadership in digital health.

Chapter 2 describes a case for Ecosystems which are gaining ever-increasing importance in digital business environments. New digital business models implemented using digital platforms heavily rely on ecosystem network. Thus, it is worthwhile investigating what role ecosystems play in the process of the digital transformation of companies. This chapter provides a theoretical background of the ecosystem research concerning digital trends, such as digital transformation, digital platforms, and digital service innovation.

Chapter 3 explores the demands that digital service innovation places on those stakeholders who use the digital service through three lenses: workers and workplace, digital literacy, and, finally, digital education. The chapter commences with an overview of digital service innovation. The impact of digital services generally is explored from the perspective of the workforce and the future of work developing an argument for the need for ongoing upskilling in the digital space.

Chapter 4 addresses a conceptual review of the methodology to define strategies and supports for carrying out effective, efficient and innovative decisions and strengthen management and leadership, given the dynamic changes that are taking place in the current turbulent environment. The main catalyst is the digital revolution and its implications in a new management style. The functions of anticipation and early assessment of threats, risks and opportunities, constitute the central functionality of the new strategic approach.

Chapter 5 explains the emergence and adoption of digital technologies that provides organizations with affordable, easy-to-use and easy-to-learn technologies that opens line-of-business (LOB) managers to participate in information technology led innovations. Such participatory actions by the LOB-managers in information technology (IT) innovations at the department levels will lead to the creation of new roles like the 'chief digital marketing officer.' While the overall outlook for innovation becomes positive through such expanding contributions of the LOB-managers, such additions will ignite new challenges.

Chapter 6 reports that digital era has catalyzed the contemporary workforce and workplace. Employees are now required to develop skills that may be outdated in a matter of years, and consciously aware that their society is evolving rapidly. This study conceptualizes the characteristics of a digitally empowered worker to encompass awareness, creativity, agility, and a positive learning orientation. In doing so, the authors consider the way in which an authentic leader and authentic may influence the development of the digitally empowered.

Chapter 7 examines the impact of mandatory Corporate Social Responsibility (CSR) expenditure targets of firms on financial performance in India. Extant studies have shown using event study methodology that announcement effect of compulsory CSR expenditure targets on abnormal returns of these firms is negative. Further, it is examined on how financial performances of these firms which met or exceeded the targets (Socially responsible firms) or were not able to meet the targets (socially irresponsible firms) differ. Available evidence shows that CSR expenditure that spur digitalisation and technological innovation helps the poorest and most vulnerable which enable them to lead a healthy, productive life.

Chapter 8 investigates the Digital technologies adoption has modernized the performance of healthcare organizations to a superior level. Today's healthcare leaders in their transformation journey are seeking the challenge to adopt and adapt different strategies that will achieve effective desired outcomes. In the digital era, organizational innovation plays a critical & significant role to remove barriers that hinder the progress. An important organizational innovation factor is managerial innovation that can have an important impact to fast track the innovation process.

Chapter 9 provides an analysis for Digital Entrepreneurship for immigrants (DEI) through the lens of entrepreneurship research theories and practices. The authors argue that the current focus of digital entrepreneurship (DE) research has centred on successful digital entrepreneurs with proactive innovation strategies and high-performance outcomes, largely being well-known brands. On the other hand, immigrants who are engaging in digital entrepreneurship have been neglected in current related literature. This chapter presents a conceptual framework for combined immigrant and digital entrepreneurship study. The authors propose future research themes for DEI drawing from in-depth results and implications of Chinese immigrants for adopting digital entrepreneurs in New Zealand and Alibaba e-commerce platforms in China.

Chapter 10 illustrates E-procurement systems which represent a digital transformation of procurement function through streamlining and automating procurement process using emerging digital and advanced analytics tools. These systems require a rigorous implementation process to attract their successful use. However, despite having a rich body of literature on e-procurement systems focusing on various issues, little is currently known on how buying organisations generally undertake implementation process of their chosen e-procurement systems to link with suppliers.

Chapter 11 explores Digital health which is fast becoming an integral part of healthcare services. Research evidence suggests that digital health can benefit stakeholders involved in healthcare including patients, and care providers. As digital health continues to integrate into routine healthcare, practitioners may require new knowledge, skills, and competencies to make the best use of it and to be able to communicate with an increasingly digitally-enabled consumer. While much effort has been made to systematic education and training relating to digital health which is an important aspect in developing the digital health workforce.

Chapter 12 articulates that the emerging technologies have an undeniably significant impact on the hospitality industry. Hotel owners and general managers face many challenges nowadays to try to compete with their competitors in an ever-increasingly competitive market. Many hotels are quick to adopt some of the emerging technologies to maximize brand visibility, gain operational efficiencies, and enhance customer experiences. The digital transformation is well underway. Hotels need to keep up with technological changes; otherwise, they are taking risks being left behind. This chapter encompasses the digital transformation of the hospitality industry to examine the impact of it on operational competence and organizational performance in an emerging country like Taiwan.

Chapter 13 demonstrates the development of digital government leads to the digital transformation of public administration. This requires a mix of competences to design, implement and manage digital government endeavors and the resulting transformations and to use new digital tools. However, researchers agree that digital government competences are under studied. Moreover, the rare existing studies focus on singular aspects of digital government competences such as technical skills. Thus, this chapter aims to propose a synthesis of the extant literature about the digital government competences and how they are approached by researchers.

Chapter 14 rationalises that Cloud computing represents a popular and frequent notion in media and scientific research. However, in post-transition countries, there is still low offer of cloud computing solutions and users have not still accepted this concept besides its broad spectrum of benefits and this creates a challenge for digital businesses. The goal of the chapter is to examine the possible reasons why, despite high media attention, there are not many examples of using cloud-computing solutions by companies in post-transition countries, using Croatia as a case study.

Kamaljeet Sandhu
University of New England, Australia

Acknowledgment

I would like to acknowledge the help and contribution of all people, more specifically, to the authors and reviewers that took part in the review process. Without their support, this book would not have become a reality. All the authors have made valuable contribution by making many revisions to their chapters, presenting their interesting research to the readers of this book.

Reviewers provided continuous and important feedbacks to the authors for improvements in their chapters, coherence, and content presentation of chapters.

I would like to personally thank all the authors and reviewers who have worked very hard to complete this book successfully. At every step we worked together as a highly coordinated and focused team.

In this journey we became very good and close colleagues, and this book has brought us all together from different parts of the world, and we have a valuable working relationship and I look forward to working with this great team again in another interesting book project to come soon.

I would like to personally thank the staff at IGI-Global, development editors, who have provided the highest level of service and have been very helpful and supportive in this research at each stage of the book development process, without their persistent help this publication would not have been possible.

I would like to thank my family, including my father and mother who passed away leaving beautiful memories and the importance for higher education and new learning that was ingrained in my very early life and which later was built into my ability to provide this advanced scholarly work to the advancement of new scientific knowledge.

This book is a tribute to my dear mum for all her love, support, belief, kindness, and trust, that I have received which transformed me into an author for scholarly research. I would not have achieved this book success without it.

Kamaljeet Sandhu
University of New England, Australia

Chapter 1
Leadership to Advance Innovation for Digital Healthcare Transformation

Mohan Rao Tanniru
https://orcid.org/0000-0002-9584-0090
University of Arizona, USA

Youmin Xi
Xi'an Jiaotong-Liverpool University, China & Xi'an Jiaotong University, China

Kamaljeet Sandhu
https://orcid.org/0000-0003-4624-6834
University of New England, Australia

ABSTRACT

Complexity theory argues for bounded instability to allow organizations to run operations at a regular speed while also allowing them to explore innovations at a faster speed in support of digital transformation. HeXie management theory uses a mix of systems engineering and holism to argue for a theme around which empowered employees can explore and couple the dividends from such explorations to the organizational vision and mission. Authors integrate these two theories and multiple leadership processes (administrative, enabling, and adaptive) around four guiding principles: alignment around theme, dynamism of employees, transitiveness of dividends, and adaptiveness to support organizational growth and capacity building. These principles are used to discuss how digital leadership has guided healthcare transformation both inside and outside a hospital in multiple use cases, thus providing insight for thought leadership in digital health.

INTRODUCTION

The competitive environment that organizations are facing today is influenced by choices for the adop-

DOI: 10.4018/978-1-7998-2799-3.ch001

tion (Rogers 1995) of different modern technologies and evolving customer expectations that can have an impact on business processes. New models require leadership for digital business transformation which are continuously being modernized to address gaps in the delivery of customer value created to challenge new digital competitors online having superior technologies for instant service delivery with minimum or no human interactions. The essential notion of 'time-based service management' which was a critical factor in earlier studies (Rockhart 1982, 1979; Rockart and Van Bullen 1986) for business, is now based on service on demand by customers and instant delivery of services via the internet in 'real time'. Therefore, organisation are now competing for instant delivery of services in real time which is a critical deciding factor for the customers- how and where such services can be used, which also provides flexibility in customer choices. The generic term "organization" is used to refer to both for-profit and not-for-profit organizations, as well as both product and service organizations. The traditional use of a top down approach to developing strategies that address vision and mission, and filtering these down to lower tiers to influence policy and operational level decision making, is too slow to address the rapid changes occurring today outside the organizational ecosystem. In fact, academic and practitioner discussion on leadership is calling for alternative approaches to developing strategies. These approaches include bringing cross-industry thinking (e.g. bringing a new CEO from a different industry segment to infuse fresh ideas (Sylvers, 2019)) and broadening the performance metrics from near term to longer term (e.g. move from shareholders to stakeholders, including customers and suppliers (Ramey, 2019)).

Both these approaches implicitly recognize the need for leadership today to address fast-changing customer expectations and the multitude of external competitors who are willing to address these expectations. This demonstrates that markets are driven by customers who are very knowledgeable and aware of different information to make decisions. In this context leadership is becoming an essential factor for modern digital environment, in which a technology is used to create different business models to address customer expectations and drive growth in new and existing markets (Chan, 2018). In fact, the role CEOs and CDOs (Chief digital officers) is being debated as organizations engage in digital transformations to address the complex market dynamic (Siebel, 2017, Tumbus et al, 2017). Much of the discussion on digital transformation (i.e. organizational transformation in the digital age) is focused on organizations build agility to adapt for changes that can quickly improve inner business structure and streamline process delivery. Such agility may come from creating two different organizational models (one to run at a regular speed and the other to run at a faster speed) (Bossart et al, 2014), using best practices (Ivancis et al 2019)). using a mix of internal and external resources to design and develop services to address value gaps (Lusch and Nambisan, 2015), or even rethinking and shaping the organization around the value it creates or co-creates with its customer using evolving business models (Weiland, et al, 2018). The premise behind all this thinking is the need to balance the organization's effort to seek stability and growth with the instability that arises when the changing ecosystem from outside the organization demands continual innovation and adaption to create value for the customer.

The unpredictable and chaotic environments (volatile, uncertain, complex, and ambiguous: VUCA environments) of today call on organizations to build resiliency (Kantur & Say, 2012). Complex systems enhance their capacity and build resiliency by changing their behaviors or strategies to respond to environmental problems or internal demand for performance improvements (Holland, 1995). Biological, environmental, and social systems (e.g. slime mold, ant colonies, immune systems, brains, markets, and companies) are complex adaptive systems that have evolved over time by adapting to changes in their environment (Johnson, 2001). The commonality among these systems is the large number of interacting agents responding to stimuli and exhibiting a response behavior learned from experience (Holland,

1995). Researchers of Complexity theory argues that the interactions of organisms or agents (e.g. semi-autonomous entities such as atoms, molecules, biomolecules, organisms, processes, people, groups, firms, industries, etc.) determine the nature of their adaptive behavior as they react to changes and influence their future state in an environment that is forever evolving, and at the same time the dynamics of business model is undergoing rapid transformation that embeds modern technologies into the lifecycle of a business process (Ferber, 1999, p. 9, Axelrod and Cohen, 1999).

Research have shown that leaders who are focused on solving a problem and making an opportunity to make a problem into a business have a very high success rate (e.g. internet startups solving problems). Another approach is to recognise a problem (which a problem presents), and then to grow a business out of it, linking different innovative technological products into it. Such an approach is believed to be highly effective as businesses venture into the problem areas to find solutions and also setting up industries and newer opportunities around the problem, that can attract the best leadership talent of scientist, researchers, practitioners, and professional, which can also be good for the community, customers, businesses, and also to create employment. Initially business needs to provide leadership to kick start such projects and new start-ups, that can export such expertise to other countries.

Hence the important research question being studied are: What is the best way to articulate *the role of a leader within a leadership process* as organizations exhibit emergent behaviors like those of complex adaptive systems? Are today's modern leadership have advanced characteristics to steer the modern business(es) towards a sustainable Digital business model? In this chapter we will study two different leadership philosophies based on their unique settings, derived from complexity theory and HeXie management theory, and their synthesis to answer the research question.

The next section will look at prior research on both complexity and HeXie management theories and compare these in order to develop a framework for thought leadership for organizational transformation. The third section focuses on one type of leadership needed to address digital transformation (organizational transformation in the digital age), called digital leadership, and develops a model to support this leadership. Section four uses this model to illustrate digital leadership in multiple healthcare use cases. The last two sections identify areas for future research and provide some discussion along with concluding comments.

BACKGROUND

Today's complex market dynamic is seeing changes constantly with unpredictability and non-linearity, and organizations need to leverage relationships and network interactions among several actors from multiple ecosystems to adapt to these changes (Shipton et al, 2008). Different leadership styles are called for to support activities used for facilitation, adaptation, and capability building to address uncertainties (Uhl-Bien et al, 2008), while also helping organizations innovate, promote quality outcomes, and recognize interrelationships (Uhl-Bien et al, 2011). Leadership (as opposed to leaders) is seen as emerging to connect people and ideas (Lichtenstein et al 2006). Three leadership-process behaviors – *administrative, enabling, and adaptive* – are identified as needed to help organizations facilitate innovations for exploration and support their adaptation, and over time to improve organizational capacity to operate and grow in a state of "bounded instability."

Figure 1. Leadership Processes and Activities Derived from Complexity Theory

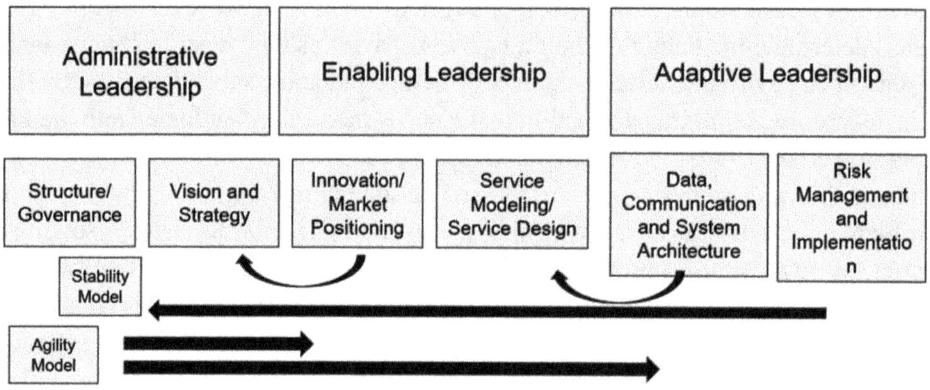

The energy expended in this journey, the adoption of new innovations as part of the adaptive leadership process, is influenced by the structures, routines, rules, and norms (Schein, 2010). Specifically, changes in the

- *Contemplated business and technology architecture* proposed by the innovation – routines, rules, and norms associated with the new innovative services designed to create value.
- *Contemplated people and processes* – network of interactions and relationship behaviors among actors who must institutionalize change.
- *Engagement of the role of internal and external actors* – that guide or constrain the needed behaviors
- *Views and opinions of heterogeneous actors* – whose conflicts need resolution (Uhl-Bein, 2008, Agarwal et al, 2008).

In summary, the three leadership processes make exploration of innovations deliberate and make organizations proactive as they move between stability to instability to address challenges and explore new opportunities that support growth.

Figure 2. Leadership constructs from HeXie Mgmt. Theory

HeXie management theory views Eastern and Western cultures and wisdom as having complementary principles that can be integrated to address today's management challenges. It blends the methodologies of system engineering and China's holism and practical wisdom to solve problems by addressing the complex issues organizations face in uncertain environments (Cao et al, 2011; Xi et al, 2019). Instead of stating it as a philosophy and mechanism, HeXie management theory evolved over time to develop constructs such as HeXie themes, He and Xie principles, and HeXie coupling, all framed under HeXie leadership to address today's global challenges (Xi- et al, 2012).

Figure 1 shows the leadership processes and activities derived from the complexity theory, and Figure 2 identifies the leadership constructs from HeXie management theory. Before we compare these two approaches and synthesize the concepts discussed within them, it is important to discuss how these activities and constructs were developed.

Table 1. Leadership differences seen through the complexity and HeXie management theory lens

Administrative Leadership	**Xie Principle**
Vision setting, strategy formulation, and developing an organizational structure and governance are used to run business at a regular speed using a "stable" organizational model	The Xie principle uses rules and propositions to reduce uncertainty through scientific design, the optimization of organizations using institutional structure, processes, and communication, and a coordinating architecture to improve operational efficiency and decision effectiveness.
Organizations deliberatively move to instability using agile organizational model and test new ideas by leveraging enabling leadership process activities.	The Xie principle proactively creates a "theme," or a region of "bounded instability," for those operating under the He principle to explore and innovate.
Administrative leadership may develop a broader organizational strategy for developing innovative ideas by enabling leadership.	The theme creates an innovative ecology, i.e. an environment within which a certain degree of risk-taking behavior and innovative culture can flourish.
Enabling Leadership	**He Principle**
The activities here include *innovative idea generation and positioning these ideas in the market* as value co-creators, and *modeling and designing products* in support of these ideas, using advanced technologies when appropriate.	Problem solving is viewed in the anticipated or future context around the "theme" developed.
These activities are supported by an agile organizational model with a structure and governance that allows for the use of internal and external partners, including customers.	Such a theme can help support the phased development of goals and problems that must be solved by the organization over a specific time period.
Innovative ideas are viewed through the lens of organizational strategy, but at times it may differ, calling for reconciliation between *enabling* and *administrative* processes.	The "themes" provide certain degrees of freedom for exploration along potential organizational ecosystem dimensions (e.g. customer, technology, regulation, etc.) which may see volatility.
Adaptive Leadership	**HeXie Coupling**
Adaptive leadership includes a set of activities needed for organizations to change to support adoption and diffusion of innovations, evaluated as plausible by the enabling leadership.	HeXie management theory defines *coupling* as a construct to connect innovations explored to the *themes* and identify several types of dividends.
Assessing the impact of these changes on business and technology architecture, people and processes, and relationships with external partners may lead to several options for adoption.	These dividends include sharing of ideas and symbiosis of thought and exploration by breaking organizational boundaries; support growth by achieving integration with existing operations through adaptation; build resiliency to future changes from reflection, discourse, and knowledge sharing.
It can also lead to *enabling* and *adaptive* leadership processes to reflect on the change contemplated and potentially revisit the time and scope of the innovation for adaptation.	Viewing exploration along multiple dividend dimensions allows for organizations to look for near term benefits to support growth and longer-term benefits to build an innovation culture and organizational capacity.

The activities discussed under each leadership process in Table 1 were developed by one of the authors over four years (2014-2017) as part of a digital leadership class taught for graduate students in the US, France, and China. This project-based course evolved from an IT-enabled innovation course in 2014 (21 projects by working graduate students in the US) to a digital transformation course in 2015 (21 projects by 60 graduate students in France and the US) to a digital leadership course that used service modeling and balance score card concepts in 2016 (76 graduate students in France; 82 graduate, 80 undergraduate and 14 executive MBA students in the US; and 21 undergraduate students in China). Currently, as discussed in Figure 1 shows the activities within the three leadership processes are part of a 10-step digital leadership methodology discussed in a management information systems book (Pomffyova, 2018). In parallel, the second author has evolved the constructs underlying the HeXie management theory over several years, beginning with an in-depth case analysis of three different organizational leadership styles that varied in scope: centrally directed, distributed coordination, and entrepreneurial (Cao et al, 2011; Xi et al, 2019), and since adapted to leadership in education (Zhang et al, 2018).

The rest of the section compares the leadership processes derived from complexity theory and HeXie management theory using Table 1 and highlights their differences.

Figure 3 provides a synthesis of these two theories as thought leadership in support of organizational transformation. Note that the "themes" connect administrative and enabling leadership, and "dividends" couple adaptive leadership with administrative and enabling leadership processes. Based on this synthesis, a new model for leadership to support of digital transformation is developed in the next section.

MODEL FOR DIGITAL LEADERSHIP

The rapidly evolving technology landscape means that organizations need to rely on several external technology actors and their resources to quickly address the growing customer demand for digital services (i.e. information technology services to create value). The 'time' to deliver a 'digital service' is a significant factor that can have an 'impact' on customers perceptions about the quality of the digital service (Hadeel and Sandhu, 2021). Given the frequent and varying scope (radical and incremental) with which advanced technologies are used to design and develop digital services, it is likely that not

Figure 3. Thought Leadership for Organizational Transformation

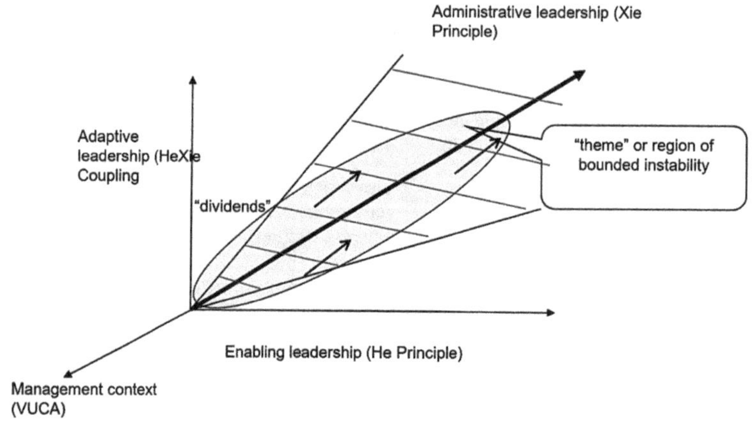

all such explorations will lead to successful adoption. Many will help organizations build their technical competencies and organizational capacity to address changes in the future. Also, given that many digital service innovations are designed to support changing customer expectations, some of these innovations can lead to developing new business models. Some are smaller in scope to retain current customers, and others may be radical to alter an organizational strategy. The complexities in understanding customers expectations from digital services is still evolving as new technologies are being embedded into the business model and new processes by organisation (Hadeel and Sandhu, 2021).

The focus of this section is to develop a conceptual model that explicates the role of multiple leadership processes discussed in the previous section for the innovative design of digital services to meet changing customer expectations. At the foundation of this model are the four guiding principles that help connect the roles leaders need to play within each leadership process as organizations move between instability and back to stability, thus answering the research question. While complexity theory was discussed specifically in the information systems development (Benbya et al 2006), the focus here is on viewing digital service innovations through an organizational leadership lens.

Model

The model proposed in Figure 3 integrates the diverse and rich perspectives from complexity theory and HeXie Management theory and arrives at four guiding principles: alignment, dynamism, transitiveness, and adaptiveness. These principles, in principle, connect the three leadership processes. The administrative leadership (Xie principle) articulates the bounded instability region around "themes," thus allowing digital leadership (i.e. enabling leadership to support digital transformation) to explore innovations and evaluate them along multiple "dividends." This leads to adaptive leadership to institutionalize some innovations to support growth and others to support organizational capacity building. The development of "themes" for exploring innovations that can frame innovative digital services for broader digital strategies. For example, improvements in technology, customers, regulation, etc. can provide dimensions to the theme around which digital strategies can be developed, and specific customer services can be explored as digital services. Similarly, dimensions such as demand side (value creation to support customer), operations (value creation to improve efficiencies or effectiveness), or supply chain (value creation by reducing costs or improving quality) can also frame digital strategies. This will align digital services that will improve internal performance metrics. Some digital strategies can be modest or extreme in scope (Dawson et al, 2016), thus dictating the degree of change the organization is seeking. Figure 4 provides the guiding principles for digital leadership, and the focus of each is discussed next.

Alignment

When organizations recognize the benefits of a new technology in creating value for a customer directly (through external services) or indirectly (through internal operational improvements), they are motivated to explore these for potential adoption. Larger explorations contribute to significant changes but occur less frequently than small explorations (Hershey, 1985). Such explorations, needed to address disruptive forces in the market (e.g. enterprise systems), need policy level discussions between those in the administrative and digital leadership, even if they are framed within the digital strategies previously articulated, as they can shift the way the organization operates in the future. However, smaller or incremental explorations are not widely known to all involved, and the value derived from such explorations

Figure 4. Guiding Principles for Digital Leadership

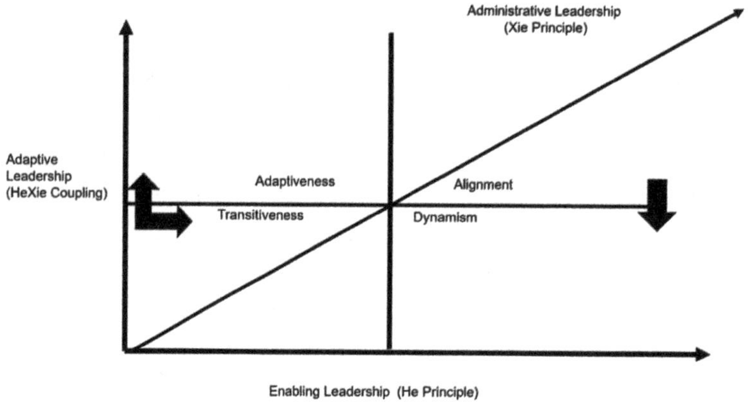

tends to be more localized, and they need visibility to gain traction for both exploration and adoption. However, such smaller explorations occur frequently to address changes in the customer or technology ecosystem. Therefore, by having espoused digital strategies derived from the theme, it will be easier to give visibility to smaller explorations. For example, a hospital that establishes theme and digital strategies around "improvements in patient satisfaction or reductions of cost attributed to patient care," will help those involved in small explorations that use technology to improve patient or emergency room operations gain visibility and rationality for their efforts. In the digital transformation journey, many hospitals have challenges to maximize the acceptance & use of digital services, due to the high cost of technologies, employees' resistance to change, complex processes, lack of leadership support, training, and support services (Hadeel & Sandhu 2021; Sandhu 2020; Venkatesh & Davis 2000; Davis 1989).

In summary, the alignment principle in the upper right-hand quadrant of the DL framework is derived from themes established, and it acts as a "bounded" region for formulating digital strategies and exploring innovative digital services, which are discussed next.

Dynamism

Complexity theory researchers argue that environmental pressure acts as energy exchange between the organization and the outside environment (Anderson, 1999). Such pressures radiate from global and domestic clients, customers, suppliers, regulatory bodies, and agencies, along with market and technological developments. They can lead to several changes in the organization. While organizations are often poised to deal with institutional forces and legitimacy constraints to remain competitive, the driving force for technological or customer service changes may come from multiple sources and call for the development of new ideas constantly (Van de Van et al, 1995). While those engaged in senior leadership can track major institutional or environmental changes, multiple sensors are needed to track changes coming from many consumer-driven technological changes, as well as constantly changing customer demands for new services. The concept of dynamism views environments as a source of information (Dutton, 1993), and such information must be used effectively to influence organizational action (Anderson, 1999). Dynamism captures the rate of change and the direction an organization is willing to undertake to influence the attitudes of those within the organization. A clear articulation of the theme and digital strategies around which explorations occur a priori helps organizational dynamism, as it allows those involved in innovation explorations to direct both the rate and direction of change, thus empowering them to seek resources and partners to rapidly design and develop new digital services for evaluation.

In summary, dynamism, shown in the lower right-hand quadrant, allows digital leadership to empower those involved in exploration to design and develop digital services to take advantage of opportunities that surface.

Transitiveness

The quick evaluation of any a new digital service to assess its viability is a challenge, especially when this assessment must be based on actors outside an organizational ecosystem (i.e. customers). The uncertainty of such evaluations is associated with not having an established routine that can provide a well-predicted path for defining what is considered a success for eventual adoption. Leaving older processes or business practices and moving to new ones is equivocal and uncertain (Brown, 1998) in general. It is especially risky when such implementations touch customers and have high failure visibility. While some radical explorations limit their exposure to risk by taking a structured sequence of steps (e.g. ideation to concept to development to prototype to pilot test and to implement), smaller explorations often have fewer resources and visibility within an organization and face challenges in making a case for justification of their implementation. Therefore, *transitiveness*, conceptualized as a culmination of enabling change and progress, is key to ensuring that the evaluation of innovation explorations is viewed along many dimensions. In other words, the dividends that accrue by implementing some of the innovations can be coupled either directly with the theme on tangible outcomes, or indirectly to support improvement in a process or improve the capacity of a unit to address future changes. For example, improving customer experience or understanding customer context can help organizations leverage this capacity to address customer needs in the future or from different regions or cultural contexts. Even if no explicit dividends are identified, the learning from exploration of the innovation can be documented as a part of the knowledge repository for future use.

In summary, transitiveness, shown in the lower left-hand quadrant, allows for assessing the impact of explorations along multiple dividends. If the dividends appear to support growth, they may be moved for adaptation as discussed next. Otherwise, some of these explorations may be delayed for later adoption, narrowly scoped for future consideration, redesigned for a change in the scope of its use, or even abandoned. The goal here is to recognize that explorations of innovations do contribute to support immediate needs (i.e. growth) or longer-term needs, including contribution to learning and capacity building (e.g. improved technological competencies, enhanced learning needed to explore new technologies or customer expectations, etc.)

Adaptiveness

Adaptiveness refers to the preparedness and resiliency of the organization to make the changes needed to address environmental influences. In the digital age, adaptiveness is needed along two dimensions: adaptiveness of the technology architecture and adaptiveness of organizational processes. An adaptive technology architecture needs to have a sense of predictability and accountability to accommodate laws and regulations, policy guidelines, corporate standards, best practices, etc. The organizational process adaptiveness relates to resource planning and allocation to couple the innovation with the rest of the operations. In complexity theory, such adaptiveness may occur through experimentation, openness, and improvisation (Lewin, 1999). Experimentation calls for some degree of improvisation to facilitate change, especially when there are areas of least resistance for change to be implemented and its success is propagated over time to others. Openness allows organizations to maintain an ability to plan for the unexpected and explore opportunities to implement a change. For example, an unexpected request from a customer or a new demand for a service offered by a competitor may create an opportunity to adopt a service to address this need to react quickly for a local context. While large scale implementations often have organizational visibility to support change, smaller implementations often need more improvisation.

In summary, adaptiveness, shown in the upper left-hand quadrant, calls for effective change strategies that help revise current processes and practices to support the established theme and advance organizational growth.

The four guiding principles – alignment, dynamism, transitiveness, and adaptiveness – help organizations address the demands of the VUCA environment. Digital leadership is needed to align the innovations explored to the themes and digital strategies developed by the administrative leadership (Xie principle) and work with adaptive leadership (or support HeXie coupling) to successfully support the adoption of those innovations explored which support the organizational theme. At the same time, digital leadership needs to create a culture of dynamism by empowering teams to rapidly explore new ideas and design and develop digital services for evaluation and use transitiveness to help gain benefits from these exploration efforts along different dimensions. Some of the dividends from these explorations are used to sustain a culture of dynamism and organizational capacity to continue to innovate to address changes, while other dividends are used to support the themes to sustain organizational growth. As indicated, while these principles are discussed within the context of digital leadership, many of them are relevant even when innovations are not driven by advanced technology. This generalizability is briefly addressed in the discussion under future research.

The next section will discuss how these principles may be used by a healthcare organization to demonstrate digital leadership.

SOLUTIONS AND RECOMMENDATIONS

Healthcare organizations (referred to here as hospitals or provider of ecosystems) are facing interesting challenges in the digital age as they continue to create value for patients. Healthcare has been viewed as a complex adaptive system (Mahajan et al, 2017), and research has begun to focus on how such a view can start to alter clinical practice (Martin, 2018), address service quality (Horvat et al, 2018) and influence priority setting of hospital practices (Barasa et al, 2017). Health services research has recognized the relevance of complex adaptive system (CAS) in addressing many of the challenges hospitals face inside and outside the hospital. Greenhalgh et al (2018) reports six cases that cover mental health services, respiratory conditions, medicines management, hospital-based rapid response teams, system-level accreditation mechanisms, and digital health solutions (such as video consultations, assisted living technologies, and remote monitoring) to discuss how complexity theory may be useful to analyzing complex health systems. In this section, we will first discuss digital leadership inside a hospital (in-depth use case 1) in Southeast Michigan over a six-year time period. We will then expand our discussion of digital leadership in four other use-cases as care is delivered outside a hospital. First, a note on the research methodology.

Research Methodology

The study adopts a variant of the ethnography approach, where one of the authors has spent significant amount of time studying the environment, engaged in detailed collection of data through field notes and interviews, and extensively reviewed data from multiple perspectives (Crabtree et al, 1999, Ventres et al, 1996). Table 2 summarizes some of the professionals interviewed for the study and their experiences in a hospital (use case 1). Research also suggests the use of multiple case studies can be used to provide a stronger base for theory building and testing (Yin, 1994). Deeply exploring one hospital case (use case 1) and comparing and clarifying emergent findings with other cases (discussed in use cases 2-5) can lead to generating several propositions (Eisenhardt, 1991) deeply grounded for future empirical evidence (Eisenhardt and Graebner, 2007). Details of the use cases 1 through 5, along with the methodology used, were published in prior research (Khuntia et al, 2017, Tanniru et al, 2018, Fowler et al, 2018, Plum et al, 2019, Tanniru, 2019). In addition, the research methodology integrates the two authors' experience in analyzing leadership in distinct domains (manufacturing, healthcare, and education) in two different countries, China and the US.

Table 2. Hours spent interviewing individuals involved in the digital transformation of the hospital

Interviewees	2011	2012	2013	2014	2015	2016	2017	Nature/Topics/Areas of the Interactions
Chief Executive Officer	3	2	3	5	6	9	7	Broad vision, strategy, and leadership contexts
Chief Information Officer		2	2	2	3	3	4	Business and technology architecture as well as IT enabled innovations
Chief Medical Information Officer				4	1	2	2	Clinical-IT alignment and implementation issues and risk mitigation
C-level Nurse Manager			2	2	3	2		Patient room management: people and process issues
Business Intelligence Manager					1	4	4	Data and architecture issues in the development of business intelligence
Nurse Manager				2	2	2	1	Process, workflow, and partner issues in addressing patient support and care management
Manager in charge of patient engagement area						1	2	Patient engagement and metrics on administrative and clinical performance to address healthcare challenges
Manager overseeing patient flow & scheduling						4	3	Process management
Information technology architect					3	3		Partner details and internal and external technology integration
Nurse manager in charge of fall risk and multipurpose rounding						2	3	Risk mitigation and implementation strategies and decisions related to adaptation

Digital Leadership Within a Hospital (Use Case 1)

This use case looks at how a 443-bed community teaching hospital in Southeast Michigan has over a period from 2010 to 2015 used several principles discussed in the previous section. The hospital has embarked on several advanced digitization initiatives since 2010 to achieve excellence in quality, safety, performance, and patient satisfaction. These initiatives have transformed the hospital from a traditional and well recognized hospital with clinical expertise to one of the most wired hospitals for several of its digital innovations in 2014 and 2015. This has been a by-product of the visionary digital leadership of the chief executive officer (CEO), working in close partnership with senior leadership, including the chief medical information officer (CMIO) and the chief information officer (CIO). Many of the observations made below reflect how both radical and incremental transformations have supported the hospital's vision and mission and its capacity to continue to innovate even today, despite many organizational leadership changes.

Alignment and Dynamism

The administrative leadership has set a broader vision to move the hospital to leverage technology to improve patient care, but it set its "theme" for digital innovations to accomplish two key goals: improve patient care quality and satisfaction while reducing costs associated with care. Note that while there are other environmental demands regarding regulation, etc., the dimensions of focus here are on how technology and patient ecosystems are changing the healthcare landscape. This framed digital strategies for the empowered team to explore innovations. The team is called the business intelligence or BI team and has multiple stakeholders, including the CMIO, CIO, nursing, and IT staff, with the CEO playing an important member role as and when needed in this team.

One of the early innovations was a clinical innovation called (Visensia) that helped the nursing and hospital staff monitor five vital signs of patients continuously to track anomalies in medical conditions and address these quickly. This has had an impact on mortality and morbidity of the patients (quality metric) as well as reducing costs (increased length of stay of patients if the patients must end up in intensive care units means increased costs). The CMIO and CIO played a lead role in this exploration, which used external vendors in support of its design and development.

The second innovation emphasis is for making data driven decisions and transparent to all involved for improving accountability. This is a challenge in many US hospitals where two different organizational structures (clinical and administrative) are used to support patient care, both of which report independently to the hospital board. The dashboard technology was used to align organizational metrics and clinical actions on several dimensions that impact patient care: improving patient flow and satisfaction, reducing costs associated with hospital acquired infections and patient falls, etc. Given its complexity for both IT and organizational culture, the exploration was divided into two stages. The first exploration was focused on organizational culture, and the second exploration focused on technology integration. The CEO led the exploration and evaluation of this innovation with support from the CIO, by using daily rounds of key executive staff to discuss metrics and making the dashboards display key metrics in all high-footprint areas throughout the hospital. The second phase of the innovation, the IT integration, was led by the CIO.

The third innovation was for an intelligent care system (ICS) that digitized several of the patient room services to reduce costs and improve patient satisfaction. Some of these included smart beds to reduce patient falls, RFID based gel dispensers to reduce hospital acquired infections, Pillow Talk and Votle (iPhone) to improve patient communication with nursing staff, etc. The complexity led the innovation to be divided into two stages. The first stage focused on technology integration and was led by the CIO, CMIO and nursing staff, and the second stage of nursing process integration was led by the CEO and CMIO with support from nursing staff.

Two other innovations studied were incremental in nature. One improved communication between nursing and pharmacy staff to help dispense drugs to patients faster (called the medication board). The other dealt with remote communication with critical patients discharged using wearable devices and iPads and supported by local care providers (called RSVP). Each of these was led by different team members, including nursing staff for the medication board and the CMIO and CIO for the RSVP system.

In summary, the key aspect of the dynamism associated with digital leadership is the role different individuals played in innovation exploration and the way complex explorations were divided into stages to support focused evaluation.

Transitiveness and Adaptiveness

Each of these innovations have reached varying levels of adaptation, and the dividends the organization accrued varied as their impact was assessed during the transitivity phase. The vital sign monitoring technology found limited value and is used only by one unit, but the dividend from its use has led to a variant of its use by RSVP when patients are sent home for remote monitoring. The first phase of dashboard technology has been assimilated and continues to be used and refined as organizational priorities evolve. However, the second part of the dashboard exploration, technology integration, has not continued due to uncertainties in organizational technology architecture. Yet, the dividend of the technology integration effort has increased the capacity of the BI team to adapt to changing priorities in the dashboard use, as well as its future integration to a new enterprise system planned for 2020 implementation.

The intelligence case system brought many new digital services during the first stage of innovation exploration, and the technology diffusion is widely adopted as of today. However, the second stage of exploration has seen varying dividends, and not all were adopted for use. For example, smart bed technology was discontinued after some initial success. RFID gel dispenser, along with PillowTalk and Volte Phone, were improved before full adaptation. Others have seen no significant change, and their adaptation has seen limited success. Independent of the varied adaptation rates, one of the primary dividends that came out of the evaluation phase is the nursing and IT staff's ability to look for other complementary non-digital service innovations to improve patient support and add other incremental digital innovations (e.g. digital incident reports) to improve their coupling with the theme. Both the medication board and RSVP systems have shown success with their support of the theme and were adapted and continued to grow in use. Table 3 summarizes the digital leadership in the hospital.

As discussed above, hospitals can use digital leadership effectively when they recognize that not all innovations explored may bring the same type of dividends, even when some of these may be adopted to improve clinical processes and practices. The goal is to use the explorations to develop a culture of innovation and willingness to explore new opportunities that will come along the way to develop a capacity to address futures changes, especially when some of these can be used to support patients as they move to care outside the hospital ecosystem. Such is the case with the RSVP system, and it is discussed along with others in the next section. Some of the work discussed in this section was discussed in-depth in prior research (Tanniru et al, 2018, Fowler et al, 2018).

Table 3. Digital leadership in the hospital

Innovations	Alignment w/Theme	Dynamism of Actors	Transitiveness to Assess Impact	Adaptiveness for Use
Vital sign monitor	Patient mortality and morbidity	CMIO and CIO with external vendor	Limited acceptance and used by one unit	Adopted for use by patients after discharge
Dashboard - Phase I-unit action and KPI alignment	Support tracking of costs and satisfaction	CEO with CIO support and external vendors and university partners	Evaluated to be of value	Adopted for use by organization
Dashboard - Phase II IT integration	Support data integration and analysis	CIO with BI team	Not sufficiently robust to address future needs, used with smaller scope	
ICS - Phase I IT integration with multiple vendor technologies	Need to support patient data communication with nursing staff	CIO and CMIO with nursing staff and IT staff	Evaluated to support data sharing	Adopted for sustained use to support communication
ICS - Phase II People and process integration	Need to support patient care delivery and cost reduction	CEO, Chief nursing mgr, CMIO and nursing staff	Impact varies across different services; some locally used; some abandoned; and some moved for adoption	A few of these are adapted with adjustments to people and process change
Medication board	Support faster delivery of drugs to patients	Nursing and pharmacy staff with IT staff support	Impact found to address delays	Adopted for use
RSVP	Support patient care post discharge from hospital	CMIO, CIO, external technology and care delivery partners	Impact found to be improving care follow-up	Adopted for use at two different sites

Digital Leadership Outside a Hospital (Use Cases 2 Through 5)

Given the diversity of patient populations and the number of clinical and non-clinical care providers with whom patients must interact when they leave a hospital, is difficult to customize digital services for each situation. However, the principles discussed above will apply in developing services for unique contexts. For this purpose, we present four such contexts, one of them being the RSVP system discussed in the previous section. They illustrate the potential for digital leadership, as hospitals seek to extend their care delivery by using several actors in the patient ecosystem. Figure 5 discusses four use cases, and Table 4 summarizes the digital leadership for the four use cases.

Patients Monitoring Health in Special Care Facilities (Use Case 2 in Q4)

Patients in this quadrant move from a hospital to specialty care centers such as skilled nursing facilities (SNF) or rehab centers for short-term stay (and sometimes for longer term stay) to undergo focused monitoring. These facilities often must care for diverse patient groups and do not have the adequate clinical staff and administrative support to act independently. They rely on consultation with specialists and the patient's primary care physician for certain clinical follow-up, and they send the patients back

Figure 5. Use cases two through five

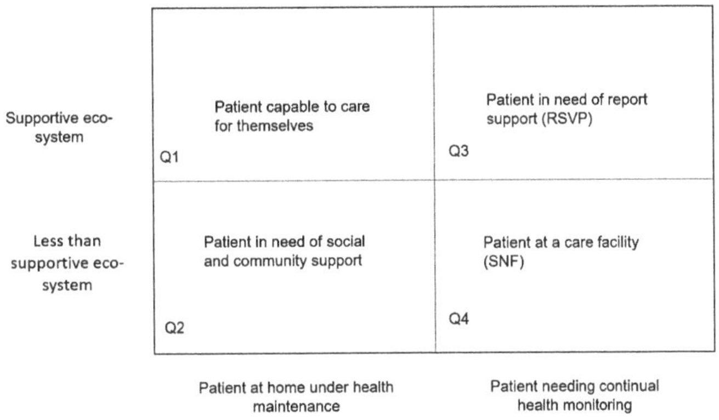

to emergency rooms if there are complications in care. Many reimbursement guidelines and penalties have started to apply to both hospitals and SNFs, thus aligning their goals to the theme of "support continuity of care with fewer readmissions." Digital leadership in use case 2 leverages an intermediary, an advanced nurse practitioner (ANP), with support from a physician as a needed bridge to address skill gaps between SNFs and the hospital for treatment of cardiac patients. This intermediary, besides reducing readmissions, has begun to improve the capability of the SNF to leverage special consults and has improved internal processes to address the needs of other patient groups and improve its quality rating (Tanniru et al, 2019).

The use of technology adoption and acceptance (Rogers, 1995; Davis 1989) referred in our case study is minimal. However, with the scope of the "intermediary" increasing as it starts to support several other SNFs, the role of tele-consults to improve communication between the staff of the SNF and the ANP along with remote monitoring of patients will continue to increase. An additional dividend is improved communication between ANP and the patient's family members. Another model to address patients who need continual monitoring for some time is being addressed by sending patients to their home and using intermediaries such as home care services to do such monitoring. Ultimately, if the broader theme is to support care monitoring in a less expensive facility closer to the patient's home or at his home, then there needs to be an alignment between hospitals and external clinical care providers, if the patient cannot effectively be engaged in their own care monitoring. This leads to use case 3 in Q3, where patients take on some of this monitoring responsibility.

Patients Monitoring Health Using Community Care Providers (Use Case 3 in Q3)

As discussed in reference to case 1, the hospital combined the capability of a patient to use a wearable and an iPad to monitor their care daily from home, with a clinically qualified community provider (e.g. an emergency management technician from a nearby fire station) visiting to report on the patient condition to the hospital staff periodically or as needed. The innovative use of local community care providers and patients, while designed to reduce readmissions by paying for some of these care provider services, also has the goal of improving the capacity of the patient ecosystem to care for its patients using local clinical care providers.

The pooling of many such patient monitoring services using nearby clinical care providers can be brought under the umbrella of facilities such as home care services discussed earlier. The diversity of patient needs and their geographical spread in some regions (e.g. patients in rural communities) can make a federated model that independently connects clinical care monitoring individuals or facilities (e.g. mobile clinics) more appealing, especially if they gain the trust of diverse patient populations with varying social and cultural needs (healthcare on an Indian reservation, supporting immigrant community, etc.). One such model that supports homeless populations is discussed in use case 4.

Table 4. Digital Leadership in Four Use Cases

Use Case Summaries	Alignment	Dynamism	Transitiveness	Adaptiveness
2: SNF in Q4: Skilled Nursing Facility - Support care delivery of patients at an external care facility	Both hospital and SNF have a common goal of reducing patient readmissions; intermediary is supported by the CMS (government)	Innovative team included advanced nurse practitioner with physician support, and SNF management	First phase led to reduced readmissions; Other dividends included improved capacity of both ANP and SNF to handle diverse patients	While the innovation has seen limited use in the hospital, ANP and physician established a new intermediary to support patients at various SNFs; SNF began to use such models for other patients
3: RSVP in Q3: Support patients who need continual monitoring at home	Both hospital and patient have common goal of reducing readmissions by providing care at patient's home; community workers act as intermediaries and are paid by the hospital	Exploration team included patient, emergency room patient, contact nurse at the hospital and community care provider (Emergency management technician)	Reduction of readmission is not fully established, but patient comfort and care quality has become a dividend to increase hospital's use to another region to test its long-term viability	Not fully adapted to support institutionalization, but is being explored to assess business model viability
4: Recuperative Care Center in Q2: Provide assistance to patients with limited support system	Hospitals and public health agencies have a common goal to reduce readmission of homeless population and provide alternatives to support broader care; charitable organizations are funding the intermediary which is a not-for-profit organization	Members from the public health agencies and non-profit organization, along with other regional centers that support work and temporary shelters	The center has shown viability and increased its capability to provide both short term clinical care and other types of rehabilitation needed to reduce homelessness	Currently, some of the providers use this service as part of their effort to reduce readmission, but it is not adopted as a general practice (as not all become eligible to use this service)
5: ASH Line in Q1: Provide counseling to patients with supportive ecosystem	Public health agencies and some providers and government support care delivery to reduce patients' addiction to smoking	Members of the intermediary and public health agencies, along with regional university center working with students and counseling staff	Research on ways to address addiction to smoking and exploring interventions such as "patches", as well as continue to build intermediary capacity to support more patients	Many healthcare providers use this service by referring patients, but it is not institutionalized as part of their treatment process

Patients Managing Health Using Community Care Providers (Use Case 4 in Q2)

Many hospitals use local non-clinical care providers to support patient needs for transportation, child-care, or education and to improve patient adherence to care plans. Examples here include supporting patients with mental health problems, sextually transmitted diseases, or diabetes (Plum et al, 2019, Tan-niru, 2019). This use case takes an unusual step to broaden the care related support from health to areas outside health that impact health. It is a non-profit organization that operates a recuperative care center and provides some clinical support for homeless patients discharged from a hospital, but it also looks to address their other needs. By working with other partners, such as county public health agencies, temporary shelters, soup kitchens, and work-related training facilities, the goal is move patients from the current condition to a state where they improve their capacity to address areas beyond health. As many social factors (Williams et al, 2008) determine a patient's ability to manage their health condition, digital leadership needs to look for the capacity building initiatives of patients and the community within their ecosystem, if the theme is to support quality of care that is more "holistic" in its approach. Such a broader theme needs to align the goals of many external clinical and non-clinical care providers, as well as other agencies within the community. This leads to our next case where the focus is on increasing the capacity of patients with chronic conditions using a public-private partnership.

Patient Managing Health Within a Supportive Ecosystem (Use Case 5 in Q1)

Healthcare providers have shown digital leadership by providing patients who leave the hospital with access to portals for sharing information, engaging in tele-consultation with physicians when needed, and using patient wearables to connect health management to their other activities, such as eating, exercising, and visiting healthcare facilities. While the goal is to improve patient access to information and improve their ability to self-manage some of their health conditions, such as glucose levels, blood pressure, and cholesterol levels, one of the challenges that continues to remain is: How can the hospital ensure patient adherence to wellness behaviors, and what explicit or implicit incentives are needed to improve such adherence?

Use case 5 discussed in Q1 is one where an intermediary is used to help patients quit smoking. While the hospital seeks the interest of a patient to quit smoking to reduce future costs and improve their quality of life, the patients need intermediaries within the community to motivate them to quit smoking. Addressing future costs for a healthcare system or a future pain for a patient is by itself is a not a motivator to improve adherence to wellness behavior. Frequently interacting with the patient and providing counseling services where the patients are might help, along with other types of incentives, such as free patches they can try.

In summary, as we moved from use cases 2 to 5, what becomes apparent is that the focus of the theme around which digital leadership explorations are aligned changes. For example, when the focus of the theme is cost reduction, the digital strategies and innovations explored inside and outside the hospital tend to focus on this theme. However, if the theme is broader in focus to address the quality of life of the patient, then innovations explored move from those that influence costs to those that will improve a patient's capacity to manage their health condition within their ecosystem. Such patient capacity can influence health care costs over the long run. More importantly, such a broader focus alters the nature of capacity building for a hospital. Hospitals need to explore innovations that help them work with multiple agencies and actors within the patient ecosystem in support of patient care. As discussed earlier, while the focus of digital leadership within a hospital is to improve its capacity to address a patient's "sickness," digital leadership outside a hospital calls for capacity building of both the hospital and the patient as they leverage multiple actors in the patient ecosystem to address the patient's "health" – thus moving from "sick care" to "health care."

FUTURE RESEARCH DIRECTIONS

We used four use cases to discuss digital leadership of care delivery models within the patient ecosystem, which can be very complex. There is high degree of variability in the social determinants that influence a patient's capacity to self-manage their care, and this poses challenge for digital leadership. For example, for patients with very limited ecosystem support as discussed in use case 4 (support of homeless populations), the focus of innovations should not be primarily on improving patient capacity to self-manage their care but rather improving the capacity of organizations such as HOPE that supports such patients. In other words, the focus of digital leadership is to support an intermediary such as HOPE by enabling it to raise funds from philanthropic organizations or to develop mobile apps that improve patient adherence to non-clinical care services. This raises the question: Do these innovations represent digital leadership of a hospital or an intermediary? On the other hand, for patients who are in a supportive ecosystem, the innovation explorations can directly focus on dividends that will help patients gain access to their care conditions using wearables, apps, and text-based alerts, thus supporting the hospital's growth strategies around the theme of reducing costs. In summary, the proximity of an organization to its customer, direct or via an intermediary, may determine the focus of digital leadership.

Similarly, if we apply the principles of digital leadership to an educational domain, digital leadership is influenced by many educational determinants of students. For example, the disciplinary content of lower-lever undergraduate students is different from that of upper level under-graduate or graduate students. Similarly, the focus of knowledge shared (structured as opposed to less structured and experiential) can vary, and so does the cognitive depth (process thinking vis-à-vis critical thinking). These determinants influence not only where the innovations should focus, but also how these are aligned with the themes set by educational institutions. If the theme is student retention and supporting students in urban campuses with challenging environmental conditions, which innovations apply and to what level of students? In other words, digital leadership that seeks alignment with a theme is looking upstream for administrative guidance, while social or customer ecosystem determinants may need to assess the dividends from innovation explorations downstream – who is to adapt these innovations. If those who adopt innovations and develop themes work for the same organization, as in the case of a single hospital case, the goal of digital leadership is to improve the hospital's capacity to grow and improve its capacity. But, if those who need to adopt innovations and who establish themes operate in different ecosystem, then digital leadership used to support explorations has to reconcile the conflicts between upstream alignment and downstream dividends.

CONCLUSION

In conclusion, thought leadership of an organization in today's market dynamic needs to not only address specific changes as they surface proactively but also build organizational capacity to explore, learn, and adapt proactively to address future changes. Using leadership concepts and constructs from complexity theory and HeXie management theory, this chapter developed four guiding principles around which the exploration of innovations can be pursued. These are alignment of exploration around organizational themes; use of organizational dynamism and a culture of innovation to explore new ideas; use of transitiveness to assign dividends from explorations to support growth in the near term as well as over a longer term by capacity building; and use of adaptiveness to institutionalize explorations that support growth and competitiveness in the near term. Using several use cases, the digital leadership needed to transform hospitals using these principles was discussed, as they transition care from within the hospital to outside the hospital.

REFERENCES

Agarwal, R., Tanniru, M., & Wilemon, D. (1997). Assimilating Information Technology Innovations: Strategies and Moderating Influences. *IEEE Transactions on Engineering Management*, *44*(4), 347–358. doi:10.1109/17.649864

Aghena W., De Smet, A. & Weerda, K. (2015). Agility: It rhymes with stability, *McKinsey Quarterly*, Dec.

Anderson, P. (1999). Perspective: Complexity Theory and Organization Science, *Organization science* (10:3), pp. 216-232.

Axelrod, R., & Cohen, M. D. (1999). *Harnessing Complexity*. New York, NY: Basic Books.

Barasa, E. W., Molyneux, S., English, M., & Cleary, S. (2017). Hospitals as complex adaptive systems: A case study of factors influencing priority setting practices at the hospital level in Kenya. *Social Science & Medicine*, *174*(Feb), 104–112. doi:10.1016/j.socscimed.2016.12.026 PMID:28024239

Benbya, H., & McKelvey, B. (2006). Toward a complexity theory of information systems development. *Information Technology & People*, *19*(1), 12–34. doi:10.1108/09593840610649952

Bossert, O., Laartz, J., & Ramsey, T. J. (2014). Running your company at two speeds, *McKinsey Quarterly*, Dec.

Brown, S. L., & Eisenhardt, K. M. (1997). The Art of Continuous Change: Linking Complexity Theory and Time-Paced Evolution in Relentlessly Shifting Organizations. *Administrative Science Quarterly*, *42*(1), 1–34. doi:10.2307/2393807

Byrne, D. S. (1998). *Complexity theory and the social sciences: an introduction*. New York, NY: Routledge.

Cao, X., Zhang, X., & Xi, Y. (2011). Ambidextrous organization in harmony: A multi-case exploration of the value of HeXie management theory. *Chinese Management Studies*, *5*(2), 146–163. doi:10.1108/17506141111142799

Chan, S. (2018). What the heck is digital transformation- Finally a definition anyone can understand? *Open Consulting Group*, May.

Crabtree, B. F., & Miller, W. L. (1999). *Doing qualitative research*. Thousand Oaks, CA: Sage.

Davis, F. D. (1989). Perceived usefulness, perceived ease of use, and user acceptance of information technology. *Management Information Systems Quarterly, 13*(3), 319–340. doi:10.2307/249008

Dawson, A., Hirt, M. & Scanlan, J. (2016). The economic essentials of digital strategy, *McKinsey Quarterly*, March.

Dutton, J. E. (1993). Interpretations on Automatic: A Different View of Strategic Issue Diagnosis, *Journal of Management Studies* (30:3), pp. 339-357.

Eisenhardt, K. M. (1991). Better stories and better constructs: The case for rigor and comparative logic. *Academy of Management Review, 18*(3), 620–627. doi:10.5465/amr.1991.4279496

Eisenhardt, K. M., & Graebner, M. E. (2007). Theory building from cases: Opportunities and challenges. *Academy of Management Journal, 50*(1), 25–32. doi:10.5465/amj.2007.24160888

Ferber, J. (1999). *Multi-Agent Systems*. Harlow, UK: Addison-Wesley.

Fowler, B. L., Johns, J., Tanniru, M. R., Balijepally, V., Roumani, Y. F., Bobryk, D., & Mitchell, K. (2018). Engaging patients through Multi-Disciplinary Rounding -The case study at a Michigan hospital. *Journal of Hospital Administration, 7*(5), 17–26. doi:10.5430/jha.v7n5p17

Ginsberg, A. & Venkatraman, N. (1992). Investing in New Information Technology: The Role of Competitive Posture and Issue Diagnosis, *Strategic Management Journal* (13: S1), pp. 37-53.

Gioia, D., Schultz, M., & Corley, K. (2000). "Organizational identity, image, and adaptive instability," *Academy of Management Review*, 25: 63–81.

Greenhalgh, T., & Papoutsi, C. (2018). Studying complexity in health services research: Desperately seeking an overdue paradigm shift. *BMC Medicine, 16*(1), 95. doi:10.118612916-018-1089-4 PMID:29921272

Gronn, P. (2002). Distributed leadership as a unit of analysis, *Leadership Quarterly*, 13: 423–451.

Hadeel, A., & Sandhu, K. (2021). PLS Model Performance for Factors Influencing Students Acceptance of Elearning Analytics Recommender. *International Journal of Virtual and Personal Learning Environments* (IJVPLE): New Perspectives in Online Education: Environments, Approaches, Authoring Tools.

Hershey, P. (1985). *The Situational Leader* (4th ed.). Warner Books.

Holland, J. H. (1995). *Hidden order*. Reading, MA: Addison-Wesley Publishing.

Holland, J. H. (2014). *Complexity: a very short introduction*. Oxford, UK: Oxford University Press.

Horvat, A., & Filipovic, J. (2018). Service quality and maturity of health care organizations through the lens of complexity leadership theory. *Journal of Evaluation in Clinical Practice, 24*(1), 301–301. doi:10.1111/jep.12789 PMID:28714561

Ivancic, L., Vukšic, V., & Spremic, M. (2019). Mastering the Digital Transformation Process: Business Practices and Lessons Learned. *Technology Innovation Management Review, 9*(2), 36–50. doi:10.22215/timreview/1217

Johnson, S. (2001). *Emergence*. New York, NY: Scribner.

Kantur, D. Z., & Say, A. I. (2012). Organizational resilience: A conceptual integrative framework, *Journal of Management & Organizations, 18*(6), November.

Kernick, D. (2006). Wanted – new methodologies for health service research. Is complexity theory the answer? *Family Practice, 23*(3), 385–390. doi:10.1093/fampra/cml011 PMID:16597669

Lewin, A. Y. (1999). Application of Complexity Theory to Organization Science, *Organization Science* (10:3), pp. 215-215.

Lichtenstein, B. B., Uhl-Bien, M., Marion, R., Seers, A., James, D., & Schreiber, C. (2006). *Complexity leadership theory: An interactive perspective on leading in complex adaptive systems, Institute for the Study of Coherence & Emergence*. Lawrence Erlbaum Associates.

Litaker, D., Tomolo, A., Liberatore, V., Stange, K. C., & Aron, D. (2006). Using complexity theory to build interventions that improve health care delivery in primary care. *J Gen Intern Med*. 21(S2:S30).

Long, K. M., McDermott, F., & Meadows, G. N. (2018). Being pragmatic about healthcare complexity: Our experiences applying complexity theory and pragmatism to health services research. *BMC Medicine, 16*(1), 94. doi:10.118612916-018-1087-6 PMID:29921277

Lusch, R. F. & Nambisan, S. (2015). Service Innovation: A Service-Dominant Logic Perspective, *MIS Quarterly* (39:1), pp. 155-175.

Mahajan, A., Islam, S. D., Schwartz, M. R., & Cannesson, M. (2017). A hospital is not just a factory, but a complex adaptive system—Implications for perioperative care. *Anesthesia and Analgesia, 125*(1), 333–341. doi:10.1213/ANE.0000000000002144 PMID:28614127

Manson, S. M. (2001). Simplifying complexity: A review of complexity theory. *Geoforum, 32*(3), 405–414. doi:10.1016/S0016-7185(00)00035-X

Marion, R. (2008). Complexity theory for organizations and organizational leadership, *Complexity Leadership*, Part 1, 1-15.

Martin, C. (2018). Complex adaptive systems approaches in health care - A slow but real emergence? Journal of Evaluation in Clinical Practice. *Journal of Evaluation in Clinical Practice, 24*(1), 266–268. doi:10.1111/jep.12878 PMID:29589876

McDermott, F. (2014). Complexity theory, trans-disciplinary working and reflective practice. In Applying complexity theory. Bristol: Policy Press; 2014. pp. 181–98.

Plsek, P. E., & Greenhalgh, T. (2001). The challenge of complexity in health care. *BMJ (Clinical Research Ed.), 323*(7313), 625–628. doi:10.1136/bmj.323.7313.625 PMID:11557716

Plum, A., Tanniru, M., & Khuntia, J. (2019). An Innovation Platform for Diffusing Public Health Practices across a Global Network, *HICSS-2019*, Hawaii, January.

Pomffyova, M. (2018). Management Information Systems, *Infotech Open,* October. https://www.intechopen.com/books/management-of-information-systems

Ramey, C. (2019). The Boardroom Sage who was into good governance before it was cool, *Wall Street, Business Edition*, Nov. 15.

Rockart, J. (1979). Chief Executives Define Their Own Information Needs. In *Harvard Business Review* (pp. 81–92). March/April.

Rockart, J. (1982). The Changing Role of the Information Systems Executive: A Critical Success Factors Perspective. *Sloan Management Review*, *23*(1), 3–13.

Rockart, J., & Van Bullen, C. (1986). A Primer on Critical Success Factors. In J. Rockart, & C. Van Bullen (Eds.), *The Rise of Management Computing*. Homewood: Irwin.

Rogers, E. (1995). Diffusion of Innovations, 4th ed. The Free Press New York.

Sandhu, K. (2020). Digital Systems Innovation for Health Data Analytics. Handbook of Research on Optimizing Healthcare Management Techniques. Hershey, PA: IGI Global.

Schein, E. H. (2010). *Organizational Culture and Leadership*. John Wiley & Sons.

Schwandt, D. R. (2008). Individual and Collective Coevolution. In M. Uhl-Bien, & R. Marion (Eds.), *Complexity Leadership Part 1: Conceptual Foundations* (pp. 101–127). Charlotte, NC: Information Age.

Shipton, H., Armstrong, C., West, M. & Dawson, J. (2008). The Impact of Leadership and Quality Climate on Hospital Performance, *International Journal for Quality in Health Care* (20:6), pp. 439-445.

Siebel, T. M. (2017). Why digital transformation is now on the CEO's shoulders. *The McKinsey Quarterly*, *4*(3), 1–7.

Stacey, R. D. (1995). The Science of Complexity: An Alternative Perspective for Strategic Change Processes, *Strategic Management Journal* (16:6), pp. 477-495.

Tanniru, M. (2019). Engagement leading to empowerment - Digital Innovation Strategies for Patient Care Continuity [October.]. *J Hosp Management and Health Policy*, *3*, 28. doi:10.21037/jhmhp.2019.09.01

Tanniru, M., Jones, J., Kazziha, S., & Hornberger, M. (2019). Strategies to Address Care Transition - A Methodology and a Case Study. *Journal of Hospital Administration*, *8*(3), doi:10.5430/jha.v8n3p38

Tanniru. M., Khuntia, J., & Weiner, J. (2018). Hospital Leadership in Support of Digital Transformation, *Pacific Asia Journal of Association of Information Systems, (PAJAIS), 10*(3).

Tumbas, S., Berente, N., & Brocke, J. (2017). Three Types of Chief Digital Officers and Reasons that Organizations Adopt the Role, MIS Q Executive, Jan., 16(2).

Uhl-Bien, M., & Marion, R. (2008). *Complexity Leadership Part I: Conceptual Foundations*. Charlotte, NC: Information Age Publishing.

Uhl-Bien, M., Marion, R., & McKelvey, B. (2011). *Complexity Leadership Theory: Shifting Leadership from the Industrial Age to the Knowledge Era, Leadership, Gender, and Organization* (pp. 109–138). Springer.

Van de Ven, A. H., & Poole, M. S. (1995). Explaining Development and Change in Organizations, *Academy of Management Review* (20:3), pp. 510-540.

Venkatesh, V., & Davis, F. D. (2000). A theoretical extension of the technology acceptance model: Four longitudinal field studies. *Management Science*, *46*(2), 186–204. doi:10.1287/mnsc.46.2.186.11926

Ventres, W., & Frankel, R. (1996). Ethnography: A stepwise approach for primary care researchers. *Family Medicine*, *28*(1), 52–56. PMID:8720228

Wieland, H., Vargo, S., Akaka, M. S., & Barbeau, B. (2018). A Unifying Perspective for the Technological, Business Model, and Market Aspects of Innovation, *Handbook on Service Dominant Logic*. Sage. *Lusch and Vargo, Ch.*, *29*, 508–521.

Williams, D. R., Costa, M. V., Odunlami, A. O., & Mohammad, S. A. (2008). Moving Upstream: How Interventions that Address the Social Determinants of Health can Improve Health and Reduce Disparities, *J Public Health Management Practice*, November 14 (Suppl), S8-17.

Xi, Y., Liu, P., & Zhang, X. (2019). Living in the Present of the Future: In Memory of James G. March (1928–2018). *Management and Organization Review*, *15*(4), 895–899.

Yin, R. K. (1994). *Case Study Research: Design and Methods*. Newbury Park, CA: Sage.

Zhang, X., Fu, P., & Xi, Y. (2018). Promoting Management Education in China Through Developing Practice-Based Management Theories: An Interview with Practitioner-Scholar Youmin Xi. *Journal of Management Inquiry*, *27*(2), 212–223. doi:10.1177/1056492617708056

ADDITIONAL READING

Adner, R. (2017). Ecosystem as Structure: An Actionable Construct for Strategy. *Journal of Management*, *43*(1), 39–58. doi:10.1177/0149206316678451

D'Innocenzo, L., Mathieu, J. E., & Kukenberger, M. R. (2016). A Meta-Analysis of Different Forms of Shared Leadership–Team Performance Relations. *Journal of Management*, *42*(7), 1964–1991. doi:10.1177/0149206314525205

Dawson, A., Hirt, M. & and Scanlan, J. (2016). The economic essentials of digital strategy, *McKinsey Quarterly*, March

Elsbach, K. D., & Stigliani, I. (2018). Design Thinking and Organizational Culture: A Review and Framework for Future Research. *Journal of Management*, *44*(6), 2274–2306. doi:10.1177/0149206317744252

Fischer, T., Dietz, J., & Antonakis, J. (2017). Leadership Process Models: A Review and Synthesis. *Journal of Management*, *43*(6), 1726–1753. doi:10.1177/0149206316682830

Kohnke, O. (2017). It's Not Just About Technology: The People Side of Digitization, G. Oswald, M. Kleinemeier (eds.), Shaping the Digital Enterprise, © Springer International Publishing Switzerland

Swan, B. H. (2017). The leadership shift—From directing people to action to inspiring people to action. *Healthcare Management Forum*, *30*(5), 237–241. doi:10.1177/0840470417718194 PMID:28929848

Tremblay, K. (2017). Healthcare @ The Speed of Thought: A digital world needs successful transformative leaders, *Healthcare Management Forum*, Vol. 30(5) 246-251 ª2016

KEY TERMS AND DEFINITIONS

Care Delivery Models: Models used to describe diagnostic and/or treatment services used by providers to address the needs of patients or healthcare consumers inside or outside a hospital.

Coupling: Similar to adaptive leadership, it allows organizations to connect dividends to help improve organizational capacity and sustain growth.

Digital Health: Stands for the use of advanced technology in the healthcare delivery and the technologies can refer to devices or services provided by healthcare professionals.

Digital Leadership: Enabling leadership that explores innovations needed to support of digital transformation.

Digital Transformation: Organizational transformation in digital age, where market demands are driven by advances in information technologies.

Dividends: Values derived by an organization through innovation exploration and these range from those that reduce organizational barriers, support a culture of innovation, or help organization meet its performance objectives.

He Principle (HP): An adaptive mechanism arising from organizational members' initiative and self-determination to innovate.

HeXie Management Theory: An integration of Eastern and Western cultures and wisdom to address complex issues organizations face in uncertain environments.

HeXie Theme: A direction organization takes by translating "objective constraints" of a situation it faces in its environment.

Leadership Process: Processes used by leaders to address problems by working together and moving the organization toward a desired outcome.

Organizational Capacity Building: Building an organization's ability to address changes in its environment using agile processes, organizational structure, and governance.

Use Case Modeling: Illustrates how an individual customer service need or decision-making situation is supported by an organization with the use of people and processes.

Xie Principle (XP): Refers to rules and propositions used to reduce uncertainty through scientific design and optimization.

Chapter 2

Fostering Innovation and Value Creation Through Ecosystems:

Case of Digital Business Models and Digital Platforms

Mario Spremić
Faculty of Economics and Business, University of Zagreb, Croatia

Lucija Ivancic
https://orcid.org/0000-0003-0491-7230
Faculty of Economics and Business, University of Zagreb, Croatia

Vesna Bosilj Vukšić
Faculty of Economics and Business, University of Zagreb, Croatia

ABSTRACT

Ecosystems are gaining ever-increasing importance in digital business environments. New digital business models implemented using digital platforms heavily rely on ecosystem network. Thus, it is worthwhile investigating what role ecosystems play in the process of the digital transformation of companies. This chapter provides a theoretical background of the ecosystem research concerning digital trends, such as digital transformation, digital platforms, and digital service innovation. To deeper understand how ecosystem postulates are applied in companies, case study findings from two companies operating in the service and manufacturing sector are presented. Moreover, the ecosystem role is observed in selected companies both in the process of innovation generation (value creation), as well as in the implemented digital business model (value capture).

INTRODUCTION

During the last decade, massive improvements in information reach, computing, communication, and connectivity, have made digital technologies key emerging technologies that can fundamentally impact

DOI: 10.4018/978-1-7998-2799-3.ch002

the business environment, which includes the impact on services, processes, business models and whole industries. In a very short time, the term *digital* became very popular. It changed the usual vocabulary of information science from information technology (IT) to digital technologies; from IT strategy to digital strategy, also introducing what we now call *digital disruption* and *digital economy*. Another notable change of course happened in technological governance in organizations (Spremic, 2017). From mainly internally oriented IT governance mode, organizations shifted to an externally focused use of digital technologies. On the one hand, the IT initiatives have become more internally focused, mainly intending to align with the current business process. Digital technologies, on the other hand, have become externally oriented, thus enabling digital services proliferation and enhancement of customer experience. These changes of organizational focus came along with disrupting the current business model, changing the organizational culture and affecting the entire business ecosystem (Ivančić, Vukšić, & Spremić, 2019).

Furthermore, ecosystems role is obtaining ever-increasing importance in digital environments (Gartner, 2017). Social networks, and other sorts of digital platforms available for customers and partners, as well as knowledge and innovation harvesting settings that include academia, are all examples of ecosystems of the digital age.

Therefore, the shift from IT to digital technologies opened new research directions, namely on how ecosystems research and applications can be correlated with digital transformation, business model innovation, and digital business models in modern organizations. Furthermore, Scharmer and Kaeufer (2013) bring an extensive analysis of current economic trends and models in the world. In its most broaden sense; ecosystem economy can be comprehended as developing a business around the intersectoral collaboration between the participant of the system, thus having in mind collective awareness of the system and actors in the surroundings. As a result, innovated business models, mostly supported by digital technologies are emerging. It can be concluded from their analysis that economic systems will become more and more oriented towards the ecosystem model.

This chapter aims to shed some light onto this forthcoming phenomenon. The main objective of this chapter is to explore how digital transformation and digitally enhanced services are fostering innovation throughout ecosystems and digital platforms. In order to do so, the research objectives of this chapter are defined as follows.

RO1: *Give a brief literature overview on the ecosystem research and related trends.*
RO2: *Explore what role do ecosystem actors play in companies.* More specifically, RO2(a): *Explore what role do ecosystem actors play in innovation processes in companies.*, and; RO2(b): *Explore what role do ecosystem actors play in digital business model in companies.*

DIGITAL TRANSFORMATION AND SERVICE INNOVATION

Digital transformation is a holistic term for organizational changes induced by digital technologies. It includes some of the well-known concepts such as business process management methods, operational backbone of enterprise systems, and change management initiatives in the process of digitally transforming (Ivančić et al., 2019). However, a distinct goal of digital transformation in companies is to foster innovation and disruption of existing business models and services, enabled by digital technologies (Kane, Palmer, Nguyen Phillips, Kiron, & Buckley, 2016).

Different digital technologies such as mobile technologies, robotics, cloud computing, data analytics, and sensors – Internet of Things (IoT) are present nowadays on the market. They can extract information from physical goods (data on sensors about the condition of a physical device), disseminate it quickly (transferring data using mobile technologies), store it in the cloud, and analyze it instantly (using big data and advanced analytics). By doing so, these technologies are integrating products, services, and processes and making disrupting impact on established business models. Although digital technologies are pervasive, accessible, and even affordable, they have some unique and distinctive characteristics. Ross et al. (2016) stressed that these technologies are widely available and affordable to all companies, and many smart applications can be easily replicated or imitated, but they can't offer sustained competitive advantage. However, it is the integration enabled by the simultaneous use of different digital technologies, rather than the useful implementation of a single one, that constitutes the basis for competition in the digital economy. Nylén and Holmström (2015) and Yoo et. al (2012) stand on that point and argued that digital technologies expanded beyond internal dimensions, penetrating the company's product and service offering. They conclude that digital technologies have some unique features, such as:

1. digital technologies are *flexible, easy to use* and *reconfigure* – for example, entirely digitalized business model enable Tesla to change the car configuration in a single software update; Netflix is using flexible cloud computing services on a *pay-per-use* basis;
2. digital technologies have certain extent of *unpredictability* and *fluidity*, or a degree of *generativity*, which stands *for technologies overall capacity to produce un-prompted change, driven by large, varied and uncoordinated audiences* (Zittrain, 2006) – for example, Apple's new operating system will enable virtual reality and augmented reality possibilities which will be perfectly aligned with hardware capabilities; intensive use of 3D printers might disrupt many industries, the initial niche market of Airbnb was to provide service for meetings and events for which hotel space was sold out;
3. digital technologies constantly *evolve* toward higher *processing capacity*, lower cost, and dramatic *price-performance improvements*.

Key essence for successful digital transformation, according to Ross et. al (2016), is that digitally matured companies are trying to integrate existing business capabilities with new capabilities made possible by various digital technologies, which is quite hard to replicate. Many different digital technologies have distinctive features, but when used strategically and holistically, they can integrate various internal and external business processes and enable product and service innovation throughout the entire ecosystem.

Due to its unique features and accessibility, the focus of implementing digital technology is no more only to improve internal operations, but also to expand internal dimensions, reach customers and external partners, innovate services, integrate processes, disrupt markets and fundamentally change industries. Therefore, it is no surprise that digital transformation has become a high priority on leadership agendas. Recent researches (Bonnet, Ferraris, Westerman, & McAfee, 2012) illustrate that nearly 90% of business leaders in the U.S. and U.K. expect IT and digital technologies to make an increasing strategic contribution to their overall business in the forthcoming decade.

While many researchers and practitioners agree on the necessity to transform the business using distinctive features of digital technologies, it is not straightforward how to do that (Lucas, Agarwal, Clemons, El Sawy, & Weber, 2013; World Economic Forum, 2018). There is no clear recipe for successful digital transformation and there are many ways how to manage the process of digital transformation. An especially important question that remains is how to include external partners in the value creation process. Kane et al. (2016) concluded that digital transformation has reached a new stage in which digital technologies aren't limited to separate digital divisions, channels or functions, but are used across organizations to support corporate strategies that address industry opportunities. Similarly, Ross et. al (2016) suggest that integrated products, services, channels, and processes can address the disruption. Moreover, they agree that it is not a separate technology strategy that matters, rather unique digital strategies that incorporate the opportunities of digital technologies.

In order to be able to generate such innovations, diverse innovation management initiatives are set in organizations. The stages of innovation process include several phases: (i) idea generation (also called ideation); (ii) idea management and collection (iii) idea selection, and; (iii) innovation implementation.

Changes in a business model that occur as a result of the innovation implementation phase can range in the scope of their disruption. Business models can be innovated in a scale of operational and incremental optimizations, up to the radical changes of the business model that are disruptive for the organization itself or the whole industry (Foss & Saebi, 2018; Ibarra, Ganzarain, & Igartua, 2018).

Digital business models leverage the opportunities of digital technologies to provide new services to customers and add value to the company. Digital business models are arranged around digital platforms. Manufacturing industries are adding services into their once only a physical product portfolio, while the service industries are finding innovative ways of communicating with customers and including them into internal business processes. Regardless of the strategic approach, both the manufacturing and service industries will incorporate their new services into a digital platform model. According to Weill (2011), a digital business model has three distinguished components: a good quality *content* to provide to customers; a finely tuned *customer experience*, and; a digital platform as a media for the previous two which integrates internal resources with the new content provided for the external environment.

So, what is exactly a digital platform (Hossain & Heidemann Lassen, 2017)? Digital platforms are mobile, or desktop applications intended for external users (customers and partners), or internal users (employees) (Ruggieri, Savastano, Scalingi, Bala, & D'Ascenzo, 2018). Digital platforms can have numerous purposes (Hänninen, Smedlund, & Mitronen, 2018; Hossain & Heidemann Lassen, 2017; Korhonen et al., 2017). They can be set up as a communication platform with the aim of exchanging messages or exchanging ideas (Kane et al., 2016). They can incorporate functionalities of a crowdsourcing platform. Besides the crowdsourcing platforms collecting funding for new products or projects, platforms can crowdsource ideas. Idea crowdsourcing can range from internal idea management platforms in companies, up to the digital spaces where demand and offer are joined, and where freelancers such as artists and programmers offer services that can be delivered in a digital format. A digital platform can also serve as a complement to a physical product by providing additional services or information about the product through its lifecycle. Digital platform businesses such as Uber or Wolt have recognized the unoccupied market niches and started successful companies. Service industries are likewise finding new value in their companies and their data. As a result, data exchange platforms and data aggregators are appearing in the market (Visconti, Larocca, & Marconi, 2017), for instance in the telecom industry. Fintech companies, on the other hand, have started to occupy the existing niches and threaten to radically change the financial industry as we know it (Thomas Dapp, 2014).

Therefore, the classification of digital platforms with respect to their strategical incentive can be made as follows: (i) digital platforms supplementing physical products, and; (ii) digital platforms introducing new services. The latter can be further distinguished into (ii.i) digital platforms supplementing existing services (digitally transformed incumbent service industries), and; (ii.ii) digital platforms introducing (new) service product on the market (digitally born companies).

ECOSYSTEMS ROLE IN DIGITAL INNOVATION

The innovated and new business models that are arranged around the digital services platform are having in mind the inclusion of the ecosystem actors (Sebastian et al., 2017; Yablonsky, 2018). The recent Gartner report even claims that platform-based service innovations would be impossible to create without the partners from the ecosystem (Gartner, 2017). This is the reason why companies must arrange their strategies that leverage digital technologies having in mind value capture mechanisms both for the company and for the ecosystem (Pellikka & Ali-Vehmas, 2016).

Ecosystems are networks of employees, customers, suppliers, partners, the public sector and other stakeholders that together create value for the organization (Ivančić et al., 2019; Lusch & Nambisan, 2015). They also act as facilitators of the transformational efforts of companies towards establishing a digital business model. Hence, it is evident that ecosystems in the digital age are compounded by both technical and social elements (Winby & Mohrman, 2018).

In the context of this chapter, ecosystems' role can be divided according to two stages in the process of value generation: value creation, and value capture. Value can be created or found, but after that, it also must be captured in order for the company to gain new profit streams, i.e. a monetization strategy must be deployed. When it comes to digital environments, value is co-created and captured through actors' interactions in the ecosystem (Pellikka & Ali-Vehmas, 2016). According to Pellikka and Ali-Vehmas (2016), every company should ask itself several questions when developing an ecosystem-oriented strategy. Firstly, the company should analyze the impact it is having on the ecosystem performance, i.e. how partners' performance can be enhanced. Secondly, the company should investigate existing collaboration models in the industry or similar to the company's innovation perspective. This could, of course, prove to be challenging since such a collaboration network can be kept hidden as strategic leverage. Thirdly, the company should reevaluate value systems in place, anticipate and propose how they can be altered in the future.

Although ecosystem actors have always been to some extent, a part of the organizational life, ecosystem importance has only recently begun to be more significantly addressed. With the proliferation of numerous digital technologies and the exponential growth of available knowledge, companies can no longer rely only on their internal resources to effectively cope with industry trends and market pressures. As a result, companies should recognize that innovation ecosystems are vital for progress and competitive advantage (Barrett, Oborn, & Orlikowski, 2016).

The first direction of the research on the ecosystems is innovation ecosystems or communities with the aim of external knowledge sourcing which would correspond to the stage of value creation. In order to fulfill digital innovation capabilities, companies reach out to the ecosystem in search of new ideas. OECD differentiates several sources of external knowledge. These include suppliers; clients; competitors; consultants; commercial labs; universities and research institutes, and; joint innovation projects (Roszkowska, 2017). In addition to these, customers are also becoming an important source of knowledge in the ecosystem. Besides customer journeys, customers' behavior and habits are taken into consideration when developing new services and redesigning existing ones (Ibarra et al., 2018). On top of that, with process digitization, customers play a more active role in organizational processes that were once intended to be strictly internal, such as claims processing in the insurance (Ivančić et al., 2019).

Collaboration with the academic community can develop and close valuable projects. On the other hand, the approach more in line with the economies of scale is also very popular. Companies participate in innovation project contest where they pose a challenge or a problem to a wider audience (usually students) seeking for innovative and fresh solutions from younger generations, which results in an abundance of project proposals. These projects do not have to be implemented completely. Certain project parts can be also recognized as beneficial when reevaluated and adapted by the company. Similarly, companies organize their own contests offering appealing prizes in financial form and/or employment.

Secondly, it can be noticed that ecosystems role stretches beyond the knowledge harvesting and innovation generation, i.e. it reaches out to the stage of value capture. Actors from the ecosystems also play an important role in new digital business models. These ecosystems are then referred to as *digital platform-based ecosystems* and are inherent to the organizations that are turning their business models into mainly digital so they can be called *platform organizations* (Helfat & Raubitschek, 2018; Saadatmand, Lindgren, & Schultze, 2019). Yablonsky (2018) notices that platform ecosystems are platform networks that are characterized by interaction and collaboration. In their study on business models based on digital platforms, Ruggieri et al. (2018) bring evidence for common features of such business models: importance of the ecosystem; network effects; business scalability; revelation of new sources of value; dematerialization of digital platforms, and; intermediating role between supply and demand.

Digital platforms for idea harvesting are incorporating ecosystem's role in both stages - value creation and value capture. Hossain and Heidemann Lassen (2017) identify several platforms for idea and technology soliciting, and knowledge transfer: problem-solving platforms; ideation platforms; co-creation platforms; online marketplaces; public crowdsourcing platforms; collective intelligence, and; freelance and microtask platforms. These platforms differ in the number of participants, types of involved ecosystems actors, the object of trading, and payment models. However, they have a common feature - reaching out to the ecosystem to expand digital innovation capabilities and find new income streams.

Ecosystem orientation in the platform economy is very well envisioned by the two-sided nature of digital platforms. The idea behind a business model of a two-sided platform is joining demand and supply. Many of the startups see themselves as a bridge for joining producers and consumers (Korhonen et al., 2017). This is done by actively bringing together two groups of users (Ritter & Lettl, 2018) - ones offering a solution, i.e. creating content, and the other ones that are seeking the solution, while the company providing the platform is the supporter in the shadow. Typical examples of these platforms are Uber, Airbnb, and Upwork. They clearly emphasize how value can be created by one side of the ecosystem and captured thought the interactions between two ecosystem parties on the platform. These platform organizations are not only benefiting from connections with the ecosystem, rather they could not even exist set apart from the ecosystem.

If observing socio-economic trends, it can be inferred that today's digital economy is actually a platform economy, and if related to increasing ecosystem orientation of the society - an ecosystem economy (Gartner, 2017; Hänninen et al., 2018; Korhonen et al., 2017; Scharmer & Kaeufer, 2013). While observing the social implications of digital communities, Barrett et al. (2016) found out that in a digital environment value is created through a combination of strategy, digital platform, and stakeholder engagement. They present a case of a digital platform that brings benefits for several parties included in the ecosystem. Benefits include not only financial value, but also reputational and platform value, and ethical and service value. They also argue that value in such digital communities embodies complex network ties of a wide range of ecosystem stakeholders.

METHODOLOGY

Case Study Methodology

The research was carried out using qualitative methods from multiple sources since cases are suitable for illuminating the phenomenology of an issue (Eisenhardt, 1989). In this research, it implies that the cases represent a new type of digital service innovation through the ecosystem and digital platforms integration. Two comprehensive case studies were conducted with the aim of exploring how ecosystem postulates are applied in companies. The explorative case study design presented in Yin (1994) was employed. It is suitable for answering the questions of how or why a certain phenomenon occurs, and for obtaining first-hand and in-depth understanding (Yin, 1994). Many previous authors have appraised the case study as a methodological approach in business and information science since it enriches the theory and brings valuable intrinsic insights of the organizational setting (Benbasat, Goldstein, & Mead, 1987). Accordingly, the case study as a methodology has been extensively used in information science research (see for instance: Sandhu, 2008; Sebastian, Ross, et al., 2017; Suša Vugec et al., 2019). Hence, a case study design was chosen as a suitable methodological approach for addressing the research objectives of this study, since it aims at gathering first-hand insights into how ecosystem integration is used in companies in two stages of value generation.

Purposeful Selection

An important factor for the case selection was researchers' access to the information since ecosystem ties and use of digital technologies are strategic issues of the competitive advantage that companies are not keen to reveal. Cases were selected from two different industries. On top of that, chosen companies are well-established market players and hence act as representatives of their industry. One of the companies is chosen from the service sector (telecommunications), and the other one from the manufacturing sector, in order to be able to inspect differences and similarities between the sectors regarding the objectives of the research. Although the service sector can find it easier to introduce digitally innovated services, it was noted that manufacturing industry is also introducing service-based business models, along with the traditional product-based ones (Bosilj Vukšić, Ivančić, & Suša Vugec, 2018; Liere-Netheler, Packmohr, & Vogelsang, 2018). Therefore, the cases were chosen according to the following criteria: (i) implementing digital service innovation through the ecosystem and digital platforms integration; (ii) being representatives of service and manufacturing sector, and (iii) researchers' access to the informants.

Data Collection

Firstly, in-depth interviews were conducted with top-level managers in charge of digital transformation in companies (Chief Executive Officer or Chief Digital Officer). Interview protocol semi-structured questions and prompts were employed to ensure reliability (Yin, 1994). The questions were based on the work of Kane and coauthors (Kane et al., 2016), adapted to suit the research objectives. Questions covered a wide range of business areas, such as digital transformation, innovation management, business process management, IT management, ongoing projects, and strategic plans. Secondly, we scanned the Internet and official web pages of companies for information about their digital innovation projects and collaboration with stakeholders. Thirdly, some of the information that was considered to be relevant for the analysis was gathered from the interviewees by e-mail at a later date.

Data Analysis

Since the research on ecosystems topic is still in its infancy, this paper employs explorative case studies to address the research objectives expressed in the Introduction. We followed analysis coding and case study protocol outlined by Eisenhardt (1989), including the steps of within-case analysis, and cross-case analysis of patterns. First, the interview recordings were transcribed. After that, transcripts and case information have been analyzed to find and mark examples of ecosystem integration using Word as a tool. Then, examples of ecosystems integration were rearranged into case study stories with the purpose of presenting the material coherently. In the end, a cross-case analysis was conducted in order to find patterns of ecosystem integration. Employed coding and data analysis protocol is common for researches using interviews as a data-gathering method (e.g., Sebastian, Ross, et al., 2017).

CASES

Case Study A: Telecom Industry

The first case study company (hereinafter company A) presented in this chapter is a telecommunications provider, founded in the nineties. It is a part of an international European company. Company A has over 3,500 employees, while the parent company counts for more than 200,000 employees in total. Like the entire telecom industry, company A is experiencing change necessity from legacy business models due to declining yields in the core telecom business. The company is seeking new business models and revenue streams. In this process, business model innovation results in entering niches that were previously not inherent to the telecom business. Company A is also an attractive partner in the ecosystem because of two reasons. Firstly, it has access to a large number of users, and secondly, it can gather big data about customers' behavior that is valuable for other companies to analyze.

Customers are very important stakeholders in telecom ecosystems, so it is no surprise that surveyed company A finds them as one of the key partners for value creation through business model innovation. According to the surveyed managers, *the collaboration with partners is essential for digital transformation.* As discussed in previous sections, simultaneous and integrative use of different digital technologies in a holistic and strategically focused way enabled this company to track the groups of mobile users, capture valuable anonymous data about their behavior (current location, movement, behavior, interests, purchasing habits, expenditures), store and analyze it internally and offer personalized services through business model innovation. Examples of customer induced business model innovation include changing the opening hours of brick-and-mortar stores according to individual customers' preferences (working habits/patterns) or innovating business model of B2B partners (tourist agencies, event organizers, banks, retail stores, real estate market tracking agencies).

As surveyed executive pointed out *data gathered from the ecosystem facilitates innovations and digital business models introduction,* company A was able to develop many useful services such as TO GO application which provides individual customers with location-based marketing message to their mobile devices when they are near the retail store they are interesting to visit or shop. Messages included information about products offered, shopping opportunities, sales, discount coupons, etc. However, it turned out that mobile users do not want to receive such text messages, but B2B partners were delighted with the service, which helped them develop their business and focus their marketing efforts. To be able to offer B2B partner valid anonymous data sets about customer location and preferences, company needed to integrate back-end transaction systems (CRM) with service innovation, or as Ross et. al (2016) argued: *integrate existing business capabilities with new capabilities made possible by various digital technologies*. In that way, company offer innovative service to B2B partners by selling them valuable data sets (for example, data sets sold to tourist offices), which enabled them to operate more efficiently (for example, where to place ATMs in crowdie tourist centers, or in what language and where to put a directions for different groups of tourists).

B2C segment of the ecosystem was covered by aligning social networking technologies in the telecom business model, thus enabling individual customers to communicate with the company through social media channels. In that way, the company responded to individual users' complaints about long waiting times to talk to call center operators and crowded retail stores. On the other hand, the older population is still using traditional communication channels to the full, so their habits limited business model innovation that aims at completely replacing the traditional channels of communication with social network channels.

One of the recent developments and service innovation include smart city infrastructure, which enables B2B partners to innovate their business models. By utilizing distinctive features of various digital technologies (mobile, IoT, cloud, data analytics), company A offered IoT-related innovations and a wide range of new digital services to B2B and B2G partners, mainly in the smart city environment. Recent examples include trash can sensors in Dubrovnik where sensors are tracking when the trash cans will be full and alerting utility workers, IoT-based smart parking solutions in Dubrovnik (first in the world), where IoT sensors are embedded in every parking lot, which enables parking lot providers many innovative and efficient services, ranging from internal operational efficiency and smart applications which show empty parking lots and help manage traffic, especially in peak times, or entire administration for city of Rijeka.

In that way, by holistic, strategically focused and simultaneous use of different digital technologies, company is innovating its own business model and offering innovating services for various partners through entire ecosystem (examples from B2C, B2B, B2G segments), or as company executive stressed that *collaboration with suppliers, customers, universities and high schools is important. Their ideas, suggestions, wishes and demands induce innovations which result in new business models introduction.*

The company is using digital technologies to enhance operational efficiency and improve internal business processes as well. An example of such practice is the digitalization of student contracts. Approximately 1000 students work for the company on the student contract deal. Incentives for digitalization were external (students were asking for the digitized and simplified process) and internal (to increase efficiency and automate the "low value" process). The project resulted in a very simple solution, which is using common tools such as Word and Excel and digital signature. But, major changes happened in the internal procedure and organizational change: if there is no complaint or reported problem with a student contract through newly established internal digital service in three days, it is considered valid. It is then automatically forwarded to the business process owner to digitally sign the contract and forwarded to an external supervising public authority (public student organization). In that way, employees are motivated to be more proactive and autonomous in solving problems and taking responsibility.

Digital transformation and IT transformation in Company A are occurring in parallel, thus providing improvements in operational efficiency and external reach throughout the entire ecosystem. To be able to engage customers and become more agile, the company opted for the "one CRM for all" solution, instead of a large number of old, legacy and inflexible middleware systems. The new backbone system will be modular, flexible and cover all customer touches (especially digital technologies such as chat-bots and similar). Small but agile teams are participating in the development of such systems, and all groups of customers and clients are included. *By ensuring that ecosystem representatives are included in the development of key systems and services, the company fosters business model innovation and value creation through the entire ecosystem.* As learned from the literature review, the ability of the company to simultaneously, but holistically and strategically use various digital technologies through an ecosystem is a must for successful digital transformation. Besides, big data and data analytics are the keys to business development and every-day work while introducing biometrics into services and current solutions is considered. Finally, the existing technology enables new business models introduction (e.g. offering of banking services, insurance services) thus resulting in the disruption of other industries on the market.

Case Study B: Manufacturing Industry

A second case study in this chapter is developed around the company (hereinafter company B) which is a heavy manufacturing company. Its production is mostly designated for export. Also, company B is a global market player in its niche. It was founded in the seventies and has over 300 employees. The company has been investing in robotization for several years. Automatization of production lines is the main motivator for constant staff education, employment of young and competitive engineers, and collaboration on academic projects. Recently, they started developing a digital platform that would connect their partners and suppliers in a digital ecosystem. The platform will more bring together more closely actors in the ecosystem, include partners in the production process, and shorten the supply chain time from order, through production, to the delivery.

Company B has a unique business model in a way that every product needs to be adjusted or adapted to specific clients' needs. According to CEO of company B, *ecosystem (clients) induces innovations and digital business models introduction.* Specific customer request in a niche market makes the entire process starting from product design, planning, and construction to production very long and complicated. On the other hand, clients expect efficiency, which implies that products must be delivered strictly on time or, even ahead of schedule, or in a shortest possible time. In a case of exceeding deadlines, losses are extremely high and can expose the company to several risks, namely reputation risk and financial risk. Therefore, one of the executives' priority objectives is to optimize and digitalize the production process in order to respond to the client's time requests and enable efficient time to market. This means that the incentive for change and product or process innovation is coming from the customers' side. Additionally, as a manufacturer, company B needs to foster operational efficiency, while at the same time, being able to demonstrate both process- and product-based innovation.

A change requirements coming from customers and the necessity to adjust the final product to the specific need of every single client has induced company B to shift from mass production mode to more flexible and agile, and quick and efficient. The shift to more personalized, agile and quick production mode makes executives think of process innovation. As noted previously, specific customer needs make the entire process, starting from product design, planning and construction to production, very long and complicated, and process innovation seems to be more than welcomed. To do so, company B engages as many employees they can, from managers, technicians, and engineers to communicate and work together on how to innovate every step of the complex manufacturing process. By introducing process innovations, technicians and engineers communicate directly with CNC machines, thus managing the production process and facilitating product-based innovation at the same time. These digitally enhanced process innovations are enabling them to produce more (efficiency increase of 30%) with the same number of employees, which boosts operational efficiency. In that capacity, constant demands for process optimization and efficiency lead to innovations.

It is not just the clients who are pushing the company to innovate, but *the suppliers too are crucial partners in creating value throughout the ecosystem.* Company is developing the software to keep track of raw materials and product stocks. This software will also automatically start the procurement process when supplies are under the prescribed limit. The software will start the procurement process by sending suppliers a request, and purchase and delivery will be automatically managed in the case of regular suppliers.

For company B applies: *buyers as key participants of all phases of the products' lifecycle.* Thus, the digital platform under development will enable buyers to order standardized products and parametrize specific parts to suit their needs. In that way, buyers will also be able to launch the production process directly over the digital platform. Additionally, the digital platform will be able to recommend a technical specification of a product that is most similar to buyers' needs. The company is planning to use the digital platform to offer post-sale services such as maintenance, non-stop communication with buyers over the modern communication channels and quick reaction to their questions, efficient diagnostics and failure elimination. In that capacity, buyers and suppliers will be engaged in the ecosystem through digital platform services aiming to support the entire Product Life Management (PLM) cycle.

As noted by the CEO of company B: *ecosystem induces quality control digitalization.* Manufacturing process heavily relies on efficient quality assurance and quality control. Low level of errors, low or no waste of raw materials, and low numbers in product complaints and recalls, affect operational efficiency and company's key performance indicators (KPI's). The main objective is to manage every step of subsequent activities in order to prevent unnecessary waste of raw materials and facilitate quick but efficient final product assembly. This is achieved by an enhanced use of digital technologies in quality control, including sensors, computer vision, and machine learning, sophisticated software able to follow quality changes in an organizational culture like 360° system where every employee evaluates its subordinated, superiors and colleagues, and surveying and interviewing employees about quality control process and management. The company's objective in this area is also to introduce lean production in the next two years.

Company B is using a specific business model which is predominantly driven by specific clients' needs, which is also a trigger for learning and innovation. Turbulent environment and changing requirements from customers make executives, technicians, and engineers constantly looking for new solutions with the same infrastructure (machines, robots, plants). For example, if producing a product using *n* machines, *m* robots, and other resources, and that final product is no longer competitive, a company should have enough organizational knowledge to quickly re-organize the production process and re-schedule machines, robots and other resources to something else. *It is very expensive if you can't reconstruct and re-plan the entire production process by yourself, or if you do not have employees capable of making such a sophisticated adjustment towards agile and flexible processes*. It is for sure that human capital capable to do is the key for such a successful transformation, but is also the subtle changes in an organizational culture that are motivating employees to experiment, to make a mistake and learn from that experience to be able to come up with some innovation or big improvement.

Modern technologies introduction and professional competences development are the keys to success. The collaboration with educational institutions is important. Clients expect the company to use cutting edge technologies backed up by superb organizational and leadership frameworks. In that capacity, the company is extensively collaborating with educational institutions in order to benefit from modern concepts new technologies that can be used in designing and production. As well trained employees are the company's key strengths it is important to ensure lifelong learning and various incentives for continuous education (internally and externally organized). Therefore, company B is including educational institutions to the ecosystem to align educational programs with the industry needs. This approach ensures a sufficient number of competent experts that can foster innovation and contribute to business development. In that sense, executives are trying to manage different generations of employees. Younger ones are faster to adopt the changes and are more inclined to learn. As salaries are not the only motivation for work, the company is attracting top employees by offering them opportunities for professional and personal development, incentives to grow, learn and acquire new knowledge and distinctive organizational culture. As highly motivated experts in the field and employees are striving for new knowledge, they are eager to implement their ideas, which will result in process, product and business model innovation.

Top managers of company B agree that *operational efficiency is a top priority for every single company, especially those competing in operations and manufacturing*". Several projects of IT 4.0 implementation and digital platform development were launched aiming at achieving the ultimate goal to become *"a project-oriented organization driven by technology*. They ended up using modern digital technologies such as CNC machines, sensors, cameras, but also sophisticated software, which is able to monitor quality requirements in order to produce flexible and even customized products. Moreover, using CNC machines equipped with digital technologies eliminated the construction and product preparation phase, enabling them to meet delivery deadlines, while sustaining the requested quality assurance level. A final goal is set to introduce 5.0 technologies that will enable not only the quality control but also an autonomous reaction in case of errors in the production process. Three ongoing projects are related to (1) Supply Chain Management (SCM) system development, (2) ERP system upgrade with functionalities for quality control analytics and reporting and (3) development of communication channels with business partners, both suppliers, and buyers.

SOLUTIONS AND RECOMMENDATIONS

Case studies present practice examples of companies' ecosystem ties. However, certain issues can be addressed to more effectively manage interactions with the ecosystem. Both case companies rely on ideas coming from students, younger employees and supported by constant education of staff. However, a more systematic approach can be employed. Companies can consider establishing an innovation management system that would organize innovation generation, innovation collection, innovation evaluation, and innovation implementation in a systematic and regular manner. This can be achieved with the innovation management board which would include employees from various organizational departments, as well as the company's partners.

Company B is operating in a relatively small industrial niche. Because of that, it is constantly in contact with its partners, both suppliers and customers, and familiar with their needs, requests, and opinions. Company A, on the other hand, has many telecom services users. This presents a potential risk of overlooking customers' opinions about digital innovations. Hence, it is highly suggested that companies conduct thorough market analysis which includes customers' surveying. This procedure mitigates the risk of lost money and time invested in implementing innovations that are in the end discarded due to poor customers' acceptance.

Finally, implementing business model innovations on a regular basis can be daunting for the employees. On top of that, digital innovations come with organizational changes and changes of organizational culture. Establishing change management practices can help companies cope with changes in digital transformation.

CASE STUDY DISCUSSION

Apart from smart and strategically focused use of digital technology, case study examples showed that newly established digital platform(s) enable company A to make their key stakeholders more proactive and engaging. Whether they were B2B partners in TO GO app, individual customers in social media communications, employees and governmental bodies in digitalization projects, B2G partners in IoT or smart city projects, the digital platform enables them to create value throughout the entire ecosystem. Moreover, this means that the digital platform is a value facilitator, which enables different partners from the same ecosystem to use or make innovative services, which are creating additional value for others. For example, when an individual customer is using innovative, IoT-based, smart parking service, various ecosystem partners are exploiting some benefits, whether they are city authorities, which are capturing value by better use of infrastructure and efficient traffic management (especially in peak times, which enable them proactive planning), B2B partner like tourist agency or local tourist office, which are able to offer personalized services, or local business, which have more incentives to offer location-based marketing. In that capacity, ecosystem partners have the incentive to join digital platforms and engage in capturing and creating value. Company A, which is the facilitator of the digital platform and the entire ecosystem, is exploiting many benefits by offering ecosystem partners innovative services on a robust and efficient digital platform. But, besides these obvious value capturing incentives coming from the external environment, company A needed to make many internal changes, especially, subtle improvements in organizational culture. For example, they engaged their employees to be more proactive in business process change (student contracts digitalization project), they engage and even educate their customers (communication over social media channels vs retail shops), they intentionally create collaboration opportunities and foster agile approach (CRM project), and they are recruiting top talents.

The examples of activities, projects and the organizational culture of company B clearly show how a heavy manufacturing company is adapting its business in accordance with the requirements of the digital age. It is noticeable that the company's not only progress but also survival in the market is dependent and relying on the innovations induced by the ecosystem. If it weren't for the product and process innovations, company B could have not been able to retain a leading position in the global market. These innovations also bring the buyers a sense of business security, since they know that company B is able to quickly deliver specifically designed and adapted products. Besides the buyers' requests and the industry specifics that are forcing the company to constantly innovate, important pushes for digital innovations are coming from diverse knowledge sources in the ecosystem. These sources include younger generations of employees that strive to bring added value for the company and to find their place in the organization. They are willing to learn and don't mind questioning the old ways of doing work, hence often propose innovating ideas. Other sources for value creation include educational institutions and academia; their ideas, experts, collaboration projects, and educations they provide to the company's staff. Along the use of robotics in the production lines that is inherent to the 4.0 manufacturing, the company is experimenting with other digital technologies. This is, for instance, visible in technologies deployed for quality control and quality assurance, new supplies management system, and digital platform for the inclusion of the partners in the production process. Not only that the ecosystems' incentives for business change and process and product, innovations are coming from the ecosystem but the implemented innovations are having repercussions for the entire ecosystem. New software for tracking suppliers and automatizing the procurement process will boost the efficiency of the suppliers. With the development of the digital platform that is in progress, the company's buyers will have direct access to the company's internal processes, which makes them no longer external stakeholders, but partners in the organizational ecosystem.

Table 1. summarizes the most important case study findings according to the two stages in the value generation process; value creation, and value capture. As visible, there are specific ecosystem postulates that are mutual for both service and manufacturing companies. On top of that, even the manufacturing company B is introducing service-oriented models (Product Lifecycle Management software), which is in line with the increasing service orientation of companies happening due to digital transformation (Lusch & Nambisan, 2015). However, if looking at case study examples, it can be noticed that there are distinctive ecosystem and platform characteristics inherent for each company. This can be explained by industry differences. For instance, Company A is a telecom, in which nature is the provision of infrastructure on which other companies can develop its services. Also, its ecosystem includes B2G, B2B, and B2C segment, while company B is operating in the B2B segment. Manufacturing company B on

Table 1. Summary of case study findings

	Company A	Company B
	Model Innovations Induced by the Ecosystem (Value Creation)	
Clients/users induce innovations and digital business models introduction	• Utilizing anonymized mobile users' data for changing the opening hours of retail shops. • IoT infrastructure and support.	• Optimization and digitization of the process from design to production. • Quality Assurance and Quality Control digitization.
Collaboration with universities and educational institutions is important	• Digitization of student contracts. • Students as a source for idea generation.	• Experts' education. Knowledge source of technological trends and innovations. • Aligning the university programs with industry needs.
	Integration of Digital Platform with the Ecosystem (Value Capture)	
A digital platform creates benefits for the ecosystem actors	• IoT infrastructure. • Utilizing anonymized mobile users' data for optimizing partners' business. • TO GO application.	• Automatized procurement process. • Product Lifecycle Management software.
Digital platform functionalities must be aligned with clients'/users' needs.	• TO GO application. • Social network vs. retail shops.	• Inclusion of clients in the production process.

Source: Authors' work

the other hand is more focused on process optimization. However, new business models are emerging when integrating employing platforms' model with process digitization. Therefore, each company should analyze its ecosystem and available digital technologies to come up with innovations suitable for capturing new value. Recent research from Pellikka and Ali-Vehmas (2016) corroborates such industry and company-specific strategic considerations. In the end, although some innovations might seem brilliant, service quality and usefulness should be discussed with customers before the innovation implementation (Parasuraman, Zeithaml, & Berry, 1985), as learned from the examples of TO GO application and favoring retail shops over digital communication channels.

We can say that sampled companies have digitally grown and adapted to the new digital environment, not just because of smart, integrative and strategically focused use of different digital technologies, but also due to subtle service innovations, digital platforms which enable value-added activities throughout the entire ecosystem and, moreover, by distinctive organizational culture, which is similar to studies of Kane and coauthors (Kane et al., 2016). Digital platforms enable value creation throughout the entire ecosystem, and the ecosystem enables companies to create effective organizational culture (experimenting, collaborative, agile, and innovative).

LIMITATIONS AND FUTURE RESEARCH DIRECTIONS

This chapter is in line with the reasoning of the ecosystem economy (Scharmer & Kaeufer, 2013). We have shown how ecosystems are used in the stages of value creation and value capture in two companies that are in the process of digital transformation. The limitations of the findings presented in this chapter include the analysis of a limited number of two companies and research that is exploratory in nature. Nevertheless, companies are reputable representatives of their industry. Moreover, the service and manufacturing industry have been addressed, which brings compelling conclusions that are in good agreement with the servitization trend emerging as a result of the digital transformation of business (Lusch & Nambisan, 2015). Future researches could be devoted to quantifying ecosystem's role in companies in survey design.

According to the seminal papers of the innovation thinkers' cornerstone, analysis of the network "permits understanding the social structure as it channels the process of (innovation) diffusion" (Rogers, 2003). Accordingly, here presented reasoning on the organizational ecosystems can be translated into a network analysis approach. Therefore, future research attempts can aim at investigating particular actor interactions in the network and provide a classification of interactions based on the interaction strength and impact on the ecosystem performance. On top of that, individual adoption of innovations by customers and employees can be further discussed, possibly in quantitative studies employing the Technology Acceptance Model (Davis, 1989).

CONCLUSION

Ecosystems are organizational networks that together co-create value for their ecosystem actors. The research results from the empirical investigation revealed that ecosystem role is observed in selected companies both in the process of innovation generation, as well as in the implemented digital business models. Besides, both companies emphasized the importance of digital platforms in their business ecosystems. To present our findings, we firstly elaborated on the model innovations induced by the ecosystem and next examined how digital platforms become the new hub for ecosystem integration. Lastly, the most important characteristics of digital platforms were reported. The theoretical part of the chapter brings the systematization of ecosystem research in business and information science and its relation to digital trends. Accordingly, this research contributes to the body of literature on ecosystems and its relation with platform models which is becoming a compelling topic in information systems science.

Findings from our companies revealed that ecosystems play a crucial role both in the stage of value creation, as well as value capture. In this way, the ecosystem acts as a digital transformation capability, enabling optimal utilization of digital technologies with the aim of innovating the past business models. Thus, the digital business model in our companies was implemented through digital platforms, which allowed for new digital services and new income streams for the companies and found new value in the ecosystem. Hence, practical implications of this research steam from the systematization of ecosystems' role to examples from presented companies, both of which can be valuable as inspiration and idea source for companies to start revaluating value in their ecosystem.

ACKNOWLEDGMENT

This research has been fully funded by the University of Zagreb under the project *Digital transformation in Croatian Companies.*

REFERENCES

Barrett, M., Oborn, E., & Orlikowski, W. (2016). Creating Value in Online Communities: The Sociomaterial Configuring of Strategy, Platform, and Stakeholder Engagement. *Information Systems Research, 27*(4), 704–723. doi:10.1287/isre.2016.0648

Benbasat, I., Goldstein, D. K., & Mead, M. (1987). Strategy in Studies of Information Systems. *Management Information Systems Quarterly, 11*(3), 369–386. doi:10.2307/248684

Bonnet, D., Ferraris, P., Westerman, G., & McAfee, A. (2012). Talking 'bout a Revolution. *Digital Transformation Review, 2*(1), 17–33.

Bosilj Vukšić, V., Ivančić, L., & Suša Vugec, D. (2018). A Preliminary Literature Review of Digital Transformation Case Studies. In *ICMIT 2018: 20th International Conference on Managing Information Technology* (pp. 737–742). Retrieved from https://waset.org/Publication/a-preliminary-literature-review-of-digital-transformation-case-studies-/10009516

Dapp, T. (2014). Fintech – The digital (r)evolution in the financial sector. *Current Issues: Digital Economy and Structural Change.* Retrieved from www.dbresearch.com

Davis, F. D. (1989). Perceived Usefulness, Perceived Ease of Use, and User Acceptance of Information Technology. *Management Information Systems Quarterly, 13*(3), 319–340. doi:10.2307/249008

Eisenhardt, K. M. (1989). Building Theories from Case Study Research. *Academy of Management Review, 14*(4), 532–550. doi:10.5465/amr.1989.4308385

Foss, N. J., & Saebi, T. (2018). Business models and business model innovation: Between wicked and paradigmatic problems. *Long Range Planning*, *51*(1), 9–21. doi:10.1016/j.lrp.2017.07.006

Gartner. (2017). *Capitalizing on Your Business Ecosystems Economy: A Gartner Trend Insight Report*. Retrieved from https://www.gartner.com/doc/3759164/capitalizing-business-ecosystems-economy-gartner

Hänninen, M., Smedlund, A., & Mitronen, L. (2018). Digitalization in retailing: Multi-sided platforms as drivers of industry transformation. *Baltic Journal of Management*, *13*(2), 152–168. doi:10.1108/BJM-04-2017-0109

Helfat, C. E., & Raubitschek, R. S. (2018). Dynamic and integrative capabilities for profiting from innovation in digital platform-based ecosystems. *Research Policy*, *47*(8), 1391–1399. doi:10.1016/j.respol.2018.01.019

Hossain, M., & Heidemann Lassen, A. (2017). Q&A. How Do Digital Platforms for Ideas, Technologies, and Knowledge Transfer Act as Enablers for Digital Transformation? *Technology Innovation Management Review*, *7*(9), 55–60.

Ibarra, D., Ganzarain, J., & Igartua, J. I. (2018). Business model innovation through Industry 4.0: A review. *Procedia Manufacturing*, *22*, 4–10. doi:10.1016/j.promfg.2018.03.002

Ivančić, L., Vukšić, V., & Spremić, M. (2019). Mastering the Digital Transformation Process: Business Practices and Lessons Learned. *Technology Innovation Management Review*, *9*(2), 36–50. doi:10.22215/timreview/1217

Kane, G. C., Palmer, D., Nguyen Phillips, A., Kiron, D., & Buckley, N. (2016). *Aligning the Organization for Its Digital Future*. MIT Sloan Management Review and Deloitte University Press.

Korhonen, H., Still, K., Seppänen, M., Kumpulainen, M., Suominen, A., & Valkokari, K. (2017). The Core Interaction of Platforms: How Startups Connect Users and Producers. *Technology Innovation Management Review*, *7*(9), 17–29. doi:10.22215/timreview/1103

Liere-Netheler, K., Packmohr, S., & Vogelsang, K. (2018). Drivers of Digital Transformation in Manufacturing. In *Proceedings of the 51st Hawaii International Conference on System Sciences* (pp. 3926–3935).

Lucas, H. C. J. Jr, Agarwal, R., Clemons, E. K., El Sawy, O. A., & Weber, B. (2013). Impactful Research on Transformational Information Technology: An Opportunity to Inform New Audiences. *Management Information Systems Quarterly*, *37*(2), 371–382. doi:10.25300/MISQ/2013/37.2.03

Lusch, R. F., & Nambisan, S. (2015). Service Innovation: A Service-Dominant-Logic perspective. *Management Information Systems Quarterly*, *39*(1), 155–175. doi:10.25300/MISQ/2015/39.1.07

Nylén, D., & Holmström, J. (2015). Digital innovation strategy: A framework for diagnosing and improving digital product and service innovation. *Business Horizons*, *58*(1), 57–67. doi:10.1016/j.bushor.2014.09.001

Parasuraman, A., Zeithaml, V. A., & Berry, L. L. (1985). A Conceptual Model of Service Quality and Its Implications for Future Research. *Journal of Marketing*, *49*(4), 41–50. Retrieved from http://www.jstor.org/stable/1251430%0Ahttp://www.jstor.org/%0Ahttp://www.jstor.org/action/showPublisher?publisherCode=ama. doi:10.1177/002224298504900403

Pellikka, J., & Ali-Vehmas, T. (2016). Managing Innovation Ecosystems to Create and Capture Value in ICT Industries. *Technology Innovation Management Review*, *6*(10), 17–24. doi:10.22215/timreview/1024

Ritter, T., & Lettl, C. (2018). The wider implications of business-model research. *Long Range Planning*, *51*(1), 1–8. doi:10.1016/j.lrp.2017.07.005

Rogers, E. M. (2003). *Diffusion of innovations* (5th ed.). New York: Free Press.

Ross, J., Sebastian, I. M., Beath, C., Scantlebury, S., Mockner, M., & Fonstand, N. … BCG. (2016). Designing Digital Organizations. *CISR Working Paper No. 406*. MIT Sloan School of Management. Retrieved from http://cisr.mit.edu/blog/documents/2016/03/10/mit_cisrwp406_designingdigitalorganzations_rosssebastianbeathscantleburymockerfonstadkaganmoloneykrusellbcg.pdf/

Roszkowska, D. (2017). External Knowledge Sourcing and Innovation Processes in Modern Economic Environment. *International Journal of Management and Economics*, *53*(2), 39–56. doi:10.1515/ijme-2017-0011

Ruggieri, R., Savastano, M., Scalingi, A., Bala, D., & D'Ascenzo, F. (2018). The impact of Digital Platforms on Business Models: An empirical investigation on innovative start-ups. *Management & Marketing*, *13*(4), 1210–1225. doi:10.2478/mmcks-2018-0032

Saadatmand, F., Lindgren, R., & Schultze, U. (2019). Configurations of platform organizations: Implications for complementor engagement. *Research Policy*, *48*(8), 103770. doi:10.1016/j.respol.2019.03.015

Sandhu, K. (2008). A Qualitative Approach to E-Services System Development. In T. Sobh (Ed.), *Advances in Computer and Information Sciences and Engineering* (pp. 236–241). New York: Springer; doi:10.1007/978-1-4020-8741-7_42

Scharmer, O., & Kaeufer, K. (2013). *Leading from the Emerging Future: From Ego-System to Eco-System Economies (Croatian E)*. Planetopija.

Sebastian, I. M., Ross, J. W., Beath, C., Mocker, M., Moloney, K. G., & Fonstad, N. O. (2017). How big old companies navigate digital transformation. *MIS Quarterly Executive*, *16*(3), 197–213. Retrieved from http://misqe.org/ojs2/index.php/misqe/article/viewFile/783/468

Spremic, M. (2017). Governing digital technology – how mature IT governance can help in digital transformation? *International Journal of Economics and Management Systems*, *2*, 214–223. Retrieved from https://www.iaras.org/iaras/home/caijems/governing-digital-technology-how-mature-it-governance-can-help-in-digital-transformation

Suša Vugec, D., Pejic-Bach, M., & Spremić, M. (2019). IT Strategic Grid: A Longitudinal Multiple Case Study. In Ç. Doğru (Ed.), *Handbook of Research on Contemporary Approaches in Management and Organizational Strategy* (pp. 105–128). Hershey, PA: IGI Global; doi:10.4018/978-1-5225-6301-3.ch006

Visconti, R. M., Larocca, A., & Marconi, M. (2017). Big data-driven value chains and digital platforms: From value co-creation to monetization. *Big Data Analytics: Tools and Technology for Effective Planning*, 355–371. doi:10.1201/b21822

Weill, P. (2011). What is your Digital Business Model? *MIT CISR Research Briefing, XI*(IX). Retrieved from http://cisr.mit.edu/blog/documents/2011/09/15/2011_0901_digitalbusinessmodels_weill.pdf/

Winby, S., & Mohrman, S. A. (2018). Digital Sociotechnical System Design. *The Journal of Applied Behavioral Science*, *54*(4), 399–423. doi:10.1177/0021886318781581

World Economic Forum. (2018). The Digital Enterprise. Moving from experimentation to transformation. Geneva.

Yablonsky, S. (2018). A Multidimensional Framework for Digital Platform Innovation and Management: From Business to Technological Platforms. *Systems Research and Behavioral Science*, *35*(4), 485–501. doi:10.1002res.2544

Yin, R. K. (1994). *Case study research: Design and Methods* (2nd ed.). London, UK: Sage.

Yoo, Y., Boland, R. J. Jr, Lyytinen, K., & Majchrzak, A. (2012). Organizing for Innovation in the Digitized World. *Organization Science*, *23*(5), 1398–1408. doi:10.1287/orsc.1120.0771

Zittrain, J. (2006). The generative Internet. *Harvard Law Review*, *119*(7), 1974–2040. Retrieved from http://portal.acm.org/citation.cfm?doid=1435417.1435426

ADDITIONAL READING

Autio, E., & Thomas, L. D. W. (2019). Value co-creation in ecosystems: Insights and research promise from three disciplinary perspectives. In *Handbook of Digital Innovation* (pp. 1–21). UK: Edward Elgar.

De Reuver, M., Sørensen, C., & Basole, R. C. (2018). The digital platform: A research agenda. *Journal of Information Technology*, *33*(2), 124–135. doi:10.105741265-016-0033-3

Jacobides, M. G. (2013). Creating and Capturing Value in Your Business Ecosystems. TEDx Talks. Retrieved from https://www.youtube.com/watch?v=hxQHSyk5oyc

Jarvis, J. (2009). *What Would Google Do?* Harper Business.

Johnson, W. (2019). Your Organization Needs a Learning Ecosystem. *Harvard Business Review, July*.

Kabakova, O., Plaksenkov, E., & Korovkin, V. (2016). Strategizing for Financial Technology Platforms: Findings from Four Russian Case Studies. *Psychology and Marketing*, *33*(12), 1106–1111. doi:10.1002/mar.20945

Mars, M. M., & Bronstein, J. L. (2018). The Promise of the Organizational Ecosystem Metaphor: An Argument for Biological Rigor. *Journal of Management Inquiry*, *27*(4), 382–391. doi:10.1177/1056492617706546

Thomas, L. (2015). Creating Digital Ecosystems. TEDx Talks. Retrieved from https://www.youtube.com/watch?v=lcQ0q7cp8lU

KEY TERMS AND DEFINITIONS

Digital Business Model: A business model that leverages the opportunities of digital technologies. Existing companies can digitally transform their business in order to operate also on a digital business model. New companies (digitally born) can arise that are operating entirely on the digital business model.

Digital Innovation: Result of utilizing digital technologies to innovate a business model. Its impact can range from process optimization to more radical change of a business model.

Digital Platform: Mobile or desktop applications intended for external users (customers and partners), or internal users (employees).

Digital Transformation: A holistic term for organizational, technological, and environmental change of a company induced by, or resulting in, the industry disruption by digital technologies.

Ecosystem: Network of employees, customers, suppliers, partners, public sector and other stakeholders that together co-create value for the organization and the ecosystem.

Servitization: More radical change of a business model that includes the introduction of a new service for customers. In the manufacturing industry, products are joined with the accompanied digital platform providing a new value for the customers. Also known as service orientation.

Value Capture: The process stage in which created value can be captured so that it brings new profit streams for the company. Value can be captured through an efficient digital platform.

Value Creation: The process of finding a value. It is assumed that this value has been previously hidden or unexploited.

Value Network: The result of a digital transformation where the main holders of value are no longer business processes or a value chain, but rather a value network. It includes ecosystem ties arranged around mutual interests and/or digital platform.

Chapter 3
Capability Demands of Digital Service Innovation

Jo Coldwell-Neilson
https://orcid.org/0000-0002-3602-8334
Deakin University, Australia

Trudi Cooper
https://orcid.org/0000-0002-4758-5881
Edith Cowan University, Australia

Nick Patterson
https://orcid.org/0000-0003-4565-3614
Deakin University, Australia

ABSTRACT

This chapter explores the demands that digital service innovation places on those stakeholders who use the digital service through three lenses: workers and workplace, digital literacy, and, finally, digital education. The chapter includes an overview of digital service innovation. The impact of digital services generally is explored from the perspective of the workforce and the future of work developing an argument for the need for ongoing upskilling in the digital space. The skills required in the age of digital disruption are explored, and a definition and general understanding of what digital literacy entails is presented. Finally, the opportunities for education in digital environments are explored through three vignettes which illustrate different opportunities for upskilling and retraining. The chapter highlights adaptations required in socio-political environments, education and training, and curricula to allow digital service innovations to achieve the expected benefits.

INTRODUCTION

Digital service innovation is emerging in diverse sectors in the economy, including, retail, education, tourism, hospitality, social services, and distribution. This chapter synthesises findings from three dif-

DOI: 10.4018/978-1-7998-2799-3.ch003

ferent research projects to address questions about how digital service innovation is changing work; what types of change will be socially beneficial; what digital skills will be required by the future workforce; and how education and training can meet these needs. This chapter will explore four questions:

- how is digital service innovation changing work and workplaces?
- what are the potential socio-cultural benefits and pitfalls of digital service innovation?
- what are the implications of these changes for digital literacy? and
- how is education and training being developed to meet various digital literacy needs of the future?

RESEARCH METHODS

The research presented here builds on a recent research project on digital skill acquisition conducted by Coldwell-Neilson (2016) combined with digital discernment research by Cooper (2019) and with Patterson's extensive experience designing, building and teaching in online and MOOC environments. Three vignettes are presented which illustrate contrasting education and training skills that can be used to develop of various digital service skills.

Central to this chapter is the research conducted by the Coldwell-Neilson (2016) on future digital literacy requirements. A twenty-first century perspective of digital literacy was developed through an extensive literature review, which included a scan of existing digital literacy definitions and frameworks, as well as a review of industry expectations of the impact of digital service innovation on skills and capabilities. This was validated through a series of workshops with academics from higher education as well as through conference presentations. This led to an updated understanding of digital literacy as presented in the following sections.

Cooper conducted a research project to integrate digital discernment skills into an undergraduate degree, reported in vignette 3. Barnett and Coate's (2005) curriculum framework with Threshold Concepts (Meyer, Land & Bailie, 2010) was used to develop an initial curriculum for digital discernment for youth work and community work students. An action-learning methodology was used to fine-tune curriculum implementation changes through four iterations, as reported in (Cooper & Scriven, 2017). A cyclical inquiry process of dialogical questioning, planning, observation, information gathering, reflection and re-questioning, learning and re-planning was used as a reflexive research method to document and theorise process, and to improve delivery strategies. Digital discernment problems were identified and strategies to address each of these were developed and trialled.

BACKGROUND

Digital Service Innovation is driven by the need to deliver new solutions rapidly to a wide audience through digital technology. Digital Service Innovation is often assumed to be motivated by the pursuit of economic efficiency and the use of ICT to reduce workforce cost (Green, 2017) producing savings that are passed on to consumers as lower prices, or through increasing profit. Nylén and Holmström (2015) contend that better user experience is at least as important as price competition.

Digital Service Innovation brings about digital disruption. This has both positive and negative outcomes. Digital services, the electronic delivery of information and data over the internet, provides many benefits to employees, employers and consumers if it provides easier access to such services. Digital technology has enabled flexible working, including working from home and, in some industries, flexible work hours so that work can be better balanced with other life commitments or constraints. This flexibility enables some people who were previously excluded from the labour market to participate, such as those with disabilities or people who live in remote locations or where there are few local employment opportunities (Green, 2017). Claims have been made that the automation of more tedious jobs allows workers to have greater autonomy and creativity in their working lives (Vom Brocke et al., 2018). New forms of work, such as 'e-lancing', crowdsourcing and social media influencing, have enabled individuals to directly access their markets digitally (Green, 2017).

Less positively, some digital service innovation has had negative outcomes for some groups in society. For example, digital service innovation has increased automation in clerical, retail and warehousing sectors, resulting in the loss of low skill, full-time, entry level jobs (Green, 2017). Remaining low-skill work has become casualised, and the precariat, the social class characterised by precarious employment, has emerged as a new class of workers (Standing, 2011). Services that are 'digital by default' at first point of contact (Green, 2017) where the user does not interact with another person, have not always delivered the positive user experience that Nylén and Holmström (2015) have suggested is necessary. This is especially problematic where the service is offered by an entity that the customer cannot avoid interactions with, for example, government services such as Centrelink. To avoid inequity, those using digital services need to have access to appropriate technology as well as the skills and capabilities to use them efficiently, effectively and safely (Compaine, 2001).

Technical and human barriers to digital service innovation have been researched by others, (for example, Green, 2017) and the effects of digital services innovation on employee experiences of work have also been explored (for example, Gregg, 2011). However, there is a consensus that more research is required into digital skills in the workplace (Gekara, Snell, Molla, Karanasios & Thomas, 2019). Concern is also widely expressed in the literature about the inadequacy of digital skills education and training (Gekara et al., 2019). Green (2017) contends that rapidly changing digital technologies have placed greater responsibility on individuals to maintain their skills to ensure employability. He contends that the shortening life cycle of digital skills makes this onerous because it necessitates ongoing learning and upskilling. Green found that businesses were not good at articulating the digital skills required and did not put resources into upskilling or training in digital literacy for employees (Green, 2017). This was affirmed by Gekara et al. (2019), who claim that late technology adopters expect employees to have digital skills and resist investing in upskilling their staff. However, they found that early adopters appreciated that continuous upskilling is necessary, and they were more likely to provide training (Gekara et al., 2019).

DIGITAL SERVICE INNOVATION AND THE FUTURE OF WORK

The impact of automation and digital technologies on the future of work is mediated by the socio-political systems in which technologies develop. Futurists and novelists have been making predictions about automation, the future of work, and social change, since the first industrial revolution, with varying degrees of success. Some futurists have assumed that the low-regulation free-market neo-liberal socio-political system that is dominant in many Western liberal democracies will be unchanged by the introduction of technologies. For example, Tyler et al. (2019) assert that space tourism will be a future job, without considering how mass space tourism might be affected by other policy considerations such as the climate crisis and the need to limit carbon emissions. Others have assumed that technological change will have profound social and political impacts that may be dystopian. Science fiction writers (for example, Brunner, 1968) have explored potential social and political consequences of unregulated technology. Other futurists, such as Bill Gates, Elon Musk, and Stephen Hawkins have focused upon what regulation and policies are needed to prevent adverse socio-politico-cultural outcomes of technology and to ensure that technology meets human needs rather than being driven by economic needs. This section addresses digital service innovation and the future of work. The authors begin from the premises argued in the recent ACOLA report (Walsh et al., 2019), that:

- adoption of digital innovations will have social, political and cultural consequences;
- political choice can be made about implementation of technologies that affect the future of work; and
- these choices ought to be informed by social and ethical values rather than economic gain for the few.

If this occurs, regulation will modify how technologies are adopted and also how the socio-economic system operates, at least to some extent. The following example illustrates some important aspects of this position.

In 1930, Keynes, the economist, wrote about the future of work. He was writing just after the 1929 Wall Street stock market crash and at the beginning of the Great Depression. The Great Depression occurred during the second industrial revolution (Coldwell-Neilson & Cooper, 2019) when mass production had increased output and productivity. The economic recession caused mass unemployment in Europe and the USA, and positive feedback loops meant this was ended only by the Second World War. In 1930, Keynes had this to say about work in 2030:

...for the first time since his creation man will be faced with his real, his permanent problem-how to use his freedom from pressing economic cares, how to occupy the leisure, which science and compound interest will have won for him, to live wisely and agreeably and well...Yet there is no country and no people, I think, who can look forward to the age of leisure and of abundance without a dread. For we have been trained too long to strive and not to enjoy ...For many ages to come, the old Adam will be so strong in us that everybody will need to do some work if he is to be contented. We shall do more things for ourselves than is usual with the rich to-day, only too glad to have small duties and tasks and routines. But beyond this, we shall endeavour to spread the bread thin on the butter-to make what work there is still to be done to be as widely shared as possible. Three-hour shifts or a fifteen-hour week may put off the problem for a great while. For three hours a day is quite enough to satisfy the old Adam in most of us! (Keynes, 1930).

Keynes contended that people have been encultured to need work, suggesting for the next hundred years from the time he wrote that this would continue to be the situation. By 2030, he predicted that the standards of living for ordinary people would have increased eightfold (this is probably an underestimate based upon Piketty & Goldhammer, 2015). After this time, he predicted an age of leisure, and a change of socio-cultural values away from a focus on making money and towards a focus on how to live well. Meanwhile, he suggested work should be shared to ensure that everyone had access to the sense of purpose that work provided. At the time of writing, eleven years remain before Keynes' 'age of leisure'.

The quotation illustrates that the social impacts of automation and digital technologies in part depend upon socio-politico-cultural decisions. The standard working week has reduced slightly from a 45-hour working week common in 1930s Britain, to 37.5 hours in contemporary Australia, and most other developed economies. A four-day week (30-hour worked but paid the same as 37.5 hours) was recently trialled by Perpetual Guardian in NZ. Against expectations the evaluation found that output did not decline and workers reported less stress and an improved work-life balance. A key to gaining these results was involving the workforce in the job redesign to enable a four-day week (Coulthard Barnes & Perpetual Guardian, 2018). A similar trial was run at Microsoft in Japan. Paul (2019) reported that Microsoft saw a 40% jump in productivity. There have been various trial projects of Universal Basic Income (UBI) schemes that guarantee people a weekly payment irrespective whether or not they work to address wage loss from automation (Piketty, 2015). UBI has been proposed as a policy response to structural unemployment caused by automation and digital technology (Weller, 2017) that would provide income to displaced workers. The preliminary evaluation of such a program implemented in Finland indicates that participants experienced less stress and better health, but their economic participation was no different from the control group (Charlton, 2019). If there is structural unemployment, the lack of improvement in economic participation might be expected. Apart from this, there is little evidence that changes to socio-cultural values needed for an age of leisure have occurred. There is no evidence that political and economic systems have adapted to ensure that those with little or no work have an adequate standard of living. Keynes's account was premised upon the assumption that the benefits of automation would be shared across the population, but in Australia there has been no increase in the real value of the Newstart allowance for 25 years for example (Lewis, 2019), so this has not occurred for those who are unemployed.

More recently, technology entrepreneurs Bill Gates and Elon Musk predicted that ultimately robots will replace humans in all work (as reported in Ghosh, 2018). Gates suggested this could be addressed by taxing robots, with the income generated used to fund more jobs for people working in education and social services, meeting currently unmet social and human needs. Musk supports payment of UBI to workers displaced by robots. They both predict that unless the socio-economic system adapts there will be greater wealth inequalities (Ghosh, 2018). A declining middle class in the USA (and growing inequalities) provides some evidence that this is already occurring (Pew Research Centre, 2016).

According to a recent OECD policy report (OECD, 2018), based only upon existing technologies, one in seven (14% of existing jobs) "are highly automatable and a further 32% could face substantial change" (p. 1). Manufacturing jobs, agricultural jobs and routine jobs with low skills are at most risk. Service sector jobs at risk include jobs in courier and postal services, land transport and food services (Nedelkoska & Quintini, 2018). The OECD report also identified that young workers were more at risk than middle-aged workers who have established careers, because automation will primarily affect entry level positions (OECD, 2018). According to Nedelkoska and Quintini (2018), older workers' jobs were affected, but less so than young workers.

The relationship between automation and age is U-shaped, but the peak in automatability among youth jobs is far more pronounced than the peak among senior workers. (Nedelkoska & Quintini, 2018, p. 8)

They conclude that under-employment is currently most likely to manifest as youth unemployment, and to a lesser extent unemployment of older workers, who effectively move into earlier retirement, albeit on a lower income. It should be noted, however, that recent policy in many OECD countries has progressively raised the retirement age to save the cost of pension payments, so these effects pull in opposing directions. The OECD policy report also concluded that risks to employment from automation varied considerably between countries, depending upon their economic structures.

Many jobs will change rather than disappear. Technology will replace discrete tasks, and humans will augment technology (Business Council of Australia, 2017). Other jobs have changed substantially but use similar core skills. For example, in media and communications, the number of jobs in traditional print media journalism has declined, but social media analyst positions have emerged. Least at risk were professional jobs and social work/social care. However, even in aged care, digital service innovation is using technology to facilitate people to 'age in place' by remaining in their home rather than entering institutional care (Yang, Miao & Shen, 2015; Rantz, Skubic, Miller & Krampe, 2008). In this context, digital technologies change where and how services are delivered rather than displacing workers. This potentially contributes to increased social well-being, the goal suggested by Keynes and Gates. Yang et al. (2015) concluded that for digital service innovation in aging-in-place to succeed, implementation would have to consider many different factors including "technical, economic, organizational, societal and user readiness aspects" (p. 572). This is confirmed by the ACOLA report (Walsh et al., 2019).

The OECD policy report (2018) conceded that not all jobs that could be automated would be, and that other jobs would emerge. The report stated there was considerable uncertainty about the likely rate of change and whether jobs created would compensate for jobs lost. According to Price Waterhouse Cooper (PwC) (as reported by Ghosh, 2018) many jobs will be lost to automation within the next 10 years, but artificial intelligence will create more jobs than it will displace. The PwC report further suggests that healthcare and education will be growth sectors, as will professional scientific and technical, information and communications, and hospitality (Ghosh, 2018). However, it also predicts that more jobs will be lost than created in manufacturing, retail, transport, public service and construction. In the longer term the OECD policy report also concluded that there would be a diminishing number of job-types that in future could not be completed by robots (OECD, 2018). The ACOLA report (Walsh et al., 2019) suggested that priority for automation should be given to jobs that are "dirty, dull, or dangerous and difficult" (p. 31).

Workplace training will be essential to ensure that the workforce re-trains so that workplace skills align with employment vacancies (Nedelkoska & Quintini, 2018). However, this will require more targeted training than is currently occurring as Nedelkoska and Quintini (2018) found that workers whose jobs were most at risk from automation were least likely to receive training:

Workers in fully automatable jobs are more than three times less likely to have participated in on-the-job training, over a 12-months period, than workers in non-automatable jobs. (p. 9)

Policies that support adult learning will be essential for re-training and up-skilling workers whose jobs have been automated (Nedelkoska & Quintini, 2018).

IMPACT OF DIGITAL SERVICE INNOVATION ON SKILLS

In 2015, the OECD estimated that improving written communication, maths, problem solving and digital literacy could increase productivity by up to 7% in countries such Australia (OECD, 2015). More recently, the Foundation for Young Australians (FYA, 2017) have suggested that "traditional, linear career trajectories are rapidly becoming an antiquated notion" (p. 3) and that most will experience a portfolio of careers in their working lives. FYA go on to suggest that by 2030 workers will spend more time: learning skills on the job; and on creative and critical thinking, with a need to use more maths and science skills. Workers will need highly developed communication and interpersonal skills and develop an entrepreneurial mindset. These predictions have significant implications for young people entering the workforce, but they also have ramifications for those already in the workforce. Reinforcing the OECD's 2015 expectations, skills currently in high demand are for digital literacy, critical thinking, creativity and presentation skills (FYA, 2017, p. 21).

In a post in the Adobe Blog, Gay (2019) suggests that digital literacy is "fast becoming an important job skill at all levels of an organization". Gay further suggests that 57% of executives are calling for workers to have 'soft' digital skills as well as the more common 'hard' skills, distinguishing here between the skills that make up digitally literacy and those that are specific to a profession. In his blog post, Gay makes the distinction that "digital literacy is not to be confused with a digital lifestyle, which refers to passive use of digital platforms". He brings our attention to work undertaken by the International Society for Technology in Education (https://www.iste.org/standards/for-students), who have identified seven attributes that contribute to digital literacy. These are:

1. Empowered learner
2. Digital citizen
3. Knowledge constructor
4. Innovative designer
5. Computational thinker
6. Creative communicator, and
7. Global collaborator.

These attributes align well with the skill set that the FYA (2017) have suggested will be essential skills for young people. However, as automation and digital service innovation in particular, continue to infiltrate into the workplace, what are the implications for our current workforce?

Meeting the changing demands of workplace skills requires action at a number of levels. Employers need to take at least some responsibility for upskilling their workforce, particularly for those who are most likely to be impacted negatively by automation. Employees need to be proactive in maintaining their skill set, particularly digital literacy, as these are the skills required to interact with automation. Education providers at all levels are not exempt. They have a responsibility to ensure that students and graduates have appropriate skills to be effective contributors to the workforce.

Digital literacy skills have become an essential skillset, alongside language and numeracy. But the question is often asked, and rarely answered, what is digital literacy? Employers expect it of their employees and those stepping into the workforce often think they are digitally literate. Unfortunately, there is a history of evidence to suggest that this is not the case if the evidence from studies of higher education students is to be believed (for example Kennedy, Judd, Dalgarno & Waycott, 2010; Nataraj, 2014; Martin & Roberts, 2015; Coldwell-Neilson, 2018). Without a universal understanding of what we mean by digital literacy it is difficult to understand how anyone can adequately prepare for a job of the future. We have presented one view of digital literacy above, from the International Society for Technology in Education. There are many other interpretations of what digital literacy entails, many relating to skills that need to be developed within an education context. The earliest definition was coined by

Paul Gilster (1997) who described digital literacy as “the ability to understand and use information in multiple formats from a wide range of sources when it is presented via computers” (Gilster, 1997, p. 1). This understanding of digital literacy has evolved, but the information literacy focus has persisted for more than 20 years, particularly in higher education, despite digital technologies developing rapidly and becoming ubiquitous. Digital technologies have moved from being a ‘closed shop’ where large installations where kept behind closed doors that only the computer staff could access, to the majority having more computing power in their pockets in the guise of mobile phones. The activities of daily living are regularly facilitated through mobile digital devices but, as mentioned earlier, this does not imply that those using the devices are digitally literate.

Coldwell-Neilson (2018) has defined digital literacy as the ability to identify and use technology confidently, creatively and critically to effectively meet the demands and challenges of living, learning and working in a digital society. Being digitally literate is recognising that the evolution of technology and digital media is not static, that the associated skill set needs to be as flexible and expanding, as well as transferable across technologies and disciplinary boundaries, to meet the demands of this dynamic environment. Digital literacy is a mindset and an attitude, not just a skill set. This definition recognises that our digital literacy skills are evolving over time and that different sets of overlapping skills, capabilities and understanding are required in the various aspects of our lives. Ensuring our digital literacy skills continue to be relevant entails an exercise in lifelong learning.

Drawing on the work of JISC (2014), the European Commission (2006) and others, Coldwell-Neilson (2018) has developed a framework of digital literacy that underpins the definition stated above. Digital literacy incorporates seven elements that all contribute to the seven attributes of a digital literate as suggested by the International Society for Technology in Education:

1. ICT proficiency
2. Information and media literacy
3. Information and data management
4. Creation and innovation
5. Communication, collaboration and participation
6. Security and privacy, and
7. Digital identity and wellbeing

Each of these elements provide the foundation to: develop an understanding of how digital technologies work and build confidence in using them; develop the agility and flexibility to engage with, and manoeuvre a rapidly changing digital environment; actively develop skills to understand the modern media world to enable critical engagement with the environment; develop the skills to be able to recognize when information may not be reliable at best or fake at worst; develop the skills and capabilities to be a responsible digital global citizen; and develop the skills and capabilities to harness the power of digital technology for the betterment of yourself, your community and more broadly, the world we live in.

There has been, and continues to be, a significant increase in the number of roles that will require higher levels of digital skills beyond digital literacy. A useful four-band classification of the digital skills required in our workforce has emerged as part of the UK Digital Skills Taskforce (Select Committee on Digital Skills, 2015). This report indicates that in the context of the workforce approximately 7% of workers will not require any digital skills (the report refers to these workers as digital muggles). Digital citizens will make up 37% of the workforce. These workers will need to be able to use digital technology purposefully and confidently to communicate, find information, purchase goods and services and interact with government services. Digital workers, making up 46% of the workforce, will need the ability to evaluate, configure and use complex digital systems. Elementary programming skills such as scripting will be required for these types of activities. Finally, digital makers will need skills sufficient to build the digital technologies that digital workers and citizens will be using. This group will make up 10% of the workforce. This classification suggests that the majority of the workforce of the future will required quite sophisticated digital skills, underpinned by a strong foundation of advanced digital literacy skills and capabilities.

DIGITAL SERVICE INNOVATION AND TRAINING

A recent Digital Pulse report (Deloitte Access Economics, 2019) indicates that the "highest policy priority for the digital economy is skills development" (p. 3). The report also suggests that increased investment is required to develop workers' foundational literacy, numeracy and digital skills supported by flexible learning programs that target mastery-based learning; and recognition of short courses as formal credentials. This finding is supported by McKinsey's report, which indicates that approximately 14% of the global workforce may need to change occupational categories due to digitization, automation and advances in artificial intelligence in the workplace (Illanes, Lund, Mourshed, Rutherford & Tyreman, 2019).

With the advent of new digital services, the requirements and needs for professionals to be digitally literate has increased and will continue to increase as the workplace evolves technologically. It is suggested that without the skills required to use digital tools, individuals will be left behind not only in their work lives but also possibly in their social lives (Chase & Laufenberg, 2011). Higher education is a key stakeholder in upskilling the workforce and its success or failure may be integral to, and a powerful indicator of, the knowledge-producing and talent-attracting capacity of the country (Hazelkorn, 2018). This section will focus on the training that will be required to meet the demands of the digital workforce. Who should be responsible for upskilling the workforce? What are effective ways of delivering training to and for individuals in the workforce? Two vignettes are presented which demonstrate the effective use of free online education; the first aims to upskill digital literacy skills and the second provides opportunities to gain micro-credentials that measure and certify skills developed within the workforce.

Technology is shaping the future of work, for example, through the proliferation of websites and mobile applications that change the ways people interact with services:

- providing digital access to large, global organisations such as Uber and Amazon;
- Australia Post's Keypass identity card for those who do not have a suitable way to identify themselves online;
- cloud storage solutions such as Dropbox and Google Drive; and
- activities associated with daily life such as making appointments, online banking applications or accessing government services.

These innovations have potential benefits for some businesses, and sometimes for service users and for society, however, not all technology has had beneficial social effects. Technology in everyday life, provides entertainment and opportunities to connect to others through social media such as Instagram and Facebook, and through dating applications. However, it has also enabled 24 hour a day access to gambling (Cherry, 2015), which may have adverse social consequences for some people.

In the work context, the digital era has seen a significant change in the way work is organised, for example, academic work in universities. Academics now routinely use digital devices such as computers, laptops, tablets and mobile phones, and also other digital technologies and software packages to support our research and teaching, for example:

- audio visual equipment is used in the classroom to support face-to-face teaching, including projectors and computers; presentation software such as Microsoft PowerPoint is used and lectures are recorded for students to access;
- delivery of learning resources to students is via a learning management system which also supports online discussions, provision of class lists, submission of assignments and notification of feedback and grades;
- Microsoft Outlook, Teams, Skype, Zoom and virtual meeting points are used for communication with colleagues across multiple locations;
- virtual private networks to access secure institutional intranet when working off-site;
- word processors to prepare electronic documents;
- project management tools such as Trello, Teams and Project to manage work activities and for student group work;
- Microsoft SharePoint and wikis to gain access to policies and procedures, curriculum documents, or to simply share documents with colleagues;
- other web applications for uploading student final grades, uploading exam scripts, managing tutorial and seminar groups;
- searching internal and external library databases to source information;
- online databases for research funding opportunities; and
- using reference management tools to curate all the information collected.

Capacity building to effectively use these technologies is often just-in-time, by trial and error, or relies on informal peer supported learning or the autodidactic capabilities of individuals, which often rely on digital resources. Strategies employed by staff vary according to circumstances. These approaches are sometimes formally supplemented by courses or by individual support, provided on an 'on-call' basis (helplines or mobile expert staff).

As suggested in the previous section, building flexible digital skills provides a starting point for training. Learning to use a specific tool or package may be adequate in the short term, but sometimes an in-depth transferable understanding can be useful in the longer term. In depth digital skill development can help people to transfer skills between technologies. In conjunction with Coldwell-Neilson's (2018) definition of digital literacy, awareness is required of the context in which digital skills are applied. Jones and Hafner (2012) suggest that although digital literacy relates to operating digital technology, it also relates to the ability to transfer affordances and constraints of the technology to different circumstances and contexts. However, training to use digital technologies is often dependent on the context and discipline or profession. Further, the skills associated with digital literacy may change as digital technologies advance (Coldwell-Neilson, 2017). Therefore, digital skills development needs to be addressed through lifelong learning. Coldwell-Neilson (2019) suggests that digital literacy is a mindset and attitude, not just a skill set. Providing individuals with the ability to elastically adapt from one technology to another without a steep learning curve will help them to adapt to all the requirements in the workplace of the future.

Digital literacy and digital education could be developed progressively from primary school through to high school and into tertiary education. Hulten and Ramey (2018) indicate that education provides the foundational infrastructure of literacy, numeracy and general information which contributes towards the functioning of an advanced society as well as its economy. In Europe the European Commission identified digital competence as one of eight key competencies for lifelong learning (European Commission, 2006). More recently the Australian Government developed a national policy through the National Science and Innovation Agenda (Australian Government, 2015) which was implemented by ACARA (2015).

Although these initiatives are aimed at addressing the digital literacy capabilities of young people, the workplace of the present and future will require almost everyone to have digital literacy skills. One approach to improving digital literacy skills more broadly is for digital literacy education to be provided free or at an affordable price through tools such as Massive Open Online Courses (MOOCs), where the reach can be global through simple delivery mechanisms such as smart phones. One example is 'Becoming Career Smart', which is offered on the FutureLearn™ educational platform. It has reached over 36,000 people in over 40 countries. It is estimated that 5 billion people have a smart device and over half of these are smart phones, making options such as this kind of education much more feasible. Taylor and Silver (2019) found that a relative high percentage of adults in emerging economies report owning a smart phone; for example, in South Africa 60% of adults own a smart phone, in Brazil it is also 60%, in Mexico 52% and Nigeria 39%.

MOOCs are not the only solution. There are problems, for example, high attrition when learners join courses and do not complete. There are a number of factors which contribute to this including work and life commitments, or the fact that it is free results in learners not committing fully to completing the course (Koller, 2013). The low completion rates in MOOCs is termed the 'retention funnel' and can be concerning in traditional academic environments when the completion rate may be only 5% (Koller et al., 2013). If the content is not being utilised in a 'free' educational opportunity, is this wasteful? Further research by Bralić and Divjak (2016) shows that as a tool learners were fairly satisfied with this kind of approach, but they did have difficulties such as the language barrier, needing prior knowledge or the required workload in their life to complete it. The learners also appreciated the fact that MOOCs were self-paced and provided the opportunity to assess their knowledge on a regular basis (Bralić & Divjak, 2016). The following vignette illustrates an example of a collaboration between Deakin University and Australia Post that utilised a MOOC to deliver basic digital literacy education to the general public.

Vignette 1: Australia Post Massive Open Online Courses

Massive Open Online Courses have a variety of forms and purposes. Short, introductory courses are often used as 'tasters' to entice the student to go further and complete a formal qualification. Alternatively, they deliver a specific learning opportunity to fill a perceived need. Here we present one particular MOOC which was designed to benefit the general community by increasing awareness of digital business tools, provide an opportunity for the general public to gain the skills to enable access and use of the business tools with the aim of increasing access to and adoption of the digital services. This is a relatively new approach to educating a community about a service and providing them the means to build the skills to use it. However, it also presents an insight into the potential for failure when education is not the core business of, and the target audience is external to, the organisation.

During 2018, Deakin University and Australia Post (the Australian national postal service) entered into a commissioned project to develop two courses in the MOOC to boost the general public's digital literacy understanding and capabilities. This was driven by Australia Post's implementation of a suite of digital services that the general public could access. In order to take full advantage of these services, Australia Post realised that users would need some basic digital literacy competencies which were likely to be beyond the capabilities of many of their customers.

Information about the 'Digital Discovery' program was advertised on Australia Post's website and the courses were delivered via the FutureLearn™ platform. The courses are centred around the vision of 'building your confidence online' and covered topics such as web browsing, social media, shopping online, using digital tools and cyber safety. The two courses were delivered to the public for free and each ran for a period of 2 weeks.

Participants were guided through the learning, which consisted of short readings and videos which explained the concepts and provided instruction of how to implement them in commonly used digital devices such as mobile phones and tablets. The participants were able to post questions in a discussion board, providing a means of peer support. They also completed a short quiz at the end of each week to allow them to assess their progress and a final test at the end of the course which was designed to provide Australia Post with some indication of how successful the course was at achieving its aims. The delivery platform for the project centres around social learning (where peers communicate and help each other to learn, not purely through the lead educator), so feedback could be collected from the learners' postings in the online discussion boards; comments highlighted elements the learners were doing right which they were commended on, and identified what they were doing wrong, which then provided an opportunity for further learning.

Individuals and community groups were encouraged to participate in the program by mentoring the learners or providing supported environments for the learners by delivering the program in libraries, community centres or other environments where there is a need of improved digital skills and confidence. Guides to support mentors and community organisations were made available via the Digital Discovery page.

No formal evaluation of the courses has been undertaken to date, but evidence has been collated from the discussion boards at each stage of the course which indicate that throughout the courses the participants are learning about important elements of digital literacy including: the basics of digital literacy; using digital services for solving problems in their daily life; becoming braver when using the internet and associated technologies; and coming to the realisation they know more than they thought they did. In essence, building their confidence in the digital world.

Although this approach of free digital education to help increase the capabilities of business customers is well founded, Australia Post's initiative did not live beyond the first offering of each of the two courses. This was due to a number of factors. Australia Post had difficulty in proving a return on investment for the MOOC. As FutureLearn™ owned the student data, there was no information to justify how many of the participants were in the course out of curiosity compared to the number who were there to gain digital literacy skills. There was no indication of how many of the participants in the first course went on to complete the second one. Australia Post engaged with some public libraries to run the courses as a community offering, thus allowing them to gain some real insights in the value of the courses. However, the courses were designed for wholly online delivery and were not best suited to a blended style of delivery. Overall, Australia Post was unable to justify the cost of running an advertising campaign to increase the awareness of, and participation in, the courses by their customer base.

The training and education space with respect to digital service innovation and digital skills is a rapidly evolving and useful area as we will see through the second vignette which study illustrates how certification of skills developed in the workplace, through professional credentials, have benefits for the learner.

Vignette 2: Deakin Professional Practice Credentials

Training and education are changing, particularly at the postgraduate level. How we certify or measure someone's skills in the workforce is also changing, with greater recognition that learning does not only take place in the classroom but does so to a large extent in the workplace. For many employees, their professional learning has occurred on-the-job, with no formal recognition of their capabilities. This situation is not uncommon in the ICT industry for example.

Deakin University has developed an innovative method of credentialing an individual's skills and knowledge that have been gained through work and experience. Credentials provide an authenticated and validated method for an individual to gain formal recognition of their capabilities, enabling these to be warranted to employers. Credentials do not require any formal learning but provide the means for capabilities to be formally assessed and recognised.

Credentials are offered at a number of levels aligned with Australian Qualifications Framework (AQF) levels 7 (Bachelor's level), 8 (Gradate Certificate/Diploma level) and 9 (Master's level), and focus on what are generally termed professional (or soft) skills. The skills that can be credentialed currently include (see Deakin University, 2018):

- Employability skills such as communication, digital literacy and team work;
- Technical knowledge such digital marketing, data analytics and risk management; and
- Leadership such as adaptive mindsets and leading and developing people.

The credentials have been developed in consultation with industry experts and professional bodies to ensure they are recognised in the workforce. These are benchmarked against industry skill frameworks and AQF. The ICT related credentials for example have been developed in collaboration with ICT industry experts, the Australian Computer Society, and with reference to the Skill Framework for the Information Age (SFIA).

Having selected the relevant credential, the applicant is guided through the development of a portfolio addressing the credential's assessment criteria:

1. Selecting examples and evidence that "demonstrate the criteria and dimensions of autonomy, influence and complexity of the credential" (Deakin University, 2018);
2. Prepare a reflective testimony that brings together the examples and evidence into a narrative which addresses the credential's assessment criteria; and
3. Undertake a video testimony where the applicant has to answer a series of questions relating to the portfolio and the skill being assessed. The questions are designed to validate the evidence and authenticate the individual as the person claiming the experience.

The outcomes of each stage of the process are collated and sent to two assessors – one academic and one industry based - who determine if the evidence justifies the award of the credential. This approach is an Australian-first and allows individuals to measure their professional skills but also achieve a micro-credential that is recognised and can be used towards academic certification. The micro-credentials can be used towards recognition for prior learning towards a number of Master-level degrees offered by Deakin University for example, shortening the study period considerably.

Vignette 3: Digital Discernment for Youth Work and Community Work Students

Digital discernment skills are required in youth work and in many other social service professions. These skills enable youth work students to conduct good quality research during their university studies and are also valuable after students graduate. A role of youth work is to support young people's participation in democratic institutions. To do this, youth workers need to be able to recognise propaganda and disinformation disseminated through social media and to help young people to develop an awareness of how false information may be propagated. Information discernment skills are also necessary for youth workers so that they can support young people to recognise financial and romantic internet scams in daily life. Digital media have also become an integral part of young people's social world, and, in response, digital youth work has become blended with face to face youth services, (Cooper, Tierney, Brooker & Sutcliffe, forthcoming). This means that youth workers need various types of digital knowledge and skills to use platforms ethically, effectively and appropriately.

Based upon prior teaching experience staff identified five components of information literacy and discernment problems that needed to be addressed:

1. insufficient information retrieval/search strategies,
2. misapprehensions about the purposes of university education,
3. limited strategies to evaluate information critically,
4. uncritical trust in personal experience, and
5. attitudinal resistance of some students to examining materials challenged their existing beliefs (Cooper, 2019).

The first component required students to gain information retrieval skills. This was relatively easily taught in collaboration with the university subject librarian. The second component required students to accept that a purpose of university education is to enhance their capacity for critical thinking, problem-solving and self-directed life-long learning. This concept of education was made explicit but was resisted by some students (Cooper, 2019). The last three components required students to engage with different aspects of digital discernment and critical thinking skills. These skills were the most difficult to teach because they required students to develop judgement about the quality of information from internet sources and in everyday life, the relative validity of different forms of evidence, and how language and argument can be used fallaciously to mislead. The process of teaching information discernment was embedded in disciplinary knowledge and developed and reinforced across the whole degree. The author argues that digital discernment skills should be an embedded and integral part of academic education for all graduates irrespective of discipline (Cooper, 2019).

This third vignette illustrates the importance of digital literacy and digital discernment in a non-technical profession.

SOLUTIONS AND RECOMMENDATIONS

It is clear that digital literacy is now a foundation skill in an environment of increasing digital service innovation. Education at all levels has a responsibility to ensure students are adequately prepared to engage with the digital environments and have the capability to develop their digital literacy skills to maintain relevance to future technology developments. However, not all responsibility lies with education. Employers have a responsibility to ensure they understand what digital literacy skills are required in their domain of influence and to support their employees to gain and extend their digital literacy skills aligned with these requirements. Opportunities for retraining will be an ongoing requirement as digital disruption continues to impact jobs of the future.

FUTURE RESEARCH DIRECTIONS

The three vignettes presented in this chapter demonstrate a few of the ways educational opportunities can be delivered to an audience. Further work is required to explore their relevance and reach and to ensure the opportunities meet the needs of the stakeholders. The dynamic nature of digital technology innovation and use suggests that developing an understanding of digital literacy and the associated skills and capabilities is a moving target that also requires ongoing exploration.

CONCLUSION

The introduction of digital service innovation is occurring in many fields. Digital literacy skills are becoming necessary in all professions and in everyday life. If Keynes, Musk, Gates and others are even partly correct, increased automation will have significant social impacts, including the possible replacements of full-time secure employment, for example in the public service, by less well-paid casual employment in the private sector (for example in the aged care sector). There is a strong likelihood that there will be a mismatch between some displaced workers' aspirations, their skills, and the jobs available. There is likely to be an increase in leisure-time (voluntary or involuntary) for many. How socio-political systems respond to this will have an impact on the types of work available, and whether displaced workers are able to fill the new roles. Appropriate social policy is needed to ensure that all workers, and in particular displaced workers, receive adequate support to retrain and that job creation occurs in locations that are accessible to displaced workers, or that support is provided for people to re-locate. A reduction in working hours may also need to be facilitated, to better distribute available work. Young people in particular will need extra support to gain employment, and older workers need policy that recognises that, for some, retraining for a new job may not be an option. Increased automation will ultimately change how people see the place of work in their lives, for example telecommuting and flexible work hours are becoming more common and boundaries between home and work are becoming less clear cut, as was the case prior to the first industrial revolution. Policy is also required to ensure wealth from automation is shared. Broader socio-culture values need to change to ensure that those who are displaced by automation and cannot retrain or cannot find work, have socially useful work or other ways to find purpose and gain respect (Walsh et al., 2019).

The increasing capability demands of digital service innovation will result in:

- The need for socio-political dialogue and decision-making about how to prioritise the roll-out and uptake of automation (for example, 'dirty, dangerous, dull and difficult' tasks); how to share the benefit of automation, and how to compensate those who have less work than they would like.
- Employers having a responsibility to provide opportunities for their workforce to upskill and retrain;
- Individuals having a responsibility to ensure their skills and capabilities are aligned with the current and future demands of the workplace;
- Changes to curricula in higher education across all disciplines; and
- Education institutions having an increase focus on providing appropriate learning opportunities to enable graduates to maintain and improve their hard and soft skills, through flexible, adaptable and innovative learning opportunities.

ACKNOWLEDGMENT

Support for this study has been provided by the Australian Government Department of Education and Training (Fellowship FS16-0269). The views in this study do not necessarily reflect the views of the Australian Government Department of Education and Training.

REFERENCES

ACARA. (2015). *National Assessment Program: ICT Literacy Years 6 & 10 Report 2014*. Australian Curriculum, Assessment, and Reporting Authority. Retrieved from http://www.nap.edu.au/verve/_resources/D15_8761__NAP-ICT_2014_Public_Report_Final.pdf

Australian Government. (2015). *Embracing the digital age. National Innovation and Science Agenda*. Retrieved from http://www.innovation.gov.au/page/embracing-digital-age

Barnes, C. & Perpetual Guardian. (2019). White paper - The four-day week: Guidelines for an outcome-based trial - Raising productivity and engagement. Auckland, New Zealand: 4 Day Week Global Ltd.

Barnett, R., & Coate, K. (2005). *Engaging the Curriculum in Higher Education*. Maidenhead: Open University Press and McGraw Hill Education.

Bralić, A., & Divjak, B. (2016). Use of MOOCs in Traditional Classroom: Blended Learning Approach. In *Proceedings of 9th EDEN Research Workshop*, Oldenburg. Retrieved from http://www.eurodl.org/index.php?p=special&sp=articles&inum=9&article=766

Brunner, J. (1969). *Stand on Zanzibar*. New York: Ballantine.

Business Council of Australia. (2017, October). *Future-proof: Protecting Australians through education and skills*. Melbourne, Victoria. Retrieved from http://hdl.voced.edu.au/10707/440976

Charlton, E. (2019, February 12). The results of Finland's basic income experiment are in. Is it working? *World Economic Forum*. Retrieved from https://www.weforum.org/agenda/2019/02/the-results-finlands-universal-basic-income-experiment-are-in-is-it-working/

Chase, Z., & Laufenberg, D. (2011). Embracing the squishiness of digital literacy. *Journal of Adolescent & Adult Literacy*, *54*(7), 535–537. doi:10.1598/JAAL.54.7.7

Cherry, M. A. (2015). Beyond Misclassification: The Digital Transformation of Work. *Comparative Labor Law & Policy Journal*, (3): 577–602. Retrieved from https://heinonline.org/HOL/P?h=hein.journals/cllpj37&i=614

Coldwell-Neilson, J. (2016). Office for Learning and Teaching (OLT) National Teaching Fellowship: Unlocking the code to digital literacy, *Australian Learning and Teaching Fellows*. Retrieved from https://docs.education.gov.au/system/files/doc/other/fellows_2012-2016.pdf

Coldwell-Neilson, J. (2017). Digital Literacy - A Driver for Curriculum Transformation. In *Proceedings of the Higher Education Research and Development Society of Australasia Conference, HERDSA '17*, Sydney, Australia.

Coldwell-Neilson, J. (2018). Digital Literacy Expectations in Higher Education. In *Proceedings of Open Oceans: learning without borders. 35th International Conference of innovation, practice and Research in the use of Education Technologies in Tertiary Education.* Deakin University, Australia.

Coldwell-Neilson, J. (2019). What is Digital Literacy [video file]. Retrieved from https://video.deakin.edu.au/media/t/0_80y0wiqq

Coldwell-Neilson, J., & Cooper, T. (2019). Digital literacy meets industry 4.0. In J. Higgs, W. Letts, & G. Crisp (Eds.), Education for employability (Volume 2): learning for future possibilities. Boston, MA: Brill Sense. pp. 37-50.

Compaine, B. M. (Ed.). (2001). *The digital divide: Facing a crisis or creating a myth?* Cambridge, MA: MIT Press Sourcebooks. doi:10.7551/mitpress/2419.001.0001

Cooper, T. (2019). Calling out 'alternative facts': Curriculum to develop students' capacity to engage critically with contradictory sources. *Teaching in Higher Education*, *24*(3), 444–459. doi:10.1080/13562517.2019.1566220

Cooper, T., Tierney, H., Brooker, M. R., & Sutcliffe, J. (2020, in review, for publication). Theorisation, Models of Youth Work and Digital Technologies: Opportunities and Challenges. In Z. Zaremohzzabieh, S. Ahrari, S. E. Krauss, A. A. Samah, & S. Z. Omar (Eds.), The Handbook of Research on Youth Work in a Digital Society. Hershey, PA: IGI Global.

Cooper, T., & Scriven, R. (2017). Communities of inquiry in curriculum approach to online learning: Strengths and limitations in context. *Australasian Journal of Educational Technology*, *33*(4), 22–37. doi:10.14742/ajet.3026

Deakin University. (2018). Professional Practice credentials. Retrieved from http://www.deakin.edu.au/ppc

Deloitte Access Economics. (2019). Australia's Digital Pulse: Developing the digital workforce to drive growth in the future. A report prepared for the Australian Computer Society. Retrieved from https://www.acs.org.au/content/dam/acs/acs-publications/Digital-Pulse-2019-FINAL-Web.pdf

European Commission. (2006). Recommendation of the European Parliament and the Council of 18 December 2006 on key competences for lifelong learning. *Official Journal of the European Union, L, 394*(10). Retrieved from https://eur-lex.europa.eu/legal-content/EN/TXT/HTML/?uri=CELEX:32006H0962&from=EN

Foundation for Young Australians. (2017). The new work smarts: thriving in the new work order. *Foundation for Young Australians (FYA)*. Retrieved from https://www.fya.org.au/our-research/#report-downloads

Gay, A. (2019). How digital literacy affects the modern workforce. Adobe Blog. Retrieved from https://theblog.adobe.com/how-digital-literacy-affects-the-modern-workforce/

Gekara, V., Snell, D., Molla, A., Karanasios, S., & Thomas, A. (2019). Skilling the Australian workforce for the digital economy. *National Centre for Vocational Education Research (NCVER)* Adelaide, SA. Retrieved from https://eric.ed.gov/?id=ED594704

Ghosh, S. (2018, July 18). Artificial intelligence will create as many jobs as it destroys, according to PwC. *Tech Insider*. Retrieved from https://www.businessinsider.com.au/pwc-artificial-intelligence-wont-destroy-all-jobs-2018-7

Gilster, P. (1997). *Digital literacy*. New York, NY: Wiley Computer Pub.

Green, A. E. (2017). Implications of technological change and austerity for employability in urban labour markets. *Urban Studies, 54*(7), 1638. Sage.

Gregg, M. (2011). *Work's intimacy*. Cambridge, UK: Polity.

Hazelkorn, E. (2018). Reshaping the world order of higher education: The role and impact of rankings on national and global systems. *Policy Reviews in Higher Education, 2*(1), 4–31. doi:10.1080/23322969.2018.1424562

Hulten, C. R., & Ramey, V. (2018). *Education, Skills, and Technical Change: Implications for Future U. S. GDP Growth*. University of Chicago Press. doi:10.7208/chicago/9780226567945.001.0001

Illanes, P., Lund, S., Mourshed, M., Rutherford, S., & Tyreman, M. (2018). Retraining and reskilling workers in the age of automation. McKinsey & Company. Retrieved from https://www.mckinsey.com/~/media/McKinsey/Featured%20Insights/Future%20of%20Organizations/Retraining%20and%20reskilling%20workers%20in%20the%20age%20of%20automation/Retraining-and-reskilling-workers-in-the-age-of-automation.ashx

JISC. (2014). Developing digital literacies. Retrieved from https://www.jisc.ac.uk/guides/developing-digital-literacies

Jones, R. H., & Hafner, C. A. (2012). *Understanding digital literacies*. London, UK: Routledge. doi:10.4324/9780203095317

Kennedy, G., Judd, T., Dalgarno, B., & Waycott, J. (2010). Beyond ndatives and immigrants: Exploring types of net generation students. *Journal of Computer Assisted Learning*, *26*(5), 332–343. doi:10.1111/j.1365-2729.2010.00371.x

Keynes, J. M. (1930). *Economic possibilities for our grandchildren. Essays in persuasion* (pp. 358–373). New York: W. W. Norton & Co.

Koller, D., Ng, A., Do, C., & Chen, Z. (2013). Retention and Intention in Massive Open Online Courses: In Depth. Retrieved from https://er.educause.edu/articles/2013/6/retention-and-intention-in-massive-open-online-courses-in-depth

Lewis, C. (2019). Does anyone in Australia think Newstart is adequate? *Welfare*. Retrieved from https://www.crikey.com.au/2019/04/03/newstart-federal-budget/

Martin, A. M., & Roberts, K. R. (2015). Digital Native ≠ Digital Literacy. *Principal*, *94*(3), 18–21.

Meyer, J., Land, R., & Bailie, C. (2010). *Threshold Concepts and Transformational Learning*. Rotterdam, The Netherlands: Sense Publishers. doi:10.1163/9789460912078

Nataraj, S. (2014). The need for an introductory computer literacy course at the university level. [IJBMER]. *International Journal of Business Management and Economic Research*, *5*(4), 71–73.

Nedelkoska, L., & Quintini, G. (2018). Automation, skills use and training. *OECD Social, Employment and Migration Working Papers No. 202*. Paris, France: OECD Publishing.

Nylén, D., & Holmström, J. (2015). Digital innovation strategy: A framework for diagnosing and improving digital product and service innovation. *Business Horizons*, *58*(1), 57–67. doi:10.1016/j.bushor.2014.09.001

OECD. (2015). *Students, Computers, and Learning: Making the Connection, PISA*. OECD Publishing; doi:10.1787/9789264239555-

OECD. (2018). Putting faces to the jobs at risk of automation. Policy brief on the future of work. Retrieved from https://www.oecd.org/employment/Automation-policy-brief-2018.pdf

Paul, K. (2019). Microsoft Japan tested a four-day work week and productivity jumped by 40%. *The Guardian*. Retrieved from https://www.theguardian.com/technology/2019/nov/04/microsoft-japan-four-day-work-week-productivity

Pew Research Centre. (2016, May 11). America's shrinking middle class: A closer look at changes within metropolitan areas. *Social & Demographic Trends*. Retrieved from https://www.pewsocialtrends.org/2016/05/11/americas-shrinking-middle-class-a-close-look-at-changes-within-metropolitan-areas/

Piketty, T. (2015). *The Economics of inequality*. Cambridge, MA: President and Fellows of Harvard College. doi:10.4159/9780674915565

Piketty, T., & Goldhammer, A. (2015). The Economics of Inequality. Cambridge, MA: Harvard University Press. doi:10.4159/9780674915565

Rantz, M. J., Scubic, M., Miller, S. J., & Krampe, J. (2008). *Using technology to enhance aging in place. Smart Homes and Health Telematics. ICOST 2008*. Berlin, Germany: Springer.

Select Committee on Digital Skills. (2015). *Make or Break: The UK's Digital Future*. London, UK: Authority of the House of Lords.

Standing, G. (2011). *The Precariat: The New Dangerous Class*. London, UK: Bloomsbury Academic.

Taylor, K., & Silver, L. (2019). Smartphone Ownership Is Growing Rapidly Around the World, but Not Always Equally. Retrieved from https://www.pewresearch.org/global/2019/02/05/smartphone-ownership-is-growing-rapidly-around-the-world-but-not-always-equally/

Tytler, R., Bridgstock, R., White, P., Mather, D., McCandless, T., & Grant-Iramu, M. (2019). Jobs of the future. Burwood, Victoria, Ford Motor Company of Australia & Deakin University.

Vom Brocke, J., Maaß, W., Buxmann, P., Maedche, A., Leimeister, J. M., & Pecht, G. (2018). Future work and enterprise systems. *Business & Information Systems Engineering*, *60*(4), 357–366. doi:10.100712599-018-0544-2

Walsh, T., Levy, N., Bell, G., Elliott, A., Maclaurin, J., Mareels, I., & Wood, F. (2019). The effective and ethical development of artificial intelligence: An opportunity to improve our wellbeing. Melbourne, Australia: Australian Council of Learned Academics (ACOLA).

Weller, C. (2017, February 14). Elon Musk doubles down on universal basic income: 'It's going to be necessary'. *Money & Markets*. Retrieved from https://www.businessinsider.com.au/elon-musk-universal-basic-income-2017-2?r=US&IR=T

Yang, Q., Miao, C., & Shen, Z. (2015). Digital services innovation for aging-in-place. In *Proceedings of 22nd IEEE International Conference on Industrial Engineering and Engineering Management* (IEEM). Singapore. DOI: 10.1109/IEEM.2015.7385711

Chapter 4
Strategic Intelligence Management and Decision Process:
An Integrated Approach in an Exponential Digital Change Environment

Manuel Antonio Fernández-Villacañas Marín
https://orcid.org/0000-0001-7397-6395
Technical University of Madrid, Spain & Consulting M&M Planning and Project Management, Spain

ABSTRACT

This chapter addresses a conceptual review of the methodology to define strategies and supports for carrying out effective, efficient, and innovative decisions and strengthen management and leadership, given the dynamic changes that are taking place in the current turbulent environment. The main catalyst is the digital revolution and its implications in a new management style. The functions of anticipation and early assessment of threats, risks, and opportunities, constitute the central functionality of the new strategic approach. Its impact is especially notable in the field of strategic decision making, which must be fully timely, rigorous, transparent, and credible in the face of the growing demands of customers and the public. The use of strategic intelligence has become critical, and the skills for its management and application are of great relevance in terms of defining management strategies and decisions, and complementary to the use of digital technology tools.

INTRODUCTION

The dynamic changes that are taking place in the current turbulent and disruptive environment of exponential change, have motivated us to carry out a conceptual review of the methodology to define strategies and supports for effective, efficient and innovative decision making, in order to allows the strengthening of the management processes and the leadership.

DOI: 10.4018/978-1-7998-2799-3.ch004

The current world is being hit by a series of very dynamic and energetic changes. Changes occur in a turbulent and disruptive environment, are occurring at an unprecedented rate, and their patterns are very difficult to predict and adjust to a previous pattern (Mafudza, 2018). This is a real paradigm shift in the vision and projection of strategies that are seriously affected by many different events of social, economic, cultural, technological and political. Thus, the environment in which businesses, public institutions and citizens interact is substantially different from what is known, and all have to face intense changes from all areas (Fernández-Villacañas, 2016).

In this new situation, the main catalyst for change and cause of this continuing acceleration is the digital revolution, driven by the expansion of the Internet, information technology and communications as well as universal employment, which is implying in turn the transformation of our lifestyle, of the management models and leadership. So, we are today intertwined in a process full deployment of disruptive technologies that are profoundly changing the business world, the decision analytics practice and society in general (McDaniels, 2019). Currently, the functions of prevention, anticipation, and early assessment of threats, risks and opportunities, constitute the core functionality of a new strategic approach. It seeks to optimize individual and institutional performance in complex and uncertain environments with highly critical factors, as well as developing adjusted to reality strategies, but that can produce results of high benefit provided through effective, efficient and innovative decision-making, a new management style and a new leadership.

This impact is even more noticeable in the fields of senior management, which requires decision-making processes to be fully timely, rigorous, transparent and credible in the face of growing demands from customers and the public. In this situation, strategic intelligence has become a key element, requiring a specific methodology for its collection, processing and use (Fernández-Villacañas, 2017). It is necessary to be able to build more probable future scenarios, anticipate potential conflicts and critical situations, evaluate future threats, reduce risks, identify future opportunities, discover hidden opportunities in the most complex difficulties, create added value, etc. (Clark, 2019).

In this context, the skills acquired by managers regarding the management and use of strategic intelligence in decision-making, are of great relevance and are complementary to the knowledge of the most advanced technological tools that support decision making. Therefore, emphasizing the need for trained professionals to analyse the information visually and strategically, especially with regard to the development and implementation of scientific data (López-Robles et al, 2019).

This chapter focuses on the revised study of the concepts of strategic intelligence management and decision-making processes, within the book *Leadership, management and adoption techniques for innovation of digital services*, with a complementary approach to the field of methodologies and of digital tools, and proposing new approaches to the turbulent environment of exponential change. After an introduction that allows the reader to understand the new situation described above, the chapter is divided into four parts. In the first part, the concepts of strategic intelligence are examined. In the second part, the systemic approach and organizational behaviour are reviewed as sources of significant doctrine to guide the organizational response to the exponential and digital change of the environment. The third part focuses on the decision-making process using strategic intelligence. Finally, the fourth part develops the analysis of the strategic decision and the management processes applied.

RESEARCH METHODOLOGY

The research has been developed under an exploratory or interpretive research methodology, to have a better understanding of the problem posed without obtaining conclusive results. We have used a systemic approach with a systematized deductive process (Top-Down), guided by empirical evidence which is derived from the author's experience in the field of leadership, management and digital innovation, and the qualitative interaction with different organizational structures, both public and private (Bottom-Up).

The literature reviewed has been used to synthesize the essential concepts and approaches on the problem investigated, reviewing all this with a level of detail that is considered sufficient and looking for its original aspects to guarantee objectivity and avoid interpretative biases.

CONCEPTS ON STRATEGIC INTELLIGENCE

Data, Information, and Intelligence

To start with, the most basic types and the foundation of strategic intelligence are data, information and intelligence. First, in general, data refers to basic descriptions of things, events, activities and transactions recorded, classified and stored, but are not organized according to any specific meaning. They can be numeric, alphanumeric, figures, sounds and images. Secondly, the information corresponds to the data, which has been structured, and the level of quality has been estimated, so that they have meaning and value for the receiver. This actor interprets the meaning and draws conclusions and implications. Finally, for several years there has been an increasing demand in the areas of senior management for the decision-making process to be thoroughly effective, efficient, timely, rigorous, transparent and credible as a response to the growing demands of citizens, clients and public opinion in a broad sense (Fernández-Villacañas, 2016).

In this scenario, the information resource has not only piled up, but has also progressed as a support for decision making, with a specific methodology for its production, processing and distribution, as previously established objectives, in order to build future scenarios and thus anticipating possible conflicts and crisis situations. In this way, we are able to assess future threats, reduce risks, identify early warning indicators and identify future opportunities, which are the basic elements of what we call intelligence. Its use leads to obtaining a competitive advantage in any field in which you work through the improvement of management and leadership (Noe et all, 2017).

Business Intelligence

Business intelligence is initially described as a management tool used to manage and enrich business information and to produce knowledge and intelligence updated for operational and strategic decision-making (Seitovirta, 2011). The concept refers to the information and knowledge that describes the business environment, the company itself and its status in relation to its markets, customers, competitors and economic issues, as well as the process of producing ideas, suggestions, and recommendations for management and decision makers (Pirttimäki, 2007). Business intelligence tries to identify the information needs and process the data and information collected in a useful and valuable managerial knowledge and intelligence which, by obtaining greater knowledge of the company itself and its external environment, improves proactive decision making, planning and business strategy.

Currently, business intelligence has evolved into an activity that collects and analyses all the immeasurable information and knowledge about competitors, customers, markets, technologies and broad social trends, using and exploiting all the resources of the digital environment in real time (Richards et al, 2019).

Competitive Intelligence

Competitive intelligence is the process by which organizations collect actionable information about competition and the competitive environment and then apply it to their planning and decision-making processes combining signals, events, perceptions and data in patterns and visible trends with respect to the competitive and business environment (Fleisher & Bensoussan, 2007). The primary use of competitive intelligence is to help a company assess its competitive and market conditions (Pirttimäki, 2007).

Competitive intelligence can be seen as a progression from unprocessed inputs to completed outputs. Beginning with scattered fragments of raw, basic data that are later organized by competitive intelligence professionals into information. Such information is converted into intelligence when it is converted into a useful format for a decision maker's intelligence needs. The goal of competitive intelligence can be to proactively detect opportunities or threats; eliminate or reduce blind spots, risks and/or surprises; and reduce reaction time to competitive and market changes. The goal is to ensure that decision makers have accurate and up-to-date information about the organization's competitive environment, and a plan for using that information. Competitive intelligence contributes to the foundation on which strategies and tactics are built, evaluated and modified (Fleisher & Bensoussan, 2007). It crosses and superimposes other functions, particularly those associated with marketing, planning and strategy. Competitive intelligence centres on competitor information and aims to ease tactical decision-making. Nevertheless, it has to be used in strategic decision making too.

Currently, competitive intelligence conceptually maintains its meaning but it has experienced a strong catalyst in its ability to act, exploiting all the digital resources of the environment and in turn providing a great capacity to understand the needs of customers and understand behaviour of competitors. These are actions prior to the use of the new Marketing 4.0 approaches in global markets (Kristanti & Darma, 2019).

Strategic Intelligence

Many years ago, Henry Mintzberg defined the strategy generically as "*the pattern or plan that integrates the main objectives and policies of an organization and, in turn, provides a coherent sequence of actions to be taken*" (Mintzberg, 1994). The concept has evolved and different opinions about strategic intelligence are expressed as a polyhedral and polysemic concept, which are based on the aspects to be highlighted and the approach (Palacios, 2018). Therefore, it is the result of a simple and complex assembly and analysis of information to reach conclusions about a specific problem that will be resolved in relation to other problems or as part of a final resolution process. It is also a form of organization that has been designed to collect information that becomes a useful product for decision-making for a user to solve a current or future situation. Ultimately, it is also a specific methodology whose objective is to reduce the levels of uncertainty that exist in a given circumstance to reach a particular strategic decision, thus opening viable alternatives that guarantee a greater probability of succeeding in achieving the established objectives. In other words, the objective of strategic intelligence is to collect, analyze and disseminate information that helps decision making at a strategic level (Fernández-Villacañas, 2016).

Traditional literature has always maintained that the objective of strategic intelligence is to understand where a company is headed and to know how it can maintain its long-term competitiveness in light of future challenges and changes (Thierauf, 2001). It facilitates early warnings (Gilad, 2004), serving as a radar to alert the company of environmental threats and opportunities (McGonagle & Vella, 1996), and seeking to make the best strategic decisions to maximize the success of a company (Liebowitz, 2006).

The term strategic intelligence refers to the needs of high-level decision makers and focuses primarily on proactive activities. It is noteworthy that strategic intelligence can support strategic management, especially by contributing to the collection, analysis and distribution of information. As they discover that the higher the level of decision-making, the more consolidated the information becomes and more conclusions and suggestions are added.

Traditional literature has not adequately assessed the needs for strategic information, considering that strategic information frequently appears as weak signals of qualitative data, the content of which may not be obvious but should be collected as separate fragments. In addition, it is impossible to determine information needs. As a result, this type of information cannot be processed in the same way as explicit quantitative data. In our opinion, we must assume that all levels of intelligence activities can contribute to strategic intelligence. Business intelligence produces the type of information used in long-term strategic decision-making. This concept involves the entire relevant framework of a company, not the company alone, including elements of the external environment, such as competition, industry and market information.

Although Mintzberg considered competitive intelligence as a synonym for business intelligence (Mintzberg, 1994), a few years later, competitive intelligence began to be understood as an organizational function modifying the scope between the broader area of the intelligence business and the more limited version of the competition analysis (Fleisher & Bensoussan, 2007). We believe that competitive intelligence is part of business intelligence and extends its reach far beyond competitive intelligence to also encompass competitive market information. In our opinion, strategic intelligence is a global concept that covers the signals of all levels of intelligence: business intelligence, competitive intelligence and competition intelligence. Figure 1 illustrates the relationship between the concepts.

Strategic Intelligence Cycle

The strategic intelligence cycle is a circular and repeated process to convert data into useful intelligence to meet a user's goals. It consists in a set of phases that occur sequentially and cyclically, in which the tools and elements that participate in the generation of knowledge to support a user's decision processes, interact with the different areas of influence, in a dynamic interface. Such phases of the intelligence cycle are those of planning, collection, processing and analysis, dissemination and exploitation, and feedback (Fernández-Villacañas, 2017).

- Planning: The first stage of the intelligence cycle consists of planning intelligence actions. This translates into determining the needs of the client, establishing requirements and developing a plan, to generate intelligence from certain information in a correct way, there must be a definite objective and must specify what is sought.

Figure 1. The relationship between the different levels of intelligence
Source: Seitovirta (2011)

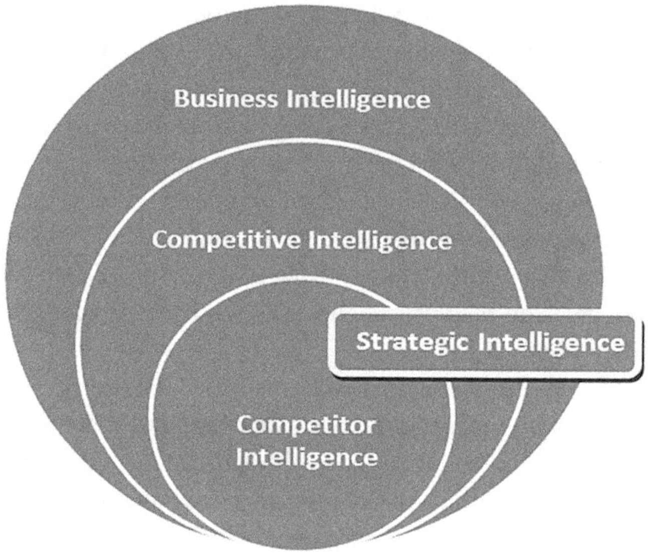

- Collection: The second stage will be the one related to the collection of data, both inside and outside the organization. It will proceed to a first classification and validation of the information collected, and its initial linking. It will be necessary to determine where (sources) and how (procedures) the data and information will be acquired (Clark & Lowenthal, 2016). Likewise, all possible options for the collection of information must have been previously determined, in view of the goals sought and once the intelligence needs have been assessed.
- Processing and analysis: The third phase will be the transformation of the information collected into something that can be easily used during its analysis, turning it into intelligence through its interpretation, a task that will depend largely on the capabilities and experience of the analyst. All reports produced must meet the goal or need defined during the planning phase.
- Dissemination and exploitation: The fourth stage will correspond to that in which the intelligences will be disseminated, providing its user with the finished intelligence product. If its users could not access in a timely manner, or could not use it, then everything acted would be useless and the established goal would not be met, resulting in its exploitation being unfeasible.
- Feedback: The final stage will be the one corresponding to the evaluation and control of the process. This can be done through progressive feedback and evaluating whether the process has satisfied the needs of the clients, or retaking it if there have been failures of any kind for which the intelligence cycle could not be successfully closed.

From all these activities, it is the analysis that entails complex individual and organizational cognitive processes. In this regard, the task of analysis is to apply rationality in that frequently ambiguous, complex and challenging that is relevant to strategic decision-makers. This requires a solid repertoire of analysis methods, tools and techniques, most of them covered in the field of data science and powered by digital technologies, to aid in finding answers that guarantee the maintenance of your organization.

Figure 2. Iceberg model – "above", "below" and "on" the waterline
Source: Fernández-Villacañas (2016)

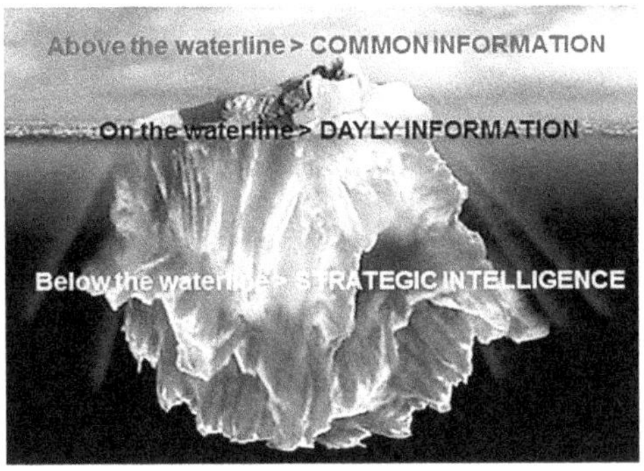

Present and future competitiveness, interpreting adequately the various scenarios, discovering what your competitors are aware of and adopting appropriate measures. Analysis is the phase of the intelligence process in which more value is created, interacting with the generation of data and information on the past, current and future situations, as well as on the options that are available (Fleisher & Bensoussan, 2007). In this way, it is also possible to understand the strategic direction that the organization wishes to take, where it needs to go, how to achieve it, and how to know if it has reached that destination.

Currently, big data and business analytics have a great impact on companies, since the revolution of networks the platforms, people and digital technology have changed the determining factors of innovation, management and competitiveness of the companies, the fuel of the future (The Economist, 2017). All this leads to generating a very valuable knowledge for organizations, which also demands a new style of more technological leadership in strategic intelligence systems. The growing expansion of the available data is a recognized trend throughout the world, while the valuable knowledge that emerges from the information comes today mainly from the digital processes of intelligence analysis (Larson & Chang, 2016).

The core activity of intelligence analysts focuses on properly interpreting the complex, ambiguous and challenging problems that are critical in assuming the decisive processes that affect. For this purpose, it is necessary to have adequate leadership, and the methodology and digital technologies that allow questions to be answered about the present and future performance of organizations. All this results in a meaningful interpretation that is based on the exploitation of the data and the significant information available, taking into account that it is increasingly necessary to resort to the intuitive distillation of managers' experiences when the environment is fundamentally volatile and uncertain (Hodgkinson & Sadler-Smith, 2018).

Leading organizations are very similar to sailing on a ship through treacherous waters. Icebergs appear in the form of competitors, but only a small fraction of the iceberg can be seen on the waterline. The most valuable and often harmful information can be located below the waterline. However, the fact is that most organizations do not receive intelligence below the waterline. Strategic intelligence provides intelligence above and below the waterline. On the waterline, there is information that your competitors do not care what they see as press releases, website ads, industry newsletters or analyst reports, etc. But below the waterline there is information that your competitors will not want you to see, such as price details, lists of sales prospects, strategic partners, marketing programs, top sales staff, R&D programs, reseller channels, acquisition objectives, competitive threats, patent applications (Fernández-Villacañas, 2016).

SYSTEMIC APPROACH AND ORGANIZATIONAL BEHAVIOUR

The explanatory variables of the current problems are strongly related, since the relations surpass the states, they correspond to very dynamic phenomena, behave atypically and are resistant to align with generalist policies, evident and simplistic, are causal and not accidental, since cause and effect behaviour changes over time and are very difficult to extrapolate in a long-term perspective. So what do we do to deal with them, if the use of classic and conventional tools is no longer effective? It is necessary to test other tools, concepts and theories to change behaviour in a structured way and generate events and results in accordance with an integrated, holistic and systemic environment, characterized by the interactions of its components and their non-linearity. The systemic approach provides a coherent internal framework for classifying and evaluating structures and organizations (Campbell & Draper & Huffington, 2019), being essentially useful in the field of strategic intelligence and decision-making processes.

Along with the systemic approach, we need to use another important area of study, organizational behaviour and leadership, which analyses the effect that individuals, groups and structures have on human behaviour within corporations to create positive organizations (Macdonald & Burke & Stewart, 2018). It is an interdisciplinary field that involves sociology, psychology, communication and management. Organizational behaviour and leadership complements organizational theory, which focuses on organizational and intra-organizational issues, and complements human resources studies, which focus more on everyday business practices. Therefore, organizational behaviour and leadership is the study of how people interact within groups. Normally, this study is applied in an attempt to create organizations that are more efficient. The main idea of the study of organizational behaviour is that a scientific approach can be applied to the administration of workers, and is used to maximize the performance of individual group members (Robbins, 2009).

There are many different models and philosophies of organizational behaviour and leadership. Areas of research work include improving job performance, increasing job satisfaction, promoting innovation, motivation and encouraging leadership. In order to achieve desired results, managers can embrace different tactics, including management style, reorganizing groups, changing compensation structures, and altering the way performance is evaluated (Mullins, 2005). Topics covered by micro-organizational behaviour include cognition, decision-making, learning, motivation, negotiation, impressions, group processes, stereotypes and power and influence. Macro-organizational behaviour includes organizations such as social systems, dynamics of change, markets, relationships between organizations and their environments, as well as identity in the organizational process, the way in which social movements influence markets and influence of social networks. The application of the theory and knowledge of the field of organizational behaviour can be divided into sections on personality, job satisfaction and management, leadership, authority, power, politics and decision making, especially the policies and strategies to undertake the planned designs of organizational behaviour and its change (Cash & Hartlev & Durazo, 2017).

THE DECISION MAKING PROCESS USING STRATEGIC INTELLIGENCE

The Decision Making Process

A basic but sufficiently illustrative decision definition tells us that it is simply a selection between two or more different lines of action. It contains two fundamental aspects of the decision concept that deserve to be highlighted. First, the existence of several alternatives that a decision maker can follow, and secondly, the selection of one of them that will constitute the course of action undertaken by him. According to this definition, the decision will represent a specific act that will involve the choice between alternatives at a given time. However, the decision itself is only the end of a prior process and, in a way, the beginning of a new one that will lead to the adoption of new decisions or the revision of those already taken. Obviously, this future decision-making process will include the synthesis of the contributions of decision researchers, analysts, professionals and much more, influencing the application of digital technologies, in particular of expert systems and of the collaborative artificial intelligence, in the communication and execution of decisions (Ademola, 2017).

Based on the above, we can provide a more elaborate definition according to which the term decision represents a process by which the numerous existing alternatives are reduced to one that is finally chosen. Therefore, the decision understood as a process implies a set of activities that integrate it and that lead to the final choice. Faced with the previous specific act of mere choice of alternatives, we found as we previously highlighted a complex and eminently dynamic process. These homogeneously grouped activities can be called phases of the decision process and we can distinguish three totally interrelated but completely different stages, and a fourth complementary stage (Fernández-Villacañas, 2016):

- Intelligence activities, that implies, in relation to the problem, an investigation of the environment so that conditions or sub-problems that require a decision are found, that is, situations that require new actions, which implies the exploration of the organizational environment both specific and general as well as the monitoring of internal activity. This phase is of fundamental importance, since *"The most common source of mistakes in management and management decisions is the importance attached to finding the right answer instead of the right question"* (Drucker, 1966). But *"The recognition of the difficulties of identifying and formulating a problem - and the question is formidable - is a first step in rational decision making ... unfortunately there is no formula that allows us to find the correct problem that has been to solve"* (Steiner, 1979). Once the diagnosis of the whole situation has been established and the specific problem that we intend to solve has been pointed out, we must think about the solutions that would allow it to be done.
- Design, modelling and conception activities, that consists in generating, developing and analysing the different alternatives or courses of action that can be followed to solve the problem posed in the previous phase taking into account all the factors identified as notable, making use of the process of research, deeper or more lax, deemed necessary. Experience is an important advantage here, although its usefulness can sometimes be significantly diminished by the changes that have occurred. Leadership, creativity, innovation and the digital technologies occupy an increasingly prominent place in the search for solutions. The development and analysis of these possible alternatives implies, first, to ascertain their viability (taking into account the restrictions imposed by existing resources: human resources, material resources, infrastructure, budgets, etc. as well as the limitations imposed by time, technology, legislation, customs, dominant social values, etc.). Then

the assessment both qualitative and quantitative of the consequences generated by the solutions considered attributing to each utility or value index based on the criteria or interests judged as important by the decision maker.

- Decision activities, which means the selection of the alternative among the possible ones that best contributes to the achievement of the objective pursued. This choice will be based on a certain criterion, so it would be necessary for the decision maker to specify in advance their corresponding decision criteria. The decision maker usually tends to evaluate alternative solutions with more qualitative criteria, so it is necessary to apply objective digital instruments that develop a quantitative analysis and allows a better rationalization of the decision process.
- It is necessary to add a fourth complementary stage of review activities that guarantees a follow-up of the execution of the decision taking into account the real characteristics of the organization at the time of practice. Likewise, an ex post evaluation will be necessary to introduce the appropriate corrections and, if necessary, propose the analysis of a new problem if a new induced situation is identified that requires a new decision. The monitoring of the review program is generally instrumentalized through the design of a monitor of qualitative and quantitative performance indicators, for whose management it is essential to have the most appropriate digital technology tools, as well as maintaining the necessary leadership for the implementation of the redirection actions advised by the control of said review program.

In theory, the analysis activity precedes the design, and the design activity precedes the choice. However, in practice the phase cycle is much more complex than this sequence suggests. Each stage in making a particular decision represents in itself an additional decision-making process. Therefore, the design phase may require new analysis activities; problems of any level generate new induced problems that, in turn, present their phases of analysis, design and choice. The same happens in relation to the review activity. Therefore, empirical analysis reveals all forms of interaction. Among these phases, the analysis of the problem is often influenced by a latent knowledge of the types of conceivable and acceptable solutions, and this influence is nothing more than a manifestation of the permanent interdependence that exists between the objectives pursued and the available means that characterize the practice of decision-making. Consequently, the purely rational model that is sequential and causal is nothing more than an unreal decomposition of pure rationalist ideology, which contrasts with the tendency of recent explanatory theories of organizational reality whose approach tends to show irrational practices in many aspects of management (Askari & El-Refae, 2019).

On the other hand, the most relevant issues involved in the decision process are the decision maker and the analyst. As trivial as the differences between the roles may seem, it seems appropriate to clearly this aspect, given that there are certain activities that are not attributed to one or the other. Therefore, as a rule, we will say that the decision maker, in his own leadership exercise, will be responsible for contributing and resolving all subjective aspects related to decision making, while the analyst's role will be technical and objective, analysing the problem and helping the decision maker to choose an alternative. The analyst would be the one who transfers the subjective information generated by the decision maker to the support models of digital services. Obviously, the responsibility for each decision will be the decision maker and not the analyst.

Internal Process by Which a Decision Maker Carry Out a Decision

It is substantial to study this aspect because the acceptance or rejection of certain premises will mean the acceptance or rejection of certain procedures for decision-making. Under this approach lies the recognition of the different ways of understanding the decision process and the implicit rationality that the decision maker applies to it.

Within the literature on decision-making processes, a distinction is made between; the optimal alternative, which is one that maximizes the set of objective functions and the efficient alternative which is the one for which the objective function cannot be improved without simultaneous damage to at least one of the other objective functions. The preferred alternative that replaces the optimal solution in decision problems with multiple objectives when they are not compatible with each other and the satisfactory alternative is a reduced subset of the set of feasible alternatives that manages to reach all the minimum aspiration levels established for each objective. These four types of alternatives facilitate the understanding of the two approaches that are generally considered in internal decision-making processes in the field of management: that of strict rationality and that of bounded rationality.

- Strict rationality: This approach, throughout history and mainly in the field of theoretical predicaments, has meant that every agent acts rationally when making decisions that concern him. The rational decision process involves three basic conditions: 1) The enumeration and knowledge of all available alternatives or courses of action; 2) The determination of the consequences that would result from the monitoring of each of the alternatives; and 3) The evaluation and comparison of the consequences associated with these alternatives.

The acceptance of these hypotheses results in a fundamentally optimizing behaviour of the decision maker, which obviously cannot conform to any alternative but must choose the best one. Regarding the total knowledge of all the alternatives, it can be said that, in most decision situations, partial and incomplete information is available so that each possible alternative can hardly be established. As for the consequences, these are generally known in the form of expectations rather than in certain terms. Only in case of total certainty about the consequences derived from the different alternatives can rational behaviour be possible. Finally, regarding the evaluation of the consequences that allow selecting an Optimal alternative and discarding the remaining ones, the establishment of a unique criterion that encompasses the assessment of all aspects that influence the decision is required, an issue that, given the complexity and the fundamentally subjectivist nature of such criteria makes such evaluation and comparison very difficult.

- Bounded rationality: the review of the assumptions on which strict rationality is based leads to the concept of this approach. The review arises as result of the acceptance of subjective, psychological, sociological or other components in the agent facing a decision situation. Bounded rationality is based on the idea that rationality is limited when individuals face decision-making processes, because of the usually reduced ability to monitor the decision problem, the cognitive limitations of the mind and external inflections, and the usually short time available to make the decision. Decision makers, from this point of view, act as satisfactory. They seek a satisfactory solution instead of an optimal one. Herbert A. Simon proposed this bounded rationality concept three decades ago as an alternative basis for mathematical modelling of decision making, as used in economics and social sciences, complementing the strict or optimizing rationality approach (Barros, 2010). The difficulties to follow a rational behaviour arises from the impossibility of knowing all the possible information about alternatives, determining with certainty its consequences and, therefore, objectively evaluating said information. In these situations in which the information is partial, the justification of the decision maker can only be limited, which implies the acceptance of two basic ideas. First, the need to seek information and alternatives derived from the lack of knowledge about them, which leads to an exploration of the different courses of action that can be determined. Second, sufficient satisfaction, that is to say, the impossibility of having all the alternatives, as well as the existing difficulties for the knowledge and the evaluation of the consequences, reduce the possibility of finding an optimal alternative and force the decision maker to comply with an alternative that sufficiently satisfies the performance levels established for the stated objectives. Aspiration levels are changing and vary over time, mainly based on the knowledge of the decision maker, his accumulated experience and his ability to learn.

Normally, most decisions in organizations are better suited to the limited rationality approach; while only in very few cases do they conform to the strict rationality approach. In sum, this gives way to limited rationality as result of the predictive and calculation limitations of decision makers. That is, it is not possible to know all the alternatives that the subject of the decision can count on, it is not possible to assess with certainty all the consequences that follow of these alternatives and it is not possible to obtain all the necessary information to choose, consequently, the best alternative.

Ultimately, the decision is represented by the set of actions taken at a specific time as result of the application of certain rules and policies to the particular conditions existing at that time. It is considered as a continuous process, as a series of consecutive changes based on the information flows that occur within the system defined by the organization itself and with its environment.

STRATEGIC AND MANAGEMENT DECISION-MAKING PROCESS APPLIED

The study of decision-making processes is based on the systemic conception, in which the systems are divided into three distinct parts - inputs, processes and outputs -, are surrounded by an environment and often include a feedback mechanism. A human decision maker is considered part of a system (Turban & Aronson & Liang, 2005), as shown in Figure 3.

The systemic approach assumes the optimal orientation of the decision-making process by involving the integrated consideration of all variables, which include, the analysis of possible scenarios, the recognition of the main values and attributes of all the components involved and the evaluation of possible consequences. Moreover, taking into account the external variables that affect decision makers, as well as intuition, as a decisive element, without ruling out any structural determinism that decides instead of the decision maker. From a systemic point of view, it is essential not to forget that the behaviour of each organization is determined by the routines established in them and by the goals and objectives that condition their action.

The systemic approach in the decision-making allows differentiating the symptoms from the real problems or the root cause of a problematic situation, and allows to solve said problematic situation, in an effective, efficient and innovate way, through the recognition of the network of influences that characterize the social systems. For the systemic approach, the definition of the problematic situation has to do then, with the world view of the decision makers, with the available information and intelligence, with the exploration or research system, which is the way we try to address the truth, and especially with the influence of culture and exogenous factors in general.

The importance of information in systems and decision-making is emphasized in the definition of decision as the process of transforming information into action (Forrester, 2007). The information is the raw material, the input of the decision, and once properly treated within the decision-making process, the action to be executed is found as the output. However, the implementation of the chosen activity will generate new information that will be integrated into the existing information to act as the basis for a new decision that leads to a new action, and so on.

The Strategic Decision-Making

Intelligence activities are intended to provide support for decision-making, and the analysis of information is an important part of that work. The strategy is the main link between the organizational configuration of a company and the external environment (Ungureanu & Avramescu, 2008). We have already reviewed the field of intelligence activities, but to discover how they help in decision-making, it is essential to observe how decisions are made. The objective of this part is to trace the field of strategic decision making by introducing several strategic approaches (Ramirez & Selsky, 2016).

There is a consensus that the strategy begins with a vision of what the organization wants to be in the future, which translates into the development of the specific actions necessary to achieve that vision. Strategic management has been defined as a way of leading an organization that seeks to develop

Figure 3. The system and its environment
Source: Turban & Aronson & Liang (2005)

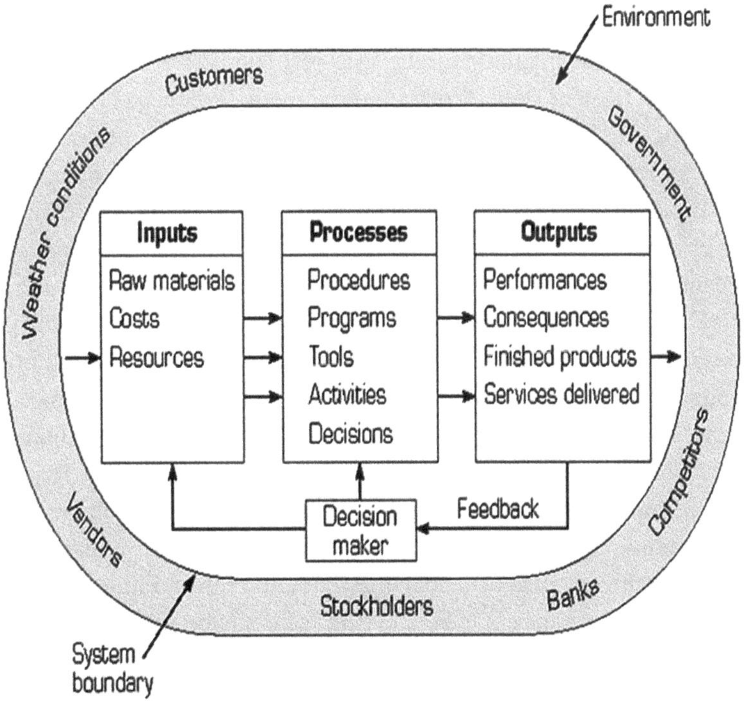

values, management capabilities, organizational responsibilities and administrative systems to connect strategic and operational decision-making (Fleisher & Blenkhorn, 2001). In the current era of digital transformation, the ability to respond quickly and strategically to unpredictable changes can determine the success or failure of the organization, which is achieved as a corporation becomes more successful in implementing innovation and Digital services. On the other hand, the ability to respond to current problems generated by a turbulent and disruptive environment will be strengthened in their culture (Ansoff et al, 2019).

The formation of strategies is usually configured as an analytical process, in which the strategic work consists in understanding the structure of the industry and seeking a position for the company that is more profitable and less susceptible to attacks (Porter, 2008). This may involve positioning the organization to deal more effectively with current competitive forces, anticipating and exploiting changes in strengths and setting the balance of these strengths to create a more favourable industrial structure for the company.

The objective of strategic decisions is to locate the company so that it can take full advantage of its competitive environment, with competitive intelligence responsible for generating this adjustment. For years, there have been criticisms about the usefulness of strategic planning in the traditional way, raising some authors with an anticipated vision of what would come next, that in the rapidly changing, accelerated and competitive world, a step-by-step strategic planning approach hinders the decision making and dynamic actions towards the required market (Fleisher & Blenkhorn, 2001). Thus, it seems necessary that instead of being carried out at certain intervals, the strategic planning process is carried out continuously (Vecchiato, 2015) using innovation and digital technologies as essential tools. Against the analytical approach of strategic planning, the design of the strategies must be emergent, surpassing the concept of analysis to become the synthesis sketch of multiple intelligence sources. It is about energizing the organization with innovative strategic thinking about the vision of the course of action that the company could take in the short, medium and long term. Planners can provide the data and help managers think strategically. They can produce the formal analyses and hard data that strategic thinking needs, but only to broaden the consideration of problems, rather than discovering the right decision. In other words, the strategy, instead of being a permanent guide for the long-term objectives of the organization, should lead to the establishment of a permanent flow of potential competitive benefits (Grant, 2016).

As a contrast to strategic planning, the strategic thinking involves intuition and creativity. The outcome of strategic thinking is an integral perspective of the company, a vision of management that is not too precisely expressed (Bratianu, 2017). This type of strategy cannot be developed in advance; it must be able to appear at anytime and anywhere in the organization and not be forced into a cyclical framework of strategic management (Fleischer & Blenkhorn, 2001). They usually arise through disorderly informal learning processes conducted by people at various levels of the organization who are profoundly involved with the particular problems in question. Strategy development requires fostering informal learning that leads to new perspectives and combinations, and strategic planning cannot be used to solve problems without judgement and intuition. Strategic planning represents a calculating management style, with the aim of reducing management power over strategy making, in a world that requires a more engaged management style where management engages people and everyone contributes to and helps to shape the direction of the company. In sum, both intuition and rationality play important roles in strategic decision-making, requiring the exercise of leadership that explains and integrates both approaches and takes advantage of the intuitive or rational characteristics of decision making in the context of the organization (Calabretta & Gemser & Wijnberg, 2017).

It is of great interest to examine the differences that exist in the minds of planners and managers, as well as the existing differences in their ways of assuming decision-making processes. Therefore, while planners may lack management authority to make compromises and may access critical information to strategy formulation and its derived courses of action, managers under pressure over time tend to encourage action over reflection and oral over written, which sometimes means they may ignore important analytical information. Therefore, planners must work as soft analysts, who aim to ask the right questions instead of finding the right answers. In that manner, they make the most complicated problems arise out of careful reflection, preventing them from being quickly closed due to rapid decisions.

For decades the managers of many companies have assumed that in the field of strategic initiative the entire organization must be involved to focus on specific corporate challenges, since it is not effective for the strategy to be planned somewhere in the top management and then communicate the rest (Hamel & Prahalad, 1989). However, in the current turbulent and disruptive environment of exponential change, strategic initiatives need to strengthen the management approach and leadership style to ensure that the launch of disconnected individual strategic initiatives becomes an integrated portfolio of initiatives. For this, five lines of action are identified that allow improving the levels of effectiveness, efficiency and innovation of the strategic initiatives of the organizations (Kunisch et al, 2019):

- Orchestrate the rhythm: Provide the necessary time in the organization to strengthen each initiative before considering the following, even assuming the risk of falling behind the competition, but energizing innovation and the necessary changes.
- Diversify objectives: A sense of immediacy must be developed so that the weak signals of the turbulent environment of exponential change allow emphasizing the need to improve instead of allowing inaction to accelerate a real crisis, diversifying objectives and achieving an initiative front that pushes the performance of the entire organization.
- Pool resources: Provide all workers with the training and level of leadership they need to work effectively and efficiently, with multidisciplinary criteria and covering all functional areas, especially those related to digital, technological and management services of analysis and decision-making.
- Configure initiative interfaces: Establish clear milestones and verify the mechanisms and interfaces necessary to track progress, making challenges inevitable for everyone in the organization.
- Foster feedback cycles through initiatives: Develop a competitive approach at all levels and in all individuals through the widespread use of strategic intelligence, feed backing each worker's efforts with those of the best employees of the organization and the competence to try to generalize the challenge at the individual level.

The Management Decision Making

An opinion of Robin M. Hogarth is very eloquent: "*Although it sounds trivial, decision making is one of the most important recurring human activities. However, there are three observations that collide me paradoxically: First, people are generally unaware of how they take decisions and often why they prefer an alternative to the others. Second, they show little concern about the quality of their own decision-making processes (although the rest's failures are quickly pointed out). Third, according to my point of view, the Scientific study of decision making has not brought the attention it deserves*" (Hogarth, 1987). Indeed, not only is it not trivial, but the importance attached to problem analysis and decision-making within management is backed by a long list of organizational phenomena that have been referred to in this work.

The classic management process raises the sequence of the planning, organization, direction, coordination and control phases, and within each of them, there is a set of decisions. Those who perform them have to decide what the organizational objectives will be, what strategies have to be put into practice, how the organizational structure should be, how the plans and budgets have to be implemented, what systems and goods and services are to be acquired, which activities should be prioritized. All of this implies the need for choice between alternatives, that is, decision-making activities. There is a great profusion in management of expressions that emphasize this approach. For example, "*The decision is the core of the entire company policy*" (Gutenberg, 1968). "*The decision is at the heart of management*" (Le Moigne, 1976). *"The management process in a global decision-making, in which the phases of which it is composed are carried out in constant interaction without being able to speak of a sequential process"* (Menguzzato and Renau, 1995).

Even when the current environment has become turbulent and time has become a very critical variable, making better decisions does not necessarily mean making faster decisions. Indeed, the business environment changes rapidly and often requires quick decisions, but this can be detrimental to the quality of decisions.

Systems are assessed and analysed in terms of two main performance measures, effectiveness and efficiency. Effectiveness is the extent to which objectives are reached. Therefore, it refers to the outputs of a system. Efficiency is a measure of the use of inputs (or resources) to achieve outputs. Peter F. Ducker suggested the following interesting and well-known manner of distinguishing between the two terms: Efficiency is doing things right, effectiveness is doing the right things (Drucker, 2002).

Ethics Decisions

No discussion about managerial decision-making would be complete without a reference to ethics since such considerations should be an important criterion in organizational decision-making (Wittmer, 2019). In our opinion, an individual decision maker who develops adequate leadership can rely on at least three different criteria to make ethical selections:

- Utilitarianism, through which decisions must be taken considering, above all, its results or consequences, providing the greatest good for the greatest number.
- The focus on rights, so decisions must be consistent and respectful of the rights of different people who are affected directly or indirectly by the decisions made.
- The reference of justice, as a supreme guarantee that decisions are always consistent with the rules of natural law and positive law.

Empirical evidence shows that ethical and unethical actions are the responsibility of both the characteristics of the individual and the environment in which they work:

- The stages of moral development are a model that allows the evaluation of a person's ability to judge what is morally correct. The greater the moral development itself, the less it will depend on external influences and, therefore, the more it will be willing to behave ethically. Those individuals who have advanced to the highest stages, give greater value to the rights of other people without worrying about the opinion of the majority, and can challenge the organizational practices that, from their personal point of view, they consider to be of incorrect moral.

- The locus of control is a personality characteristic that seeks the degree to which a person believes he is responsible for the events in his/her life (Domitriu et al, 2014). Empirical research shows that people who have an external locus of control (they think that what happens to them in life is due to luck or chance) are less likely to take responsibility for the consequences of their behaviour. However, those with an internal locus of control are more likely to rely on their own internal rules of good or evil to guide their behaviour.
- Finally, the ethical organizational environment refers to the worker's perception of organizational expectations and their organizational behaviour.

SOLUTIONS AND RECOMMENDATIONS

The great dynamic changes that have taken place in an increasingly turbulent and digital environment suggest recovering conceptually and revising the methodology to define strategies and support to make effective, efficient and innovative decisions. In this new situation, the functions of prevention, early warning and early assessment of threats, risks and opportunities constitute the central functionality of a new strategic approach, which requires the participation and participation of all workers, to strengthen management levels and leadership in organizations.

The use of strategic intelligence has become critical, and the human skills for its management and its application to the decision-making process are of great relevance, both in terms of defining management strategies and decisions, as complementary to the knowledge of thc more advanced technology tools to support this decision-making.

It is also essential to improve human cognitive abilities, both individually and in groups, in decision-making processes, aware that a large part of the decisions are not typical of the processes of strict or optimizing rationality but of those of bounded rationality or of sufficient satisfaction. The support of digital analysis systems is necessary, but not sufficient.

FUTURE RESEARCH DIRECTIONS

The field that has been analysed for the use of strategic intelligence in decision-making processes raises interesting lines of research to be developed in the field of leadership, management and techniques to be adopted for the innovation of digital services. The projection of the behaviour of human decision makers within the organizational field in the development of expert systems with artificial intelligence raises the need to search for new human-machine balances, which overcome human dysfunctionalities in decision-making, but that maintain the sensitivity and cultural coherence of organizational values.

CONCLUSION

A conceptual review of the methodology has been carried out to define strategies and supports to make effective, efficient and innovative decisions, in order to strengthen management and leadership, given the dynamic changes that take place in a turbulent and digital environment.

Firstly, strategic intelligence has been studied, as a polysemic and polyhedral term, as well as the concepts in which it develops: business intelligence, competitive intelligence, intelligence activities at different organizational levels and information analysis.

Secondly, the application of systems theory and the conceptual field of organizational behaviour has been analysed, covering aspects such as personality analysis, job satisfaction, leadership, authority, power and politics, all of which are fundamental variables that affect the decision-making processes.

Thirdly, the most significant aspects of the decision-making process using strategic intelligence have been studied in general, and the individual internal process and of leadership by which a decision maker makes a decision in particular, taking into account that acceptance or rejection of certain premises will mean the acceptance or rejection of certain procedures for decision making. Under this approach the different ways of understanding the decision process and the implicit rationality applied by the decision maker are recognized, in an affectation to the psychological process of both the individual and social environment that explain the strong tendency to move from an approach from strict rationality to another of bounded rationality.

Fourth, the essential aspects of the decision-making processes in the strategic and management field have been reviewed, reviewing from an applied conception the organizational systemic reality and the organizational behaviour, as well as the most significant recommendations to guarantee the effectiveness, efficiency and innovation of the decision processes.

Fifth, a brief analysis of ethics in the processes of organizational decision-making has been carried out, studying the implications and criteria of an organizational, individual and environmental impact.

REFERENCES

Ademola, O. E. (2017, October). *Reality of human decision-making: A leadership and management platform.* Paper presented at the meeting 9th iSTEAMS Multidisciplinary CrossBorder Conference, University of Ghana, Legon.

Ansoff, H. I., Kipley, D., Lewis, A. O., Helm-Stevens, R., & Ansoff, R. (2019). *Implanting strategic management* (3rd ed.). Palgrave Macmillan / Springer. doi:10.1007/978-3-319-99599-1

Askari, M. Y., & El-Refae, G. A. (2019). The rationality of irrational decisions: A new perspective of behavioral economics. *International Journal of E-Business Research, 17*(4), 388–401.

Barros, G. (2010). Herbert A. Simon and the concept of rationality: Boundaries and procedures. *Brazilian Journal of Political Economy, 30*(3), 119, 455-472.

Bratianu, C. (2017). *Strategic Thinking in Turbulent Times.* In C. Bratianu, A. M., Dima, & S. Hadad (Eds.), *Proceedings of the 11th International Conference on Business Excellence*, Bucharest University of Economic Studies, Bucharest ()10.1515/picbe-2017-0026

Calabretta, G., Gemser, G., & Wijnberg, N. M. (2017). The interplay between intuition and rationality in strategic decision-making: A paradox perspective. *Organization Studies, 38*(3-4), 365–401. doi:10.1177/0170840616655483

Campbell, D., Draper, R., & Huffington, C. (2019). *A Systemic Approach to Consultation. Systemic Thinking and Practice Collection.* London, UK: Routledge.

Cash, P. J., Hartlev, C. G., & Durazo, C. B. (2017). Behavioural design: A process for integrating behaviour change and design. *Design Studies*, *48*, 96–128. doi:10.1016/j.destud.2016.10.001

Clark, R. M. (2019). *Intelligence Analysis: A Target-Centric Approach* (6th ed.). Washington, DC: CQ Press / SAGE.

Clark, R. M., & Lowenthal, M. M. (2016). *The five disciplines of intelligence collection*. Thousand Oaks, CA: CQ Press / SAGE.

Domitriu, C., Timofti, L. C., Nechita, E., & Domitriu, G. (2014). The Influence of the Locus of Control and Decision-making Capacity upon the Leadership Style. *Social and Behavioral Sciences*, 141, 25 August, 494-499.

Drucker, P. F. (1966). *The Effective executive*. New York: Harper & Row.

Drucker, P. F. (2002). *Managing in the Next Society*. New York: St. Martin S. Press.

Economist, The. (2017, May 6). Data is giving rise to a new economy. Retrieved from https://www.economist.com/briefing/2017/05/06/data-is-giving-rise-to-a-new-economy

Fernández-Villacañas, M. A. (2016). The Global Strategic Intelligence Management; E-module multimedia for student's self-study. #Europehome Project, UE: European University Foundation.

Fernández-Villacañas, M. A. (2017). La inteligencia estratégica económica global: La capacidad anticipativa clave en la solución de los nuevos conflictos. In E. del Aire (Ed.), *La Nueva Economía de la Defensa en un Nuevo Orden Mundial. Reflexiones desde el Ámbito Aeroespacial* (pp. 137–163). Madrid, Spain: BBVA & Ministerio de Defensa de España.

Fleisher, C. S., & Bensoussan, B. (2007). *Business and Competitive Analysis: effective application of new and classic methods*. Upper Saddle River, NJ: FT Press.

Fleisher, C. S., & Blenkhorn, D. L. (ED). (2001). Managing frontiers in competitive intelligence. Westport, CT: Quorum Books.

Forrester, J. W. (2007). System Dynamics: A personal view of the first fifty years. *System Dynamics Review*, *23*(2-3), 345–358. doi:10.1002dr.382

Gilad, B. (2004). *Early Warning: Using Competitive Intelligence to Anticipate Market Shifts, Control Risk, and Create Powerful Strategies*. New York: Amacom.

Grant, R. M. (2016). *Contemporary Strategy Analysis: Text and Cases Edition*. Hoboken, NJ: John Wiley & Sons.

Gutemberg, E. (1968). *Economía de la Empresa. Teoría y práctica de la gestión empresarial*. Bilbao: Ed. Deusto.

Hamel, G., & Prahalad, C. K. (1989). Strategic Intent. *Harvard Business Review*, (July): 1. PMID:10303477

Hodgkinson, G. P., & Sadler-Smith, E. (2018). The Dynamics of Intuition and Analysis in Managerial and Organizational Decision Making. *Academy of Management Perspectives 32*(4), Published Online (). doi:10.5465/amp.2016.0140

Hogart, R. M. (1987). *Judgment and choice*. New York: Wiley & Sons.

Kristanti, M. V., & Darma, G. S. (2019). The Role of Marketing & Competitive Intelligence. In Industrial Revolution 4.0. Jurnal Manajemen dan Bisnis, 16(1), Januari.

Kunisch, S., Keil, T., Boppel, M., & Lechner, C. (2019). Strategic initiative portfolios: How to manage strategic challenges better than one at a time. *Business Horizons*, *62*(4), 529–537. doi:10.1016/j.bushor.2019.03.007

Larson, D., & Chang, V. (2016). A review and future direction of agile, business intelligence, analytics and data science. *International Journal of Information Management*, *36*(5), 700–710. doi:10.1016/j.ijinfomgt.2016.04.013

Le Moigne, J. L., & Bénard, A. (1976). *Los sistemas de decisión en las organizaciones*. Madrid, Spain: Tecniban, D. L.

Liebowitz, J. (2006). *Strategic Intelligence: Business intelligence, competitive intelligence, and knowledge management*. Boca Raton, FL: Auerbach Publications. doi:10.1201/9781420013900

López-Robles, J. R., Otegi-Olaso, J. R., Porto-Gómez, I., & Cobo, M. J. (2019). 30 years of intelligence models in management and business: A bibliometric review. *International Journal of Information Management*, *48*, 22–38. doi:10.1016/j.ijinfomgt.2019.01.013

Macdonald, I., Burke, C., & Stewart, K. (2018). *Systems Leadership. Creating Positive Organisations* (2nd ed.). London, UK: Routledge. doi:10.4324/9781315178486

McDaniels, T. (2019). *Four Decades of Transformation in Decision Analytic Practice for Societal Risk Management. Risk Analysis - An International Journal*. Society for Risk Analysis, Wiley Online Library; doi:10.1111/risa.13332

McGonagle, J. J., & Vella, C. M. (1996). *A New Archetype for Competitive Intelligence*. West Port, CT: Green Wood Publishing Group.

Menguzzato, M., & Renau, J. J. (1995). *La Dirección Estratégica de la Empresa*. Barcelona, Spain: Ed. Ariel Economía.

Mintzberg, H. (1994). *The Fall and Rise of Strategic Planning. Harvard Business Review*. January-February Issue.

Mufudza, T. (2018). Dynamic Strategy in a Turbulent Business Environment. In IntechOpen Ed., Strategic Management - A Dynamic View, (DOI:) doi:10.5772/intechopen.81250

Mullins, L. J. (2005). *Management and Organizational Behaviour*. London, UK: Pitman Publishing.

Noe, R. A., Hollenbeck, J. R., Gerhart, B., & Wright, P. M. (2017). *Human resource management: Gaining a competitive advantage*. New York: McGraw-Hill Education.

Palacios, J. M. (2018). The Role of Strategic Intelligence in the Post-Everything Age. *The International Journal of Intelligence, Security, and Public Affairs*, *20*(3), 181-203.

Pirttimäki, V. (2007). Business Intelligence as a Managerial Tool in Large Finnish Companies. Tampere University of Technology, Publication 646, Tampere.

Porter, M. E. (2008). *The Five Competitive Forces That Shape Strategy. Harvard Business Review.* January Issue.

Ramirez, R., & Selsky, J. W. (2016). Strategic planning in turbulent environments: A social ecology approach to scenarios. *Long Range Planning*, *49*(1), 90–102. doi:10.1016/j.lrp.2014.09.002

Richards, G., Yeoh, W., Loong, A. Y., & Popovic, A. (2019). Business Intelligence Effectiveness and Corporate Performance Management: An Empirical Analysis. *Journal of Computer Information Systems, 59*(2), 188-196.

Robbins, S. (2009). Comportamiento Organizacional. Teoría y práctica 13th Ed. Naucalpan de Juárez: Ed. Pearson Educación México.

Seitovirta, L. C. (2011). *The Role of Strategic Intelligence in Corporate Decision Making. Archive number 12600*. Espoo: Aalto University.

Steiner, G. (1979). *Strategic Planning: What Every Manager Must Know*. New York: Free Press.

Thierauf, R. (2001). *Effective Business Intelligence Systems*. Westport, CT: Quorum Books.

Turban, E., Aronson, J. E., & Liang, J. P. (2005). *Decision Support Systems and Intelligent Systems* (7th ed.). New Jersey: Ed. Prentice Hall.

Ungureanu, E., & Avramescu, T. C. (2008). Strategic Diagnosis and the Part it Plays in Decision Making. *Annals of the University of Petrosani. Economics*, *8*(2), 269–274.

Vecchiato, R. (2015). Strategic planning and organizational flexibility in turbulent environments. *Foresight*, *17*(3), 257–273. doi:10.1108/FS-05-2014-0032

Wittmer, D. P. (2019). *Ethical Decision-Making. Chapter in Terry Cooper, Handbook of Administrative Ethics* (2nd ed.). New York: Routledge.

Chapter 5
The Clash of the Titans:
CIO and LOB Engagement in IT Innovation

Sachithra Lokuge
https://orcid.org/0000-0003-4558-687X
RMIT University, Australia

Darshana Sedera
Swinburne University of Technology, Australia

Shailesh Palekar
Queensland University of Technology, Australia

ABSTRACT

The emergence of digital technologies provides organizations with affordable, easy-to-use, and easy-to-learn technologies that opens line-of-business (LOB) managers to participate in information technology-led innovations. Such participatory actions by the LOB-managers in information technology (IT) innovations at the department levels will lead the creation of new roles like the 'chief digital marketing officer.' While the overall outlook for innovation becomes positive through such expanding contributions of the LOB-managers, such additions will ignite new challenges. Especially, such roles and approaches will contest the traditional centralized technology management approaches. This chapter provides three engagement models on how the LOB-managers and chief information officers could interact harmoniously to enhance the quality of IT innovations led by LOB-managers.

INTRODUCTION

Information technology (IT) innovation is important for the survival of market competition (Lokuge et al. 2019; Svahn et al. 2017) and digital technologies have been identified as important catalysts in triggering and enabling IT innovation (Chen et al. 2015; Nambisan 2017). The advent of digital technologies such as social media, mobile technologies, analytics, internet-of-things is transforming the very nature of how organizations conduct their internal business functionalities (Ebel et al. 2016; Sedera et al.

DOI: 10.4018/978-1-7998-2799-3.ch005

2016). The application of such technologies has provided unique opportunities for departments such as sales, marketing, warehouse, finance and operations to contribute to the organization in a direct way by responding to customer needs and wants. In particular, Swanson (2012) highlights the important role of the line-of-business (LOB) managers who are becoming ever more important in initiating and adopting IT innovations. LOB managers are employees that work in departments such as sales, marketing, human resources, finance where non-managerial employees report (Hales 2005). With the advances in the organizational hierarchies, these managers are provided with more responsibilities, power, budget and control (Renwick 2003). However, the orthodox view is that the new ideas and opportunities enabled by IT in an established organization typically stem from the top management (Qian et al. 2013). For example, the traditional view is that all IT innovations are initiated, managed and executed by the chief information officer (CIO) and the IT department of the organization. A major reason for such traditional prominent view is that IT is considered as a specialized department and employees of the other departments are not required to possess IT skills and knowledge for operational activities of an organization. While organizations have valued the participation of employees in the creation of new ideas and opportunities through empowerment of operational staff and middle management leadership, factors such as complexity of IT resources and the portfolio (Chae 2014), IT knowledge of the employees (Flath et al. 2017) and misalignment of IT solutions with business strategies (Gerow et al. 2014) have prevented them taking an active role in IT innovation in organizations.

However, with the advent of digital technologies organizations can truly engage operational staff and middle managers in IT innovations. Prior literature highlights that the availability of digital technologies, ease-of-use, ease-of-learning and relative low prices of the digital technologies are encouraging innovations to be introduced at the grass-root department levels (Brinker and McLellan 2014; Lokuge et al. 2019). These innate characteristics of digital technologies (Nambisan 2017) are inspiring employees to actively take part in IT innovation activities within an organization. Such advances in digital technologies are fundamentally changing corporate competition, exerting additional pressures on organizations to adjust their business logic and processes in the digital era and increasing momentum of digitization (Ebel et al. 2016).

In order to foster IT innovations, front-line facing departments are introducing new roles such as Chief Marketing Technologist, to align relevant marketing technology with business goals, to liaise with IT, and to evaluate and choose technology providers for marketing department (Brinker and McLellan 2014). Further, according to Gartner (2016), 61% of North American organizations are increasing capital expenditures on technology at the LOB-levels. On the other hand, the success of an organization's innovativeness depends heavily on the manager's ability to adapt and progress their business models (Andries and Debackere 2007). However, this emerging phenomenon of the LOB-manager's role in IT innovations challenges the traditional view of organizations taking directions only from the CIO (or its equivalence) (Arnold et al. 2000). In this emerging model, leadership and autonomy for innovation must be given for LOB-manager by the CIO. As highlighted in their work, Augier and Teece (2009, p. 411) argue, [that] "the new world we are in requires a different breed of managers, and highly skilled employees with capacities to combine and integrate." Not surprisingly, the contemporary organizations consider leadership and autonomy to be critical success factors for innovations (Puranam et al. 2006). However, this emerging phenomenon of the leadership and autonomy for LOB-led IT innovation has received much less attention. As such, the objective of this book chapter is to investigate the engagement of the CIO and the LOB-manager in managing LOB-led IT innovations. Thus, we investigate the research question "What are the engagement models that CIO and LOB managers can adopt when initiating IT innovations?" We define engagement by following the definition of Salanova et al. (2005) as the simultaneous employment and expression of a person's preferred self in task behaviors that promote connections to work and to others, personal presence (physical, cognitive, and emotional) and active, full performances. Further, when studying IT innovations, we specifically focus on investigating the IT innovations that are led by LOB managers. Such innovations we define as LOB-led IT innovations. This book chapter investigates the leadership roles of CIO and LOB managers in managing LOB-led IT innovations, the role of the CIO in managing such innovations and the factors that are important to consider when engaging with LOB-led IT innovations.

LITERATURE REVIEW

Digital technologies have presented new opportunities and business models that dramatically changed how organizations utilized IT for achieving corporate goals (Nambisan 2013; Nylén and Holmström 2015; Sedera et al. 2016; Yoo et al. 2012). The cost effectiveness, ease of acquisition, ease of learning, ease of deployment and ease of management of digital technologies are further ascribed to the popularity of these technologies (Nylén and Holmström 2015). Even though these technologies are easy to obtain, they must be managed effectively in order to innovate (Nambisan 2013). As Penrose (1959) argues, acquisition or the owning of resources does not guarantee competitive advantage, however, "cultivating alignment between business and IT strategies could increase profitability and generate a sustainable competitive advantage" (Gerow et al. 2014, p. 1159).

In the contemporary business landscape, with the advent of the digital technologies the LOB-managers are also provided with the opportunity to innovate (Brinker and McLellan 2014; Lokuge et al. 2018). We define innovation following the definition of Damanpour and Evan (1984) as the 'implementation of an idea whether pertaining to a device, system, process, policy, program or service that is new to the organization.' Contrary to the tradition whereby the CIO and the IT department initiates IT innovations, innovations can be introduced at the grassroots functional department level as well. In accordance with the literature, the success of the IT innovations initiated at the LOB level also must concur with the strategic goals. Prior researchers have investigated such alignments from different viewpoints such as, orientation of the business strategy and IT strategy (Chan et al. 2006), IT skills, knowledge, capabilities and business strategy (McLaren et al. 2011) etc. Even though such studies consider the alignment between IT and the strategic goals, the alignment of the views and the engagement between the managers is one of the important factor that rarely discussed in prior literature.

LOBs are separate units such as marketing, sale, finance which prior research considered only as supporting business units. However, with the advent of digital technologies, LOBs also have access to a budget that allows them to acquire IT resources, manage their own IT portfolio and initiate IT innovations (Henfridsson and Lind 2014). As such, under these circumstances, the LOB-managers, most of the time, do not require to wait until they receive permission from the CIO and IT department for initiating innovations. With the access to such low budget innovations, they can initiate and launch innovations as they see fit. However, when such IT innovations are initiated, there is an impact on the IT portfolio of the organization which is looked after by the CIO and the IT department. As such, the interaction between these parties (i.e. LOB-manager and the CIO/IT department) becomes an important factor. For example, when one leader (i.e. LOB manager) is initiating an innovation that has an impact on another leader's milieu (i.e. CIO and IT department's context and responsibilities), then there is a possible clash between the two leaders. Managing such complexities is also an important aspect. Especially, in this contemporary era where LOB managers are able to initiate IT innovations due to innate characteristics of digital technologies, such complexities are likely to occur.

Therefore, it is interesting to explore how LOB-led IT innovations manage the alignment with the organizational level strategies and policies. When LOBs initiate IT innovations, it important for the CIO/IT department to make sure that there are minimal effects to the overall business processes, which ultimately have an impact on the organization. Considering the nature of the LOB-led IT innovations, there is a constant conflict between LOBs and the CIO over IT innovations which need to be constantly managed to ensure alignment. For example, any IT innovation that LOB's propose will be using one or many common IT resources in the organization. These IT resources can be the enterprise system of an organization which is organization-wide and a small change to the enterprise system will have an impact on the central database. Further, IT culture, policy and rules are usually managed by the CIO and the IT department. Also, they are responsible for any issues with the IT systems as well. Therefore, even though LOB-led IT innovations are innovative and will add value to the organization, there is a huge risk for the organization as well. As such, CIO and the IT department need to make sure to come up with solutions to mitigate any issues that they face. Based on the discussion above, in this book chapter we investigate the engagement between CIO and LOBs. Especially, it is interesting to investigate how the engagement between the CIO and the LOB-manager occurs when LOB-led IT innovations take place. To the best of our knowledge, such discourse is yet to take place and highlights a gap in the literature. While the extant studies provide a wealth of knowledge on the IT business alignment, less attention has been given in investigating the engagement of CIO and the LOB-manager when the LOB-led IT innovations are initiated.

RESEARCH METHODOLOGY

The objective of this book chapter is to investigate the engagement between the CIO and LOB-manager when LOBs are initiating IT innovations. The explorative nature of this research deemed a qualitative approach (Emory and Cooper 1991; Klein and Myers 1999; Yin 2010). To explore and to get an in-depth understanding of the phenomenon, a case study method was applied (Walsham 1993). For twelve months the researchers engaged with various organizations investigating IT projects in each case organization. In here, data was gathered about the IT projects from their inception to delivery. By following theoretical sampling, six IT projects that belong to four organizations were selected as in these projects LOB-mangers were initiating IT innovation to achieve competitive advantage. To generalize the finding of the study, multiple cases that represent diverse industry sectors and ownership structures (i.e., publicly listed and multi-national organizations) were selected.

Furthermore, all the organizations fulfilled the following criteria: (i) the organization had a dedicated CIO position and a team of IT staff that managed the organization's IT portfolio, (ii) the organization's LOBs had access to their own technology portfolio as well as a shared IT portfolio among the other LOBs, (iii) at the time of the data collection, the CIO had been in the position for at least six months, was not in the last six months of their appointment and was participating in regular meetings with the executive leadership team (e.g., the Chief Executive Officer and Chief Finance Officer) and (iv) the case organization had initiated an IT-centric project using one or more digital technologies within the past 12 months.

The snowballing technique was utilized to recruit interviewees (Myers and Newman 2007); it sought two types of informants from each organization: (i) the CIO and (ii) the LOB managers. According to Laforet (2013), when investigating a phenomenon such as innovation, observations must be made from managers/leaders who initiated or were immediately affected by the innovative endeavors for the purpose of corroboration, triangulation, and substantiation. As such, the study commenced with interviews with the CIO (leaders who focus on IT innovation), or with the individual holding an equivalent position (i.e., Chief Technology Officer or Technology Lead) was selected for the interviews. In addition to the CIO, the LOB managers and managers from each project (i.e., cases) were interviewed to obtain a closer perspective of each project. While interviewing CIOs, a closer attention was paid to adjust the interview duration to suit interviewees' emergent demands (Myers and Newman 2007). The data collection was conducted through 31 semi-structured interviews, totaling 45 person-hours. All the interviews followed the same case protocol. Each interview took between one to two hours and in most cases, follow-up interviews were conducted for clarification or due to time constraints where the CIO was unavailable for lengthy meetings in a single session. Several new probing interview questions were added based on the interviews which allowed the emergence of new themes. All the interviews were conducted face-to-face, in the English language, and were recorded and transcribed. Further, to obtain an appropriate degree of reliability, three sources of evidence such as internal documentation, general web search and interviews were used (Dubé and Paré 2003; Klein and Myers 1999). During the interviews, when the interviewee mentioned about internal documentation such as design documents, concept papers, we requested access to these documents as well. Data gathered from all these sources was used to corroborate, validate and complement the interview data.

The four organizations involved in the study were C1, C2, C3 and C4. The cases selected here are referred with pseudo-names due to the confidentiality agreements signed between the organization and the researchers. To ensure the depth of the interpretation, we selected organizations that represent different industry sectors and selected interviewees from different hierarchical levels (Klein and Myers 1999). All four organizations had an implemented enterprise system that was shared among all departments in the organization. At the time of the data collection, C1 was operating in more than two continents, while C2 – a leading producer of fruits and vegetables – had operations in Australia. C3 is in public-sector organization and C4 is a leading private insurance organization in Sri Lanka.

Following the guidelines of Klein and Myers (1999), an interpretive data analysis approach was applied. After transcription and examination of the data, two researchers developed a comprehensive case description detailing each project and summarized the narratives in a table (Walsham 1993). The two researchers initiated the data analysis process by summarizing and reviewing the interviews throughout the data collection. The initial analysis of data involved identifying different IT that were utilized in the projects, their role and the characteristics. The categorized data was then further examined, and the emerging themes were then organized to describe the engagement to the overall organization processes, procedures and strategy.

During this stage, the two researchers referred to further details such as company documents, websites, news and referred to the literature as themes emerged. In comparing the emerging themes to existing literature, the researchers adopt existing labels, concepts, or explanations when there were commonalities. Similarly, new labels, concepts, or explanations were developed when there were opportunities to extend the existing literature. For example, as we continued our analysis, we identified different engagement models possible between the CIO and LOB. Prior studies on IT innovation discuss the alignment in the firm however, missing from the existing literature was an understanding of how LOBs contribute in IT innovation and how they maintain IT business alignment. Further through the analysis, it was observed that the cases used different IT in different contexts and in initiating some of the IT innovations they did not seek the permission from the IT department. The interviews and the project documents were used to identify the sequence of events and helped to identify the outcomes of each of the projects. The data analysis process was iterative with a recursive feedback obtaining from the IT business alignment, economic theory of complementarity (Milgrom and Roberts 1995) literature, the data collected and the emerging interpretations. This iterative process continued until we reached the state of theoretical saturation. Once we had the core concepts identified, we interpreted them in the context of existing literature, and we demonstrated these empirical patterns into a conceptual framework.

SOLUTIONS AND RECOMMENDATIONS

LOB-Managers Initiate IT Innovations

The analysis of the data collected from all four organizations revealed that the LOB-managers were inspired to acquire digital technologies and initiate IT innovation in their respective departments. Here the LOB-manager took advantage of the innate characteristics of digital technology such as affordability, trialability, ease of use and ease of learning to derive solutions as per the requirements of their respective departments. Further, like the IT department, the LOBs too observed competitor behaviors, solutions of IT companies and even technologies that they could use as customers in analogues areas. These ideas and the affordability of the digital technologies led these LOB-managers to initiate such innovations.

The LOB-managers were aggressively promoting their business needs and possible opportunities to engage with digital technologies to their CIOs and IT departments through compelling business ideas. Their ideas and suggestions affected only to their departmental considerations and boundaries. Even though the LOB-managers made specific observations of the bottlenecks and inefficiencies of the current LOB business processes, some of their suggestions marked clear 'leaps' of thinking, boundaries and efficiencies. However, such LOB-led innovations are invaluable for tailoring specific IT solutions within the department. While initiating such innovations, the LOB-managers drew feedback from the CIO and IT department. This demonstrated the first point of engagement between the LOB and IT department on LOB-led IT innovations.

However, there was a general tendency to assume that it is okay to develop and deploy an application within a department to improve departmental performance, if they are within the departmental budget, creating contentions between LOBs and the IT department. However, since these innovations have a direct impact on the IT portfolio of the organizations, the CIOs did not prefer to approve any of the IT innovations without their consent. Such tensions between the two parties creates a barrier for innovation.

Further, apart from the capabilities of the digital technologies, the LOBs were inspired to innovate as they wanted to contribute to the wellbeing of the organization. They highlighted that rather than waiting for solutions to receive from the CIO and the IT department, they decided to look for possible solutions to be integrated to their departments. Since the LOBs are familiar with the business processes, bottlenecks, issues and can pinpoint the troubles they were facing, it was easier for them to come up with effective solutions. Further, the applicability of these innovations was high.

The Power Struggle and the Need to Retract Power

When the LOB-managers initiate an idea, in some cases the respective CIOs and IT departments played a substantial role in developing a prototype of the required solution. Since the necessary experience and expertise for the development of IT solutions are with the IT department, these organizations were able to launch their IT innovations faster. In these cases, the CIOs highlighted the shorter development time and effort of LOB-led IT innovations. In the iterative process of technology development, the prototype was handed over to the LOB for feedback, and testing. The LOB in turn was able to see how the proposed IT solution featured in the organizational strategy. The iterations also allowed the LOB-managers to better understand the role of corporate IT, strategy and the utility of existing technology infrastructure.

However, the major drawback the CIOs lamented about was the lack of a 'holistic' view of the organization emanating from the ideas proposed by LOB. Especially, the CIOs argued LOB mangers lacking knowledge on compliance, policy, security and creating bottlenecks as a result of stimulating part of a business process. It was also highlighted that the CIOs and the IT departments developed protypes that aligned closely with the existing infrastructure and strategies. It was identified that, when designing the prototypes, the IT department did not consider possibilities that were considered 'out the box' fearing too much misalignment. As such, in cases where the CIO and the IT department involvement are high, the integration to the existing technological platforms was prominent. The CIOs highlighted the necessity and the value of integrating the 'localized' LOB applications to the corporate IT systems and infrastructures.

Once the LOB-managers approve the technical and business architecture of the proposed solution, the IT department began implementing it. Therein, the IT department appreciated the innovation capabilities of the department – though there had been multiple points of contentions and frictions. Further, the CIO and the IT department were coming to terms with the advent of the new culture, where LOB-managers are eager to contribute to the corporate IT portfolio. Such a recognition marks the departure from a highly centralized model of managing IT to a federated IT portfolio management model. Even though this new culture was prominent, a general displease regarding the initiatives by the LOBs, as an aversion towards risk was also evident.

The CIO and the IT department were cautious about all LOB-led IT innovations, worried about IT governance, compliance and security risks. The IT leadership also recognized the opportunities and risks associated with this new involvement of the LOBs in IT innovations and made several adjustments in their management approach. For example, once a successful IT project was executed, the CIO and the IT department attempted to derive policy or guidelines that encourage re-use of corporate-wide technologies. This was observed as an attempt to consolidate a 'single corporate image' across the operational areas of marketing, sales and IT. Similarly, purchasing of new on-demand systems and infrastructure for departmental use were curtailed to a 'preferred vendor list' that gets renewed periodically.

Figure 1 shows the scope of the IT innovation and the management approach of organizations. In here the innovation scope is categorized as functional and organizational. Functional refers to applicability or the impact of the innovation to a particular department or business activity only. The organizational innovation scope refers to the applicability or the impact of innovation proposed on the organization as a whole. Such innovations go beyond a single department and has an impact on many departments or whole organization. Organizations can follow two types of management approaches: centralized and decentralized. In centralized management approach, one party is responsible for managing all activities. For example, in the traditional management approach, CIO and the IT department were responsible for initiating and managing all IT related activities. Any IT related issues and concerns had to go through

Figure 1. Scope of the innovation and the management approach

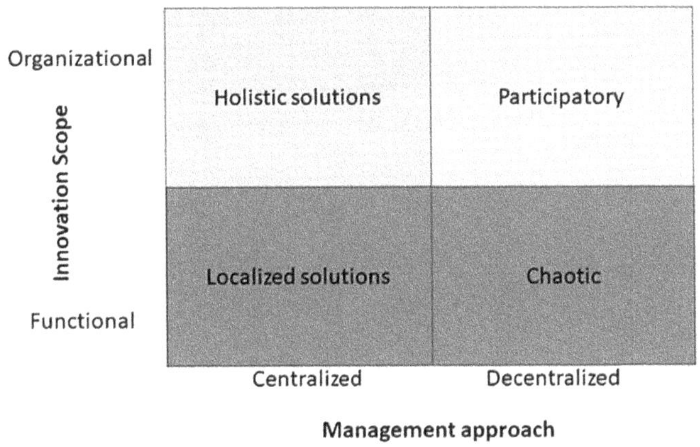

IT department first. Such management approach is known as centralized. In contrast, when the responsibilities are shared among departments, such management approach is known as decentralized approach. Usually, some organizations when they have multi-organizations around the globe, they follow a decentralized management approach to minimize the delays and make the operations effective and efficient.

As per Figure 1, when the management approach is centralized, and the innovation scope is functional the LOBs and the CIOs are focusing on localized solutions. In such scenario, the leadership of the CIO is prominent as this is centralized approach. However, when the IT solution is specially for a particular department, then the two leadership roles clash. When the management approach is centralized and innovation scope if organizational-wide, then CIO and the IT department take the leadership. Here, the LOBs will not play much of a role, but if the impact is significant, then again, a chaotic environment between the two leadership parties may occur. When the management approach is decentralized, and the innovation scope is functional, the LOB managers play the main role. From the LOB-managers point of view, it was quick and easy way of getting things done. However, the more autonomy they have, more chaotic the environment is for the CIO and the IT department. When organizations do not have proper IT policies and procedures in place, they found it very difficult to manage. As such, in such situations, these organizations started introducing new policies and procedures to mitigate risks. When the management approach is decentralized, and the innovation scope is organizational-wide the organizations saw a participatory nature in IT initiatives. However, considering different needs and requirements of different LOBs such initiatives took long time to process.

The Engagement of LOB-Manager and the CIO

It is evident that whether it is centralized or decentralized management approach, or whether it is functional or organizational-wide innovation, each scenario will have their own issues and leadership clashes. To minimize the risk and clash between the two leaders, each leader (i.e. CIO and LOB) had to play unique roles in order to make the IT initiatives successful. As depicted in Figure 2, the LOB-managers and CIOs play a distinct role in relation to LOB-led IT innovations. Further, the figure also outlines engagement between the CIO and the IT department and the LOB-manager in LOB-led IT innovations.

Figure 2. Different engagements during the innovation process

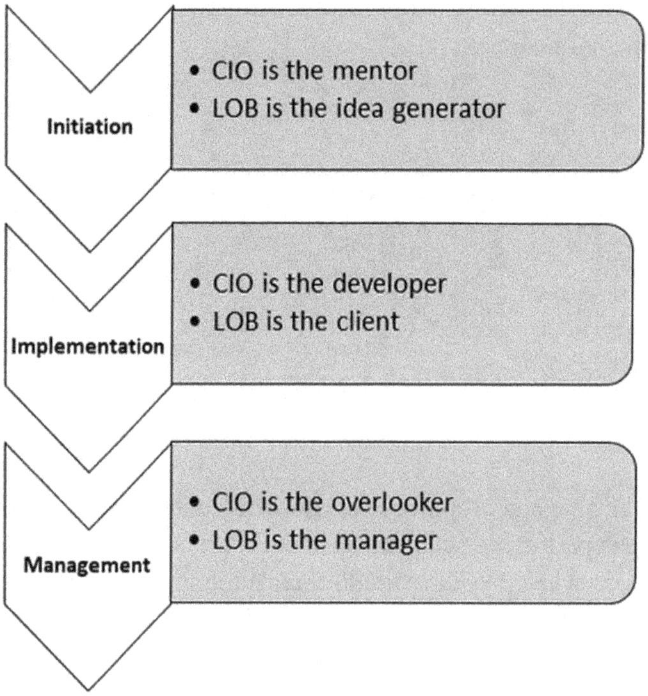

Further investigating the LOB-led IT innovations, it was observed that there are three phases where the engagement can occur: (i) initiation, (ii) implementation and (iii) management (See Figure 2). In the initiation phase, the LOB and the CIO focus solely on deriving an agreement on the proposed idea. Here, the LOBs and senior executives, negotiating the scope, management and value propositions of the proposed solution. The LOB becomes the idea generator and the CIO becomes the mentor as he has experience and expertise in IT innovations.

Second, the implementation phase focused on development of the solution, aligning the IT infrastructure (available or available to acquire) with the prototype of the proposed solution. Usually, the implementation phase involves developing a prototype. Through prototyping, a wavering agreement of the senior IT managers is obtained to increase the skills, technology architecture and processes. Here, CIO and the IT department were responsible for developing the solutions. LOB becomes the client and the CIO/IT department consult them throughout the implementation process. The organizations experienced less problems when they followed this process in LOB-led IT innovations.

Third, the management phase refers to degree to which the needs, goals, objectives, and/or structures of IT at the LOB-level are consistent with the business at the organizational level. Therein, the implemented solution is incorporated to the corporate IT portfolio and the management of the solution is given to the LOBs. However, the CIO/IT department overlooks the solution. In this phase, the leadership roles changed by CIO becoming the over-looker ad the LOB becoming the manager. Figure 2 shows the different leadership roles of the CIO and the LOB in different phases of the LOB-led IT innovations.

Even though we observed different roles of the two leaders at different phases, overall when analyzing the whole scenario, we identified prominent leadership behaviors. When observing the nature of the two leaders in LOB-led IT innovations, two types of behaviors were evident. While one leader (CIO) tries to maintain a responsible behavior to keep the policies, strategies and alignment in place, it was clear that the LOB-managers were being more responsive/agile to the environmental changes. As such, we identify two types of leadership behaviors: (i) the responsive and (ii) responsible behavior.

Characteristics of these two leadership behaviors are provided in Table 1 below. The key observants in these two leadership behaviors were innovation vs efficiency focused nature of the two leaders. While LOB is trying to innovate, the CIO was trying to balance the impact and operational side of the IT systems. While the LOBs were continuously assessing the IT systems and solutions, the CIOs were assessing the IT readiness of the organization and making sure that the organization is ready for the future and the digital era. Further, while LOBs were focused on more autonomy, the CIOs were focused on attaining IT business alignment and minimizing the impact of LOB-led IT innovations on the overall business. It was evident that the two leaders were reflecting two different characteristics in IT innovation process. The LOB manager is acting as an agent of innovation while the CIO was acting as an authoritarian. Even though in the traditional view, the CIO should act as agents of IT innovation, we identified that the LOBs too are working as agents of innovation. While CIO and IT department allow such behaviors, they make sure that the risk and the negative impact on the organization and the IT systems are minimized.

These two behaviors can be explained using the behavioral control theory perspective (Kirsch 1996). Even though behavioral control theory is much used in outsourcing literature, for explaining the LOB-led IT innovation process too we can apply this as it is also an ongoing social process which involves multiple stakeholders. Further, the success of the innovation project highly depends on the management of the CIO and the IT staff. The changes the LOB-manager and the staff carry out in LOB-led IT innovations, they need to consult the CIO to get their feedback on alignment of organizational strategies, IT resources, policies and plans. As such, in a LOB-led IT innovation project, the CIO becomes the controller and the LOB-manager becomes the controlee.

According to the behavioral control theory, a control is a mechanism that the controller sets to govern the actions of the controlee (Soh et al. 2011). However, the CIO-LOB relationships are not entirely governed by formalized organizational mechanisms. For example, the LOB-manager is not required to do IT innovations in an organization. However, the advent of digital technologies has changed their roles and the responsibilities of a manager. Even though, the LOB-managers are given the freedom to initiate new ideas, from the deployment to management phases, the CIO and the IT staff get involved. This highlights the controlling behavior of the CIO and the need to maintain the controls of an IT project. As McLachlin (1999) explains, major issues can occur if the CIO neglects to establish controls of the engagement. As such, to maintain this, during the implementation and management phases, the CIO gets more involved in the project. Therein, we identified more prominent leadership role of the CIO in the last two phases of the innovation process.

Table 1. Responsive and responsible behaviors

Responsive Behavior	Responsible Behavior
• Focus on innovation. • Continuous assessment of opportunities and challenges (recursive implementation) • High autonomous nature • Acting as agents of change and innovation	• Focus on efficiency (cost and time focus) • Assessment of the organizational readiness (e.g., culture, IT, human, skills, financial etc.) • Assessment of the strategic IT business alignment • Acting as authoritarians

In these phases, the LOB-manager requires the CIO to:

1. provide additional expertise (Poulfelt and Payne 1994) and knowledge (Dawson et al. 2010)
2. derive solutions for complex problems arise during the deployment (Kitay and Wright 2003; Poulfelt and Payne 1994)
3. provide an independent perspective
4. perform system integration (Gartner 2010).

In these scenarios, the CIO and the LOB manager are required to engage in order to compete the IT innovations successfully. When observing such activities and the nature of the leadership of the CIO and the LOB-manager, it was clear that similar to Schein (1990), three CIO-LOB engagement models were prominent in LOB-led IT innovations as well. They are:

1. Purchase of expertise – this is a detached relationship. The LOB-managers are only seeking independent perspective on their business problems.
2. The doctor-patient model – the LOB-managers seeks advices from the CIO and the IT staff for their technical and business problems. The model is focused more on diagnostic approach.
3. The process consultation model – the CIO and the IT department are facilitators providing the technical expertise on defining the business problem by offering frameworks and methodologies. The roles and responsibilities of a CIO or the IT staff are well defined, and the LOB-manager is responsible for deciding the outcome or solution, CIO or the IT staff may provide alternative solutions.

As discussed above, the CIO and the IT staff support can be used by LOB-manager in the above-mentioned models to perform a wide range of tasks in IT innovation process. Depending on the nature and the complexity of the IT innovation process, the CIO or the IT staff can select the most suitable engagement model (Kitay and Wright 2003). Even though the LOB-staff is equipped with knowledge and expertise in initiating IT innovation projects, there are three ways they can utilize the expertise of the CIO and the IT staff in their IT innovation projects. Such engagement models will allow both CIO and the LOB managers to launch their IT innovations successfully. Further, the engagement of CIO and the LOB will minimize the risk of innovation failure and the negative impact it has on the organization and the IT systems and processes.

FUTURE RESEARCH DIRECTIONS

As this is a preliminary investigation of the LOBs initiating IT innovations, there are several limitations of the study. First, the study was conducted using a small sample. To further verify these complementary engagement types, more research is needed. Specifically, a quantitative approach is more appropriate to generalize the findings. Studies that allow further variation through industry sectors, countries and other contextual characteristics can further improve the robustness and generalizability of these findings. Second, future study can add further granularity to the complementary engagement types. For example, researchers can further study the conditions for certain engagement types as well as factors that could differentially influence their successful engagement. Third, future research could investigate the leadership clash between the two leaders and antecedents and mechanisms to minimize such risks. Consideration of such factors in a future study will improve our understanding of how such factors can be incorporated into organizational learning and enhance the operational efficiency and innovation. Finally, further research required to understand LOB and CIO engagement. Such investigations can have substantial research and practical implications, specifically in identifying new success factors and outcomes of engagement.

CONCLUSION

In recent times, organizations face new challenges through the advent of digital technologies such as social media, mobile, analytics, cloud computing and internet-of-things. The participation of the LOB-managers in IT innovation process challenges the IT management structures and policies. With the background of such technology interventions, we conducted a study, using data collected from four cases to understand how contemporary organizations' CIOs engage with the LOB-managers when LOB-managers also initiate IT innovations. As such, the objective of this book chapter was to explore the engagement of the CIO and the LOB-manager in the LOB-led IT innovation process. We identify LOB-led IT innovations as IT innovations that are led by the LOB-managers in the organization. This book chapter contributes to leadership and management body of knowledge by extending the understanding of how engagement differ across innovation phases for both CIOs and LOB-managers. For example, in the book chapter we provided how the CIO and the LOB played different roles in different phases and how such roles helped them to successfully execute LOB-led IT innovations. It further conceptualized two behaviors of leaders in the LOB-led innovation process.

Further, the findings highlighted a new emerging role of the LOB-managers and yielded the conclusion that organizations are already changing their view of the role of the LOB-managers. For example, Gartner (Brinker and McLellan 2014) predicted that by 2017 each salient LOB (i.e. all functional departments, such as marketing) in all major organizations will have a designated Chief Technology Officer. Brinker and McLellan (2014, p. 83) point out that Kimberly-Clark had introduced the role of chief marketing technologist to better deliver functional requirements through the wealth of available technologies. They explain that the main objective of the chief marketing technologist's role is "to create the best technology vision for marketing" and that the chief marketing technologist will enable departments to campaign for "greater experimentation and more-agile-management of that function's capabilities" as they are "change agents of innovation." As such, this book chapter provides insights into the evolving role of the existing LOB-managers than creating a new role.

The results of the study: (i) LOB-led innovation process and (ii) the engagement types provide a commencing point for researchers to examine how different engagement types can assist organizations to complete IT innovation projects successfully. Further, the book chapter proposed organizations should sought engagements at the initiation, implementation and management phases while initiating LOB-led innovations. Also, at each phase, the role the CIO and the LOB need to play was also discussed. Such discourses will minimize the tension and the clash between the two leaders. The need for three engagement types become critical as the LOB-led IT innovations may not consider the strategy, processes and technology at the organizational level. As such, engagement at the CIO/IT department level is important for the success of the LOB-led innovations.

In addition, this book chapter introduced two leadership behaviors between the two leaders (i.e. CIO and LOB-manager). The identification of new business requirements and being agile to respond to the market needs is the 'responsive behavior' of a leader. Even though the responsive behavior of the LOB-managers were prominent in the initiation phase, the CIOs played more 'responsible role' in the deployment and management phases. The reason for having two different leadership styles between the two leaders is that even though the LOB-led IT innovation project focusses only on the department, the use of IT will have an impact on the whole organization. As such, the CIO played a more responsible role. The study also identified three engagement models between the CIO and the LOB-manager in managing innovation projects. As the future research, the authors will look further into how LOB-managers can decide on the engagement model based on the nature of the IT innovation project.

There are several implications for practitioners. The book chapter findings highlighted the emerging role of the LOB-managers in initiating, implementing and managing the LOB-led innovation process. The three engagement types provide guidance for the IT manager (e.g., the CIO) to identify the suitable engagement model for a given IT project. Further, the book chapter highlighted the collaborative role both CIOs and LOB-managers should play to successfully complete a LOB-led innovation project. It is clear through the findings that the conventional leadership approach of CIO leading IT related decision making is not suitable for the contemporary business environment. The LOB-manager's role has evolved to a more responsive role than a passive role.

REFERENCES

Andries, P., & Debackere, K. 2007. Adaptation and Performance in New Businesses: Understanding the Moderating Effects of Independence and Industry, Small Business Economics (29:1-2), pp. 81–99.

Arnold, J. A., Arad, S., Rhoades, J. A., & Drasgow, F. (2000). The Empowering Leadership Questionnaire: The Construction and Validation of a New Scale for Measuring Leader Behaviors, *Journal of Organizational Behavior* (21:3), pp. 249-269.

Augier, M., & Teece, D. J. (2009). Dynamic Capabilities and the Role of Managers in Business Strategy and Economic Performance, *Organization Science, 20*(2), pp. 410-421.

Brinker, S., & McLellan, L. (2014). The Rise of the Chief Marketing Technologist, *Harvard Business Review, 92*(7), pp. 82-85.

Chae, B. K. (2014). A Complexity Theory Approach to IT-Enabled Services (Iess) and Service Innovation: Business Analytics as an Illustration of Ies, *Decision Support Systems, 57*(1), pp. 1-10.

Chan, Y. E., Sabherwal, R., & Thatcher, J. B. (2006). Antecedents and Outcomes of Strategic IS Alignment: An Empirical Investigation, *IEEE Transactions on Engineering Management, 53*(1), pp. 27-47.

Chen, Y., Wang, Y., Nevo, S., Benitez-Amado, J., & Kou, G. (2015). IT Capabilities and Product Innovation Performance: The Roles of Corporate Entrepreneurship and Competitive Intensity, *Information & Management, 52*(6), pp. 643-657.

Damanpour, F., & Evan, W. M. (1984). Organizational Innovation and Performance: The Problem of "Organizational Lag," *Administrative Science Quarterly, 29*(3), pp. 392–409.

Dawson, G. S., Watson, R. T., & Boudreau, M.-C. (2010). Information Asymmetry in Information Systems Consulting: Toward a Theory of Relationship Constraints, *Journal of Management Information Systems, 27*(3), pp. 143-177.

Dubé, L., & Paré, G. (2003). Rigor in Information Systems Positivist Case Research: Current Practices, Trends, and Recommendations, *MIS Quarterly, 27*(4), pp. 597-636.

Ebel, P., Bretschneider, U., & Leimeister, J. M. (2016). Leveraging Virtual Business Model Innovation: A Framework for Designing Business Model Development Tools, *Information Systems Journal, 26*(5), pp. 519-550.

Emory, C. W., & Cooper, D. (1991). *Business Research Methods*. Homewood, IL: R. D. Irwin.

Flath, C. M., Friesike, S., Wirth, M., & Thiesse, F. (2017). Copy, Transform, Combine: Exploring the Remix as a Form of Innovation, *Journal of Information Technology, 32*(4), pp. 306-325.

Gartner. (2010). Market Overview: Consulting and Si Service Providers, Asia/Pacific 2010. Gartner.

Gartner. (2016). Gartner Survey Reveals Marketing Budgets Increased for Third Consecutive Year. In R. van der Meulen, & A. A. Forni (Eds.), *Newsroom,* Gartner.

Gerow, J. E., Grover, V., Thatcher, J. B., & Roth, P. L. (2014). Looking toward the Future of IT-Business Strategic Alignment through the Past: A Meta-Analysis, *MIS Quarterly, 38*(4), pp. 1059-1085.

Hales, C. (2005). Rooted in Supervision, Branching into Management: Continuity and Change in the Role of First-Line Manager, *Journal of Management Studies, 42*(3), pp. 471-506.

Henfridsson, O., & Lind, M. (2014). Information Systems Strategizing, Organizational Sub-Communities, and the Emergence of a Sustainability Strategy, *The Journal of Strategic Information Systems, 23*(1), pp. 11-28.

Kirsch, L. J. (1996). The Management of Complex Tasks in Organizations: Controlling the Systems Development Process, *Organization Science, 7*(1), pp. 1-21.

Kitay, J., & Wright, C. (2003). Expertise and Organizational Boundaries: The Varying Roles of Australian Management Consultants, *Asia Pacific Business Review, 9*(3), pp. 21-40.

Klein, H. K., & Myers, M. D. (1999). A Set of Principles for Conducting and Evaluating Interpretive Field Studies in Information Systems, *MIS Quarterly, 23*(1), pp. 67-93.

Laforet, S. (2013). Organizational Innovation Outcomes in SMEs: Effects of Age, Size, and Sector. *Journal of World Business, 48*(4), 490–502. doi:10.1016/j.jwb.2012.09.005

Lokuge, S., Sedera, D., Grover, V., & Dongming, X. (2019). Organizational Readiness for Digital Innovation: Development and Empirical Calibration of a Construct, *Information & Management, 56*(3), pp. 445-461.

Lokuge, S., Sedera, D., & Perera, M. (2018). The Clash of the Leaders: The Intermix of Leadership Styles for Resource Bundling, *Pacific Asia Conference on Information Systems (PACIS 2018)*, Yokohama, Japan: AIS.

McLachlin, R. D. (1999). Factors for Consulting Engagement Success, *Management Decision, 37*(5), pp. 394-404.

McLaren, T. S., Head, M. M., Yuan, Y., & Chan, Y. E. (2011). A Multilevel Model for Measuring Fit between a Firm's Competitive Strategies and Information Systems Capabilities, *MIS Quarterly, 35*(4), pp. 909-929.

Milgrom, P., & Roberts, J. (1995). Complementarities and Fit Strategy, Structure, and Organizational Change in Manufacturing, *Journal of Accounting and Economics, 19*(2-3), pp. 179-208.

Myers, M. D., & Newman, M. (2007). The Qualitative Interview in IS Research: Examining the Craft, *Information & Organization, 17*(1), pp. 2-26.

Nambisan, S. (2013). Information Technology and Product/Service Innovation: A Brief Assessment and Some Suggestions for Future Research, *Journal of the Association for Information Systems, 14*(4), pp. 215-226.

Nambisan, S. (2017). Digital Entrepreneurship: Toward a Digital Technology Perspective of Entrepreneurship, *Entrepreneurship Theory and Practice, 41*(6), pp. 1029-1055.

Nylén, D., & Holmström, J. (2015). Digital Innovation Strategy: A Framework for Diagnosing and Improving Digital Product and Service Innovation, *Business Horizons, 58*(1), pp. 57-67.

Penrose, E. T. (1959). *The Theory of the Growth Of the Firm*. New York: Sharpe.

Poulfelt, F., & Payne, A. (1994). Management Consultants: Client and Consultant Perspective, *Scandinavian Journal of Management, 10*(4), pp. 421-436.

Puranam, P., Singh, H., & Zollo, M. (2006). Organizing for Innovation: Managing the Coordination-Autonomy Dilemma in Technology Acquisitions, *Academy of Management Journal, 49*(2), pp. 263-280.

Qian, C., Cao, Q., & Takeuchi, R. (2013). Top Management Team Functional Diversity and Organizational Innovation in China: The Moderating Effects of Environment, *Strategic Management Journal, 34*(1), pp. 110-120.

Renwick, D. (2003). Line Manager Involvement in Hrm: An inside View, *Employee Relations, 25*(3), pp. 262-280.

Salanova, M., Agut, S., & Peiró, J. M. (2005). Linking Organizational Resources and Work Engagement to Employee Performance and Customer Loyalty: The Mediation of Service Climate, *Journal of Applied Psychology, 90*(6), pp. 1217-1227.

Schein, E. H. (1990). Process Consultation: Vol. I. *Its Role in Organizational Development*. Reading, MA: Addison-Wesley.

Sedera, D., Lokuge, S., Grover, V., Sarker, S., & Sarker, S. (2016). Innovating with Enterprise Systems and Digital Platforms: A Contingent Resource-Based Theory View, *Information & Management, 53*(3), pp. 366–379.

Soh, C., Chua, C. E. H., & Singh, H. (2011). Managing Diverse Stakeholders in Enterprise Systems Projects: A Control Portfolio Approach, *Journal of Information Technology, 26*(1), pp. 16-31.

Svahn, F., Mathiassen, L., & Lindgren, R. (2017). Embracing Digital Innovation in Incumbent Firms: How Volvo Cars Managed Competing Concerns, *MIS Quarterly, 41*(1), pp. 239-253.

Swanson, E. B. (2012). The Managers Guide to IT Innovation Waves, *MIT Sloan Management Review, 53*(2), pp. 75-83.

Walsham, G. (1993). *Interpreting Information Systems in Organizations*. Chichester, UK: Wiley & Sons.

Yin, R. K. (2010). *Qualitative Research from Start to Finish*. New York: Guilford Press.

Yoo, Y., Boland, R. J. Jr., Lyytinen, K., & Majchrzak, A. (2012). Organizing for Innovation in the Digitized World, *Organization Science, 23*(5), pp. 1398-1408.

ADDITIONAL READING

Arnold, J.A., Arad, S., Rhoades, J.A., and Drasgow, F. 2000. "The Empowering Leadership Questionnaire: The Construction and Validation of a New Scale for Measuring Leader Behaviors," *Journal of Organizational Behavior* (21:3), pp 249-269.

Augier, M., and Teece, D.J. 2009. "Dynamic Capabilities and the Role of Managers in Business Strategy and Economic Performance," *Organization Science* (20:2), pp 410-421.

Brinker, S., and McLellan, L. 2014. "The Rise of the Chief Marketing Technologist," *Harvard Business Review* (92:7), pp 82-85.

Flath, C.M., Friesike, S., Wirth, M., and Thiesse, F. 2017. "Copy, Transform, Combine: Exploring the Remix as a Form of Innovation," *Journal of Information Technology* (32:4), 2017/12/01, pp 306-325.

Hales, C. 2005. "Rooted in Supervision, Branching into Management: Continuity and Change in the Role of First-Line Manager," *Journal of Management Studies* (42:3), pp 471-506.

Lokuge, S., Sedera, D., & Perera, M. 2018. "The Clash of the Leaders: The Intermix of Leadership Styles for Resource Bundling," *Pacific Asia Conference on Information Systems (PACIS 2018)*, Yokohama, Japan: AIS.

Puranam, P., Singh, H., and Zollo, M. 2006. "Organizing for Innovation: Managing the Coordination-Autonomy Dilemma in Technology Acquisitions," *Academy of Management Journal* (49:2), pp 263-280.

Swanson, E.B. 2012. "The Managers Guide to IT Innovation Waves," *MIT Sloan Management Review* (53:2), pp 75-83.

KEY TERMS AND DEFINITIONS

Digital Technologies: New technologies such as social media, mobile technologies, analytics, internet-of-things are considered as digital technologies.

Engagement: Simultaneous employment and expression of a person's preferred self in task behaviors that promote connections to work and to others, personal presence (physical, cognitive, and emotional) and active, full performances (Salanova et al. 2005).

Innovation: Implementation of an idea whether pertaining to a device, system, process, policy, program or service that is new to the organization.

Innovation Process: The LOB-led IT innovation process was similar to any IT innovation process. Three phases were prominent in the data analysis. They were, initiation, implementation and management.

Line-of-Business (LOB) Managers: An employee that oversees other employees and operations of a business while reporting to a higher-ranking manager. For an example, sales manager is an example for a LOB manager. He is responsible for sales department functionalities, but he will be reporting to a senior executive level manager.

LOB-led IT Innovation: The IT innovations that are initiated by the LOB managers.

Responsive Behavior: The LOB-managers when initiating IT innovations were agile and looking for new ways to integrate digital technologies in initiating IT innovations. Such innovative behaviors are identified as responsive behaviors.

Chapter 6
Digitally Empowered Workers and Authentic Leaders:
The Capabilities Required for Digital Services

Joseph Crawford
https://orcid.org/0000-0002-2191-6216
University of Tasmania, Australia

Kerryn Butler-Henderson
https://orcid.org/0000-0002-6082-2108
University of Tasmania, Australia

ABSTRACT

The digital era has catalyzed the contemporary workforce and workplace. Employees are now required to develop skills that may be outdated in a matter of years, and be consciously aware that their society is evolving rapidly. This Chapter conceptualizes the characteristics of a digitally empowered worker to encompass awareness, creativity, agility, and a positive learning orientation. In doing so, the authors consider the way in which an authentic leader and authentic follower may influence the development of the digitally empowered. Authors present a model and its impact on digital innovation and digital productivity.

INTRODUCTION

The digital era has catalyzed the contemporary workforce and workplace. Employees are now required to develop skills that may be outdated in a matter of years, and consciously aware that their society is evolving rapidly. Themes of disaggregation, competition, and incentivization are driving new forms of management (Dunleavy et al., 2006). Electronic forms of human resource management have shaped the way in which organizations and their managers can influence employees to achieve their targets

DOI: 10.4018/978-1-7998-2799-3.ch006

(Bondarouk & Ruël, 2009). The rate and scale of change is unprecedented, and employees are expected to keep pace with the unfolding innovations. In fact, they are often expected to participate in shaping such innovations. Yet, for the most part, employees receive very little training to create the digital fluency skills necessary to achieve desired digital outcomes.

A digitally empowered workforce is prepared for their present needs through hard skills. LinkedIn (2018) identified cloud computing, artificial intelligence, analytical reasoning, people management, and user experience design as the top five most demanded hard skills in their annual survey of 4,000 professionals. Yet, 57 percent of senior leaders recognized that soft skills such as creativity, persuasion, collaboration, adaptability, and time management were more important than hard skills. Senior leaders globally recognize the need for their employees to be ready for the future, not just able to complete their day-to-day responsibilities.

For a workforce to be empowered, it requires three antecedent factors. The first is leaders willing to develop and empower their employees (Cheong et al., 2018). Leaders who are effective at motivating their employees can provide agency to those employees to enact their knowledge, skills, and attributes in the workplace. Leadership has evolved. The focus of behavioral research has been on the development and application of knowledge and theory in traditional contexts. The digital era has forced evolution. For example, leadership is enabled or hindered through videoconferencing (Hambley et al., 2007) and social media (Barbera et al., 2014). Complex technology enhanced environments (e.g. virtual, immersive, and augmented reality) are providing new opportunities for development, productivity, and wellbeing. Leaders are only just beginning to come to terms with what this means for their workplace, and their ability to lead.

The second is employees who are developmentally ready to be empowered. An individual's developmental readiness is based on the belief that individuals can create positive self-fulfilling prophecies about their future (Avolio & Hannah, 2008). That is, individuals who are willing to develop and better understand their own self-concept are better able to respond positively to have agency. The Organization for the Economic Cooperation and Development (OECD, 2001) identified a digital divide between those in lower and higher socioeconomic statuses. In the workplace, those in higher socioeconomic statuses tend to be in management, with a higher remuneration. This would suggest there is an existing digital divide in the workplace between managers and their subordinates (Ayanso et al., 2010). Understanding, therefore, the role of follower behaviors in this context could enable the exploration of bridging the digital fluency gap that may exist.

The third is an organizational culture or climate suitable to enable agency. Important, organizations with poor organizational climate can lead to negative long-term performance outcomes (Crawford et al., 2017). Organizations that embed digital innovation into their culture can change their market landscape (Butler-Henderson, 2014). Organizational culture is defined by many elements such as involvement, consistency, adaptability, and mission (Denison & Mishra, 1995). Earlier research defines organizational culture as the synthesis of assumptions, values, and artefacts (Schein, 1985) with adaptation to also include symbols and processes (Hatch, 1993). We do, however, distinguish organizational culture from this Chapter. While an important part of this discussion, which we encourage others to pursue, there was no scope to discuss with enough depth in this Chapter.

This chapter draws on the three antecedent factors (leaders, employees/followers, and culture) to explore the capabilities that digitally empowered employees require. Scholars are increasingly recognizing the role of behaviors and climate in successful change transitions (Martin et al., 2005), divergence in positive perceptions between leaders and subordinates (Martin et al., 2006), and digital innovation requires a consistent aptitude for change. In this Chapter, we propose a model between authentic leader behavior, authentic follower behavior, digitally empowered workers, digital innovation, and digital productivity.

The structure of this chapter has been set-out to enable the exploration of the relationship that authentic leader behaviors, authentic follower behaviors, and digital fluency have. In the next section, we begin by providing a summary of, and justify the use of, the critical review method in this chapter. In the literature review, we explore the context of digital services, and define the authentic leader and their five behaviors (awareness, sincerity, balanced processing, positive moral perspectives, and informal influence). Next, we define the authentic follower and their behaviors (psychological capacity for authenticity and positive organizational engagement). We also define the characteristics of digital fluency (creativity, social and emotional awareness, agility and adaptability, and a continuous learning orientation). Following, we synthesize the authentic leader, authentic follower, and digital fluency into a model and explain the relationships.

CRITICAL REVIEW METHOD

Critical review methods are commonly used as a 'catch-all' in semi-structured literature reviews. Take for example, a seminal critical review in organizational research (Podsakoff et al., 2003). Aside from the mention of a 'critical review' in the title, the only other place in text that presents the words 'critical review' is a title in the reference list. In the development of a typology of literature reviews, a critical review "aims to demonstrate writer has extensively researched literature and critically evaluated its quality. Goes beyond mere description to include degree of analysis and conceptual innovation" (Grant & Booth, p. 94). This literature review approach differs from systematic reviews and meta-analyzes by their clearly defined rules. Where a critical review seeks to identify the most significant pieces of work in the field, a systematic review would identify clear scoping conditions (e.g. search phrases, temporal periods, and specific journals).

The value in conducting a critical review in this piece of work is that several key areas of the literature are being examined, and codified search terms may limit the knowledge that can be extracted. In this critical review, we examine a series of disperse literatures before synthesizing their parallels and exploring their differences. The aim of which is to enable knowledge sharing across a divide. On the one hand, authentic leadership literature explores the enablers created by developing leaders (some of which are likely enablers of digital services), and on the other is the emerging literature on the digital era and the individuals seeking to operate within an innovative and evolving space. We group our critical review into the next section, with the subsequent section synthesizing the disparate parts.

LITERATURE

Today, terms such as digital disruption and digital transformation are no longer business jargon, but part of the organizational strategy. With consumers expecting to receive information from their lounge suites, any business who does not have a digital presence risks being left behind. It is the scope of that digital presence that will shape the skill requirements of the workforce. A one-person plumbing business servicing a small local area may only have an online listing on a trade website, and therefore their digital capability needs are minimal. But what are the needs of the workforce in business that offer digital services. And what attributes in both the leaders and followers will encourage digital innovation and productivity? This summary of the literature defines digital services explores the future trends that will shape workforce needs. The authentic leader and follower are introduced and the behaviors of each are examined. We conclude by delving into the literature about the digitally empowered worker.

Digital Services

Digital services are defined as "an activity or benefit that one party can give to another, that is, provided through a digital transaction" (Williams, Chatterjee, & Rossi, 2008, p. 507). They are different from normal services in that interactions with clients are, at least in part, through a digital experience. The sense of what is tangible may differ and ownership is more elusive. For example, a digital design company is hired to design a new logo for a company. They may complete all relational elements digitally with the customer. Ownership of the product, a logo, needs to be clearly defined in the service agreement, including the right for the company to modify the logo in the future. Digital services can be personalized, contextualized. and adapted in real-time to meet the client's needs. The power of the client has increased in today's market, with clients expecting services to be available anytime through one click. Client loyalty has also changed in the digital era, with clients seeking services that are client-focused and easy to access (Leimeister, Österle, & Alter, 2014). Organizations with maturing digital services share common features. These features include increased risk taking and experimentation, and investment in talent, across the workforce and leadership behaviors. Digital service employees require the behaviors of a leader to effectively engage with their clients and employees, by influencing and enabling positive outcomes. 90 percent of digitally maturing organizations are integrating their digital strategy into their organizational strategy, yet 44 percent of companies state they are not adequately preparing for digital delivery of services (Kane et al., 2016).

There are six identified megatrends predicted for the future that should shape organizational digital strategy (Hajkowicz et al., 2016). The first is an exponential and rapid technological and data growth reshaping services and the workforce. This includes significant developments in the Internet of Things, growth in interconnectivity technology shaping how we work, increases in big and small data informing business strategy, and rapid advances in artificial intelligence and automation. The second area will be where this growth impacts on changing employment markets and organizational structures. This includes increases in globalization of workforces, a change in work practices, flattened organizational structures, growth in contract or freelance workers, and an increase in co-working environments or remote offices. The third megatrend is the opportunities for entrepreneurship. With the change in organizational structures and increased opportunities for freelance work, the small business environment will see rapid growth. This will create opportunities for enterprise and entrepreneurs to drive innovation and productivity. Digital entrepreneurialism is low cost and lean, with many start-ups today established to quickly turn ideas into innovations.

Fourth, economic and population analysis shows we have an intergenerational workforce. In western cultures the average life expectancy has increased, with improvements in health and lifestyles resulting in people living longer. Whilst people are staying in the workforce for longer, participation rates are declining as people reduce their work hours. This trend will result in a higher proportion of people retired than those in the workforce, leaving many industries with an unsustainable workforce. Digital services, particularly automation of low skilled or operational jobs, will somewhat address this issue. This will result in an increasing need for higher skills-set, which is the fifth megatrend. Education and skills will have increasing importance, particularly postgraduate education. There will be a growth in science, technology, engineering, mathematics, and medicine (STEMM) areas to address industry gaps. In turn, the cost of higher education will continue to increase. Learners will seek different ways to engage in study whilst remaining in the workforce, and as such there will be an increase demand for online education delivered in more accessible formats, such as short courses, micro-credentials and massive open online courses (MOOCs). Lastly, the skills sets required for the workforce will change, due to digital empowerment. The attributes of the digitally empowered worker are addressed later in this chapter.

Education and training will become more important, particularly to upskill the existing workforce through times of automation and innovation. Digital capabilities such as digital literacy and fluency will become assets through a time of changing importance in STEMM (science, technology, engineering, mathematics and medicine). Service industries, particularly education and healthcare, will continue to drive job creation, but these workforces will shift as digital innovations alter roles and processes. In turn, the attitudes and beliefs of the digitally empowered worker will change, as will the very definition of the traditional work day. Flexible work environments will become the norm, and digital service teams will no longer be located geographically in the same location. Digitization and the intergenerational impact on the workforce will result in people staying in the workforce longer, but with decreased participation. Therefore, job transitions will shift and new job forecast models will be required.

Digital transformation of business services requires a digital capable workforce to deliver services. Digitalization has been characterized with six dimensions: interconnectedness, diminishing time lag and abundance of information, increased transparency and complexity, hierarchical removal and dissolvement of personal barriers, decision enabler and integrity enhancing, and humanizing effect (Khan, 2016). While these are not exhaustive, and the methodology is fraught, they do provide a positive initial set of ideas that we will draw on.

Authentic Leaders

Authentic leader theory was created to enable an antidote to prevailing ethical challenges of our society (Luthans & Avolio, 2003). Nested within the field of positive organizational behavior (Luthans, 2002), authentic leaders enable positive development within their followers and those around them (Avolio & Gardner, 2005). Since then, a large series of positive outcomes has been identified for followers of the authentic leader. For example, follower empowerment, performance, satisfaction, commitment, trust, engagement, and happiness (Gardner et al., 2011). Authentic leader behaviors are defined in Table 1 (Crawford et al., 2019).

The authentic leader has an important role in motivating employees, although this relationship has been underexplored in the literature. Empirical relationships exist between authentic leaders and job satisfaction (Wong & Laschinger, 2013) and work engagement (Joo et al., 2016). To some extent, it is possible to theorize that if employees are engaged in their work and they express their satisfaction with their role because of their authentic leader, then it is also likely that such a leader must do so through motivating their employees.

The digital era has caused complexity in the role of the leader, and requires careful consideration to way in which leader behaviors manifest. Motivating an employee, for example, is likely to be more difficult when a leader only sees that employee by way of video-conferencing and the annual staff Christmas party. Drawing on some of the earlier referred to digitalization characteristics (Khan, 2016), we identify the role of the authentic leader as an enabler of transparency in a complex world. Authentic

Table 1. Defining authentic leaders and their dimensions

Authentic leader	Influences and motivates followers to achieve goals through their sincerity and positive moral perspective, enabled through heightened awareness and balanced processing.
Awareness	Having insight into the behaviours of one's self, other individuals, and groups.
Sincerity	Presenting one's true self to others, honestly and openly in all relationships and with consideration for the context.
Balanced processing	The tendency to consider all relevant information available, and using this to make decisions that benefit the collective (e.g. followers).
Positive moral perspectives	The commitment to one's intrinsic ethical framework, and a willingness to subdue personal interests and ego to facilitate collective interests.
Informal influence	The ability to inspire and motivate individuals to accomplish goals of their own volition, regardless of rank or position.

Definitions quoted from multiple pages in Crawford et al., (2019b).

leaders engage in sincere interactions with others, and seek to be transparent in their behaviors (Rego et al., 2017). Complexity as an inherent element of the contemporary and dynamic environment that we are situated within, and authentic leader behaviors create the conditions to co-create a vision in times of chaos (Vitello-Cicciu, 2019).

Importantly, within a digital environment, is recognition that hierarchies are harder to sustain. Digital innovation can empower individuals, but also disempower those who leverage information power (Mansell, 2016). The growth in transparency stemming from faster access to information removes significant formal and positional power from managerial staff. In transition, coercive effects can be hindered and informal power enhanced (Powell et al., 2017). The authentic leader relies on their informally developed influence to achieve their shared vision, rather than the power granted by their position. Some leaders reliant on their position are likely to struggle during the digital transformation. Digital transformation requires more than leadership, it requires leaders with behaviors that enable them to consciously understand their changing environment and cope with the ethical and emotional pressures faced.

In discussing a positive leadership theory, it is equally important to consider the low authentic leader. Dark side (Liu et al., 2012), Machiavellian (Rego et al., 2017), corporate psychopaths (Boddy, 2011), and unethical leadership (Brown & Mitchell, 2010) theories are a few examples. Dark sides of leadership are identifiable as those who score low in their authentic leader behaviors. For example, in a typology, the Machiavellian leader is low in authenticity and high in Machiavellianism (Rego et al., 2017). Early psychological evidence focused on prescription for negative psychological conditions over promotion of positive psychological solutions (Seligman & Csikszentmihalyi, 2014). Fredrickson and Joiner (2002) describes this as positive spirals, whereby positive emotions predict the creation of more positive emotions. Likewise, the reverse is true: negative emotions predict more negative emotions (Garland et al., 2010). We recognize, therefore, that the promotion of positive leader behaviors has the capacity to mitigate the negative and dark sides of leadership. For example, bullying behaviors of leaders is possible to eliminate by high integrity, sincerity, and awareness. A leader who is aware of the impact of their behaviors on others is likely to have greater empathy and therefore be more considerate of their actions.

Authentic Followers

Followers have been commonly neglected in the leadership literature, with a constant emphasis on the follower as an extension of the leader. There have been some emerging work that has challenged the status quo, but very few significant enough to penetrate the bias towards the leader. In an evolutionary conversation of leadership, authors highlight "it is less obvious what benefits ultimately exist for followers" (Van Vugt, 2006, p. 357). Arguments of 'everyone cannot be a leader, so some have to be followers' are common in the assumptions of the literature, along with recognition that a leader requires a follower to influence. The confusion in the literature exists from an overreliance on the formal positional authority of a manager and the role of the leader. Some studies, albeit limited numbers, attempt to separate formal and perceived authority with leadership style and focus (Eglene et al., 2007). Changes in the contemporary leadership landscape have drawn attention away from positional power and authority towards informal and relational power bases (Ridgeway, 2001). In leadership theories with informal influence bases, there is importance in understanding the nature of the follower. Such an understanding should go beyond questioning how a follower is best influenced to achieve goals, but to consider how followers can be best developed, empowered, and enabled. Followers in one situation may be a leader in another (Gilstrap & Morris, 2015), so developing followers is one mechanism to develop future leaders.

In many theories of followership, engagement is a key component of consideration. In one typology, the follower goes from 'isolates' to 'dieharts' (Kellerman, 2008). The isolates are typically detached and uninterested in engaging in the organization beyond their minimum requirements. The bystander is more engaged, but usually opts to be on the sideline. Participants are engaged and supportive, but may be value-neutral in their followership. Activists are energetic and have high engagement, and diehards are heavily engaged. The challenge with this typology is the lack of recognition that followers can influence through their engagement. This model is insightful but lacks a key component, indeed it explicates the first component of an effective follower but misses the psychological capacity for authenticity.

Another popularized model of followership (Kelley, 2008) identifies two characteristics of followers: the degree to which they are independent thinkers, and the degree to which they are actively engaged in creating positive energy. In this model, there are sheep (passive and non-thinking), yes-people (positive energy, but blindly follow), alienated (think independently and negative energy), pragmatics (fence sitters, but tends to follow the status quo), and star followers (think for themselves and have positive energy). The model articulated by Kelley (2008) is more advanced than the previous, but is still limited. It recognizes the value in thinking critically and challenging the leader when they disagree, but it withholds ethics and indeed a sense of self in the followers. Followers can think critically, reflect and develop a sense of self, and engage with their organization (Crossman & Crossman, 2011). They can also be ethical, and have well defined ethical frameworks (Carsten & Uhl-Bien, 2013).

In studies of authentic followership, authentic followers have a need to be recognized and valued, with a desire to be challenged and excited (Goffee & Jones, 2006). While such notions continue to treat the follower as secondary. Mutual transparency was important however (Avolio & Reichard, 2008). The assumptions in these papers is heavily reliant on the notion that followers are subservient beings rather than demonstrative of their own independence (de Zilwa, 2016). de Zilwa (2014) defines authentic followership through the system in which they would operate. In her work, she considers positive culture and political conditions as a key element of the model. The leader-follower relationship is nested within a broader socio-political environment, and such an environment has implications for the leader and the follower alike. A careful explication of what it means to be an authentic follower is lacking however, beyond their psychological capacity for authenticity.

Authentic followers have been far less explored than authentic leaders (Agho, 2009), despite their value in the context of considering what it means to be a digitally empowered employee. Crawford et al. (2018) synthesizes the preexisting research to explain the authentic follower by psychological capacity for authenticity (e.g. de Zilwa, 2014) and positive organizational engagement (like Kelley, 2008). In their work, Crawford and colleagues (2018, p. 274) define the authentic follower by their psychological capacity for authenticity (awareness, sincerity, balanced processing, and a positive moral perspective) and positive organisational engagement:

An authentic follower is an individual who, through their capacity for authenticity and positive organisational engagement, are self-managing and follow leaders with whom they share values.

We draw on this conceptualization to consider how the digitally empowered worker shares characteristics with the authentic follower. Indeed, the literature on authentic leaders and authentic followers provides insights into the enablers of the characteristics of the digitally empowered worker, and we discuss this further in the next section.

Digitally Empowered Workers

The changing landscape in digital services described above will change the worker of the future. The worker will need to maintain engagement in constantly changing environments, largely with and through technology, in data-driven roles. The digitally empowered worker needs to be not only digitally literate (know how to work with technology), but digitally fluent. Digital fluency is described by Briggs and Makice (2012, p. 68) as "the ability to reliably achieve desired outcomes through digital technology". A digitally empowered worker needs to possess a new set of skills, knowledge, and mindset to deliver digital services. But what are the skills, knowledge, and mindset required to be an effective and digitally empowered worker? Among our critical examination of the wealth of published literature on the topic, we identify four dynamic attributes: awareness, creativity, agility, and learning orientation.

Awareness

Awareness, or consciousness, is the ability to perceive, feel, know, and understand people and events. Whilst everyone has level of self-awareness, the digitally empowered worker requires awareness of how others perceive events and the needs of others. Awareness includes a broad range of skills that can be collectively grouped as social and emotional skills. Social and emotional skills include communication and negotiation skills, interpersonal skills and empathy, leadership and managing others, entrepreneurship and initiative-taking, adaptability and continuous learning, and teaching and training others (McKinsey Global Institute, 2018). The McKinsey Global Institute (2018) predicts there will be a 26 percent increase in the number of work hours utilizing social and emotional skills. Furthermore, people management, coordinating with others, and emotional intelligence are amongst the top ten skills required by the future workforce (World Economic Forum, 2016). Interestingly, the World Economic Forum (2016) predicted negotiation skills would drop out of their top ten list as they rationalize that you cannot negotiate with a robot or artificial intelligence (AI).

Emotional intelligence is being aware of and managing your own emotions and the emotions of others, including interpersonal and communication skills (Murawski & Bick, 2017; van Laar et al., 2017). Emotional intelligence comprises emotional recognition and expression, emotions embedded and understood in decision making, awareness of others' emotions, regulation and management of emotions of self and others, and ability to function and flourish in conditions where emotional states are heightened (Gardner & Stough, 2002). Both individual and team emotional intelligence enriches creativity and team cohesion through collaborative communication and negotiation. Similarly, team emotional intelligence is connected to the organizational climate, which can influence individual and team creativity. Moreover, Barczak et al. (2010) conclude the relationship between emotional intelligence, team trust and a collaborative culture are precursors to creativity. In a digital service, emotional intelligence will enable workers to work and communicate with others in a positive and productive way. This is particularly necessary when teams are geographically dispersed and interpersonal and communication skills are through digital technologies instead of in person.

Equally, empathy is an important attribute in the digitally empower worker. Empathy is the cognitive awareness of the perspective of others and how we react to other experiences with an affective response (Colbert et al., 2016). Creativity is enhanced through information and ideas sharing, so being empathetic enables individuals to be focused on the user needs and desires. As such, empathy underpins human-centered design. Within digital services, being empathetic online and building healthy online social networks will become required attributes.

A strong social conscience also featured in the literature (Pricewaterhouse Cooper, 2017; Van Laar et al., 2017). The future digitally empowered workforce will need to have an awareness about the environment, ethics, culture, diversity, and human rights. Economic and social developments will impact innovations, and therefore the workforce must account for these when developing new ideas.

Creativity

Creativity involves forming solutions to bring one's ideas, thoughts and dreams into reality in ways that are novel and useful. It can be looking at the world in a different way to see the potential for a solution. It is creative and sometime complex, problem-solving, where innovation cannot begin without creativity. As such, the ability to be creative features strongly in the literature examining the future workforce needs (Neumeier, 2014; McKinsey Global Institute, 2018; Pricewaterhouse Cooper, 2017; Van Laar, 2017; World Economic Forum, 2016).

Creativity and innovation are often mistakenly used synonymously. Whilst creativity is the process or creation, such as thoughts and ideas, innovation is the use of those ideas to improve the way we do things, whether that be processes, products, services, or procedures. All individuals have some degree of creativity, and creativity is not limited to only innovation (Hoerl & Gardner, 2010). Therefore, creativity is a necessary attribute for the digitally empowered worker. Creativity can be encouraged in individuals by providing them with the space and time to allow people to brainstorm ideas. Continuous learning, rewarding curiosity, encouraging risk taking (within reason), and positive encouragement should be part of the cultural norm. Over time, this will improve the confidence in digitally empowered workers, leading to greater creativity.

Team culture can also facilitate or inhibit creativity, which in turns impacts innovation. Whilst individual characteristics or attributes will influence creativity, the team creativity process will influence the innovation (Somech & Drach-Zahavy, 2011). Functional heterogeneity in teams will result in diverse expertise, skills, and attributes. This richness of diversity will result in either a positive or negative impact on team creativity, depending on the team's cohesion. Furthermore, the external network of the team and individual members will further enrich this diversity as it broadens the information sources, increases members' skills, and reshapes perspectives, lending to new ideas, thoughts and dreams. Somech and Drach-Zahavy (2011) reports there is a positive relationship between personal attributes, functional heterogeneity, and team creativity. Where there is a high level of individual creativity and a high level of heterogeneity in the team, there is high level of team creativity. Individual creativity attributes are a good predictor of the level of team creativity, and therefore team creativity could be enhanced through the selection of creative personalities.

Lastly, the climate can also influence creativity. The climate informs members the expected behaviors and attitudes for the team and organization. Elements within the creative climate include psychological safety, positive team relationships, the degree of individual freedom, encouragement of creativity and setting of the vision and mission, and available resources (Denti, 2011). Denti (2011) proposes the six components of a creative climate include: i) challenging and creative tasks to motivate creativity; ii) intellectual debate to share diverse viewpoints; iii) flexibility and risk taking to push the margins; iv) top management support to challenge the norms; v) positive supervisor relations with the team to encourage creativity; and, vi) positive interpersonal exchange to create joy and togetherness. Most importantly is a climate that empowers individuals to be creative.

Agility

The ability to be flexible and quick will become important skills for the digital workforce (World Economic Forum, 2016). This includes cognitive flexibility and judgement and decision making (World Economic Forum, 2016). Individuals must be agile, demonstrating an ability to think and move quickly (McKinsey Global Institute, 2018; van Laar et al., 2017). Digital services need to be adaptable to continuously meet the demand of clients, and as such the workforce needs to be equally agile. Individuals and teams need to quickly generate new ideas, making team culture even more critical in digital services. The ability to acknowledge and manage change will be an important skill. People with adaptive leadership skills aim to understand a diversity of viewpoints, using empathy and awareness, and enhancing creativity. Those who become part of the change process will benefit the most in the digital world.

Critical thinking (World Economic Forum, 2016; Van Laar et al., 2017) and higher cognitive skills (McKinsey Global Institute, 2018) will enable the digital workforce to interpret information quicker, enabling faster problem-solving. McKinsey Global Institute (2018) forecast between 2016 to 2030 there will be a 25 percent decrease in the number of work hours requiring physical and manual or basic cognitive skills, and an increase in higher cognitive skills (9% increase) and technological skills (60% increase). The ability to think quickly, be cognitively flexible, and be part of the change process will be critical skills in the digitally empowered worker.

Positive Learning Orientation

Lastly, the digitally empowered workforce needs to be able to identify and set their own learning goals, and be open to new ways of working and learning (Pricewaterhouse Cooper, 2017; Van Laar et al., 2017). Learning orientation, or continuous learning, enables individuals and teams to expand their knowledge and skills and remain current across contemporary developments. Being agile allows digitally empowered workers to integrate new skills into their roles and strong interpersonal skills allow the translation of this information to the team. The establishment of a learning organization is critical in enabling employees to have a positive relationship with their learning experience. Indeed, authentic leaders have a significant role to play in enabling the creation of a learning organizational culture (Milić et al., 2017).

Continuous learning can take place through formal learning, social learning, or self-directed learning. As discussed above, self-actualization is an important element in continuous learning. Individuals need to be able to identify areas for growth and improvement within their own skill set and gaps within the team and organization. A culture that supports continuous learning, or a learning culture, will support individuals and teams. Self-direction and lifelong learning are critical elements in creating a workforce that is creative, socially and emotionally aware, agile, and adaptable. Leadership that supports continuous learning through demonstration and empowerment will create an environment where improvement in knowledge and skills is encouraged. The digitally empowered workforce will maintain currency of contemporary practices, enhancing cognitive agility, encouraging creativity, and heightening awareness.

SOLUTIONS AND RECOMMENDATIONS

Through the critical review of digital services, authentic leaders, authentic followers, and the behaviors of digitally empowered employees we have developed the foundation for a synthesis. We identified six megatrends that are likely to have great influence on the organizational ability to engage in digital strategies (Hajkowicz et al., 2016). These include rapid technological/data growth, where such growth will influence employment and organizational structures, new opportunities to be entrepreneurial and innovative, a more diverse employee age profile, a need for higher capabilities, and the rapid evolution of the required skills of the workforce. Each of these offers significant challenges, but underlying an effective response to each is effective leaders and effective follower behaviors.

We position authentic leaders and authentic followers as a possible enabler of greater digital empowerment. To this end, we provided a critical review of the two constructs in the context of research that parallels to digital empowerment. We also begin to examine the possible behaviors of a digitally empowered worker. These include creativity, awareness, agility and a learning orientation. At this stage of the Chapter, we believe the reader should have a grasp of the four underlying theoretical lenses we adopt in the establishment of a model of hypotheses (see Figure 1). The subsequent sections begin to explicate the nature of each of these relationships, and we continue further to explore the implications of such a model in practice.

Authentic Leader Behaviors and Digital Empowerment

The authentic leader, through their informal influence and sincere interactions with their followers, creates the conditions by which their followers are developmentally ready to engage with new information

Figure 1. The nomological network of the digitally empowered follower

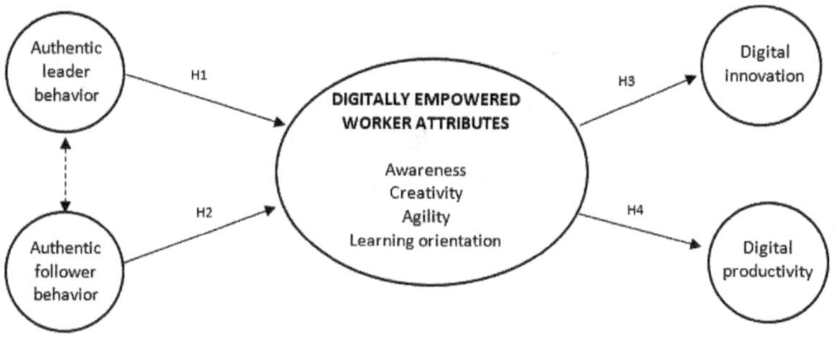

and new knowledge. In the leadership literature, developmental readiness has grown in its usage (Avolio & Hannah, 2008). Through the guise of a positive organizational culture, authentic leaders seek to create the conditions by which followers *want* to engage with self-improvement and assessment of their current conditions. Some negative forms of leadership that leverage information power or fear stifle the follower's desire to improve their work conditions and remain relevant (e.g. Blasé & Blasé, 2002). Power distance has a negative relationship with empowerment (Çakar & Ertürk, 2010), and thereby leaders who are sincere and are willing to engage in positive relationships with employees will create greater employee empowerment.

Authentic leaders can create the foundations for digital empowerment by being supportive and enable knowledge sharing, rather than harboring information. For authentic leaders, a key aim of fostering a psychological capacity for authenticity and positive organizational engagement in followers is paramount (Crawford et al., 2018). This, as discussed in the next hypothesis, is an enabler of digital empowerment. Authentic leaders have a strong influence on enabling both innovation and creativity. In the innovation and entrepreneurship literature, creativity is typically identified as a critical component of early stages of entrepreneurship (Mulgan, 2006). This includes the notion of identifying a problem, generating ideas for solving the problem, and identifying the most suitable idea to pursue. Innovation is required at the latter stages of entrepreneurship, from developing prototypes through to entering the chosen market for the first time (Anderson et al., 2014). For authentic leaders, there are established relationships with outcomes of higher individual creativity (Ribeiro et al., 2019) and individual and team performance (Luu, 2019; Shirey et al., 2019).

In digital innovation, a parallel can be drawn. The authentic leader, by facilitating an environment where individuals feel safe to take risks and explore alternate ways of operating, empowers their followers to be creative and innovative (Černe et al., 2013). When considering dark sides of leadership, evidence indicates abusive behaviors towards employees results in declines in creativity (Liu et al., 2012). To this end, a leader who engages in positive and supportive behaviors is likely to also enable creativity and innovation in their followers. For authentic leaders, the ability for employees to engage in social innovation is also important. Social innovation is creating new ideas for wider benefit and is often linked to social entrepreneurship models (Crawford et al., 2020). The ability for employees to engage in digital innovation in for-profit and for-purpose organizations is strengthened by an authentic leader creating an employee willingness and ability to learn and take calculated risks. We posit, therefore, that:

Hypothesis 1: That authentic leaders will have a direct influence on the degree to which employees are digitally empowered.

Authentic Followers as the Foundation of Digitally Empowered Workers

The authentic follower, who has a psychological capacity for authenticity and positive organizational engagement, to be aware of themselves and their environment (de Zilwa et al., 2014). By being aware of their environments, they can regularly conduct environmental scans (which includes technology changes and usage). An authentic follower uses their balanced processing to critically review options and needs to make informed decisions. This process, along with recognition of individual strengths and weaknesses creates conditions where followers are better able to engage in the use of innovative digital technologies.

In contemporary society, there is a constant bombardment of digital technologies. Some of these are exceptionally relevant, and some are not. Authentic leaders are cognizant of the strengths and weaknesses of their followers (Avolio et al., 2004). This leads to an understanding of the characteristics that might hold back a follower from being digitally empowered. For example, an older follower may have significant resistance to new piece of technology, particularly when younger employees are able to intuitively understand the new technology. The divide between digitally native employees born from the late 90s will have difficulty in empathizing with older employees who can recall times before computers, phones, and similar. These challenges, an authentic follower can overcome with sincere engagement and careful assessment of the situation. This is just one way in which an authentic follower can create the conditions for digital empowerment in their organization. We posit, therefore, that:

Hypothesis 2: That authentic follower behaviors will have a predictive effect on employee digitally empowerment.

Digitally Empowered Workers and Digital Innovation

For digital workers, their creative self-efficacy has an impact on their perceptions of personal idea generation and objective ratings of creative output. In the same study, when psychological entitlement was high, this was no longer true (Černe et al., 2019). These employees, with the kind of self-regulation and self-efficacy that an authentic follower is likely to engage in, creates the conditions for greater digital innovation. Where digitally empowered employees are empowered to be creative, innovative, moderate their sense of entitlement, and learn, we posit that they will engage in higher levels of digital innovation.

In a study of 743 employees over 93 small/medium enterprises, collective attitudes and uncertainty avoidance had an association with empowerment and innovation (Çakar & Ertürk, 2010). Therefore, employees who have awareness of their environment, are more open in conversations and interactions with others, and seek to process information in a balanced manner are likely to better engage with innovation generally. The capabilities of the digitally empowered are likely to further strengthen the application of innovation in digital settings. We propose therefore that:

Hypothesis 3: That digitally empowered workers will have higher rates of digital innovation than employees who are not digitally empowered.

Digitally Empowered Workers and Digital Productivity

Having a high level of awareness, creativity, agility, and learning orientation will better position workers in digital services compared to those who have minimal awareness beyond their immediate self, lack creativity, are rigid, and do not take part in continuous learning. To return to the digital services example earlier. A designer who creates logos is likely to work with clients across the globe. These clients will have needs that vary, and cultural norms that are diverse. Such a designer will need to demonstrate high degrees of other and situational awareness to navigate such complexities, beyond a designer that only worked with domestic clients. They would also need higher agility to manage deadlines across multiple time zones, and a positive learning orientation to acknowledge that their competitors are not two nearby design firms, but rather any design firm with a well-developed digital presence. These attributes will enable the worker in a digital environment to be more productive as they are able to develop new ideas and adapt to rapid change. Dynamic management can encourage creativity and innovation in workers to the benefit of organizations (Helfat & Martin 2015).

Hypothesis 4: That digitally empowered workers will have higher rates of digital productivity than employees who are not digitally empowered.

FUTURE RESEARCH DIRECTIONS

The implications of this chapter are extensive, but still infant in their development. Our Chapter highlights a set of implications for emerging research with significant opportunities for empirical testing. The model we propose (Figure 1) highlights the potential for a series of relationships between the authentic leader and authentic follower to digitally empowered workers, and subsequent increases in digital innovation and productivity. If this holds true, managers should seek authentic leader behavior training and development, and seek to cultivate authentic follower behaviors in their subordinates. Doing so has wide implications already discussed in the literature (see Gardner et al., 2011 for an extensive list of outcomes), and we propose that digital empowerment is produced as an outcome of authentic leaders and followers. We also believe that there may be direct effects of authentic leaders and followers on digital innovation and digital productivity, but further research is required for such a proposition.

This research, while encompassing a wide suite of literature, has its limitations and opportunities for future research. The first limitation is that our work is theoretical. While it is critical that theoretical developments inform empirical research (Crawford & Kelder, 2019), at present our hypotheses are untested outside of their logical consistency and potential for significance. We encourage scholars to continue this exploration of the role of authentic leaders and authentic followers in enabling digital outcomes (e.g. empowerment, innovation, and productivity). The second limitation is that we only reviewed one type of leader behaviors. Most evidence indicates a significant list of leadership styles, and to provide the strength of argument, and to explore in depth, we selected one that had logical consistency based on past work. This assumption, while we believe is true, could also be true of other theories of leadership such as transformational leadership (e.g. Özaralli, 2003), top management teams' theory (Ugboro & Obeng, 2000), and empowering leadership (Srivastava et al., 2006). Each of these theories have relationships to empowered employees, among others, and could also contribute to digitally empowered leadership.

Future research could examine such relationships and continue to theoretically and empirically explore the comparative effects of leadership styles on digital outcomes.

CONCLUSION

This Chapter evaluated the role of the authentic leader and authentic follower on digital empowerment and began to theorize what a digitally empowered worker might look like. Through a critical review of four underlying constructs (authentic leaders, authentic followers, digital services, and digital empowerment) we sought to develop a better understanding of each of their roles in enabling digital outcomes. In our discussion, we explicated a model of digital empowerment that views authentic leaders and authentic followers as antecedents, and digital innovation and digital productivity as outcomes. In a world that is characterized by significant growth in the use of technology, being able to understand and shape its usage for positive outcomes is critical in enabling human flourishing. We believe that authentic leaders and followers have a role in enabling the conditions of digital empowerment, and such empowerment creates the conditions that can shape and continue to innovate digitally.

REFERENCES

Anderson, N., Potočnik, K., & Zhou, J. (2014). Innovation and creativity in organizations: A state-of-the-science review, prospective commentary, and guiding framework. *Journal of Management*, *40*(5), 1297–1333. doi:10.1177/0149206314527128

Avolio, B., & Gardner, W. (2005). Authentic leadership development: Getting to the root of positive forms of leadership. *The Leadership Quarterly*, *16*(3), 315–338. doi:10.1016/j.leaqua.2005.03.001

Avolio, B., Gardner, W., Walumbwa, F., Luthans, F., & May, D. (2004). Unlocking the mask: A look at the process by which authentic leaders impact follower attitudes and behaviors. *The Leadership Quarterly*, *15*(6), 801–823. doi:10.1016/j.leaqua.2004.09.003

Avolio, B., & Hannah, S. (2008). Developmental readiness: Accelerating leader development. *Consulting Psychology Journal: Practice and Research*, *60*(4), 331–347. doi:10.1037/1065-9293.60.4.331

Avolio, B., & Reichard, R. (2008). The rise of authentic followership. In R. Riggio, I. Chaleff, & J. Lipman-Blument (Eds.), *The Art of Followership: How Great Followers Create Great Leaders and Organizations* (pp. 325–337). San Francisco, CA: Jossey-Bass.

Ayanso, A., Cho, D., & Lertwachara, K. (2010). The digital divide: Global and regional ICT leaders and followers. *Information Technology for Development*, *16*(4), 304–319. doi:10.1080/02681102.2010.504698

Barbera, P., Bonneau, R., Egan, P. Jost, J., Nagler, J., Tucker, J. (2014). *Leaders or followers?* Measuring political responsiveness in the U.S. Congress using social media data. *110th American Political Science Association Annual Meeting*.

Barczak, G., Lassk, F., & Mulki, J. (2010). Antecedents of team creativity: An examination of team emotional intelligence, team trust and collaborative culture. *Creativity and Innovation Management, 19*(4), 332–345. doi:10.1111/j.1467-8691.2010.00574.x

Blasé, J., & Blasé, J. (2002). The dark side of leadership: Teacher perspectives of principal mistreatment. *Educational Administration Quarterly, 38*(5), 671–727. doi:10.1177/0013161X02239643

Boddy, C. (2011). The corporate psychopaths theory of the global financial crisis. *Journal of Business Ethics, 102*(2), 255–259. doi:10.100710551-011-0810-4

Bondarouk, T., & Ruel, H. (2009). Electronic human resource management: Challenges in the digital era. *International Journal of Human Resource Management, 20*(3), 505–514. doi:10.1080/09585190802707235

Briggs, C., & Makice, K. (2012). *Digital fluency: building success in the digital age*. Bloomington: SociaLens.

Brown, M., & Mitchell, M. (2010). Ethical and unethical leadership: Exploring new avenues for future research. *Business Ethics Quarterly, 20*(4), 583–616. doi:10.5840/beq201020439

Butler-Henderson, K. (2017). *Health Information Management 2025: What is required to create a sustainable profession in the face of digital transformation?* Launceston, Australia: University of Tasmania.

Çakar, N., & Ertürk, A. (2010). Comparing innovation capability of small and medium-sized enterprises: Examining the effects of organizational culture and empowerment. *Journal of Small Business Management, 48*(3), 325–359. doi:10.1111/j.1540-627X.2010.00297.x

Carsten, M., & Uhl-Bien, M. (2013). Ethical followership: An examination of followership beliefs and crimes of obedience. *Journal of Leadership & Organizational Studies, 20*(1), 49–61. doi:10.1177/1548051812465890

Cerne, M., Bunjak, A., Wong, S., & Moh'd, S. (2019). Creative and deserving: Digital workers' transformation of subjective into objective creativity. Academy of Management Proceedings, 2019(1), 10070.

Černe, M., Jaklič, M., & Škerlavaj, M. (2013). Authentic leadership, creativity, and innovation: A multilevel perspective. *Leadership, 9*(1), 63–85. doi:10.1177/1742715012455130

Cheong, M., Yammarino, F., Dionne, S., Spain, S., & Tsai, C. (2018). A review of the effectiveness of empowering leadership. *The Leadership Quarterly, 30*(1), 34–58. doi:10.1016/j.leaqua.2018.08.005

Colbert, A., Yee, N., & George, G. (2016). The digital workforce and the workplace of the future. *Academy of Management Journal, 59*(3), 731–739. doi:10.5465/amj.2016.4003

Crawford, J., Dawkins, S., Martin, A., & Lewis, G. (2017). Understanding the organizational climate of unethical leadership in the Australian Football League. *The Journal of Leadership Studies, 11*(2), 52–54. doi:10.1002/jls.21525

Crawford, J., Dawkins, S., Martin, A., & Lewis, G. (2018). Conceptualising authentic followers and developing a future research agenda. In D. Cotter-Lockard (Ed.), *Authentic Leadership and Followership: International Perspectives* (pp. 271–293). Cham, Switzerland: Palgrave Macmillan. doi:10.1007/978-3-319-65307-5_11

Crawford, J., Dawkins, S., Martin, A., & Lewis, G. (2019). Putting the leader back into authentic leadership: Reconceptualising and rethinking leaders. *Australian Journal of Management*; Advanced Online Publication. doi:10.1177/0312896219836460

Crawford, J., & Kelder, J.-A. (2019). Do we measure leadership effectively? Articulating and evaluating scale development psychometrics for best practice. *The Leadership Quarterly*, *30*(1), 133–144. doi:10.1016/j.leaqua.2018.07.001

Crawford, J., Kelder, J.-A., & Knox, M. (2020). What does it take to be a social entrepreneur? Authentic leaders and their effect on innovation. In Ç. Dogru (Ed.), *Leadership styles, innovation, and social entrepreneurship in the era of digitalization* (pp. 282–310). Hershey, PA: IGI Global.

Crossman, B., & Crossman, J. (2011). Conceptualising followership–a review of the literature. *Leadership*, *7*(4), 481–497. doi:10.1177/1742715011416891

de Zilwa, D. (2016). The strengths and capacities of authentic followership. *Leadership and Organization Development Journal*, *37*(3), 310–324. doi:10.1108/LODJ-01-2014-0010

de Zilwa, D., Lapierre, L., & Carsten, M. (2014). A new conceptual framework for authentic followership. In L. Lapierre, & M. Carsten (Eds.), *Followership: What is it, and why do people follow* (pp. 47–72). Bingley, UK: Emerald Group.

Denison, D., & Mishra, A. (1995). Toward a theory of organizational culture and effectiveness. *Organization Science*, *6*(2), 204–223. doi:10.1287/orsc.6.2.204

Denti, L. (2011). *Leadership and innovation: How and when do leaders influence innovation in R&D teams?* Sweden: University of Gothenburg.

Dunleavy, P., Margetts, H., Bastow, S., & Tinkler, J. (2006). New public management is dead—Long live digital-era governance. *Journal of Public Administration: Research and Theory*, *16*(3), 467–494. doi:10.1093/jopart/mui057

Eglene, O., Dawes, S., & Schneider, C. (2007). Authority and leadership patterns in public sector knowledge networks. *American Review of Public Administration*, *37*(1), 91–113. doi:10.1177/0275074006290799

Fredrickson, B., & Joiner, T. (2002). Positive emotions trigger upward spirals toward emotional well-being. *Psychological Science*, *13*(2), 172–175. doi:10.1111/1467-9280.00431 PMID:11934003

Gardner, L., & Stough, C. (2002). Examining the relationship between leadership and emotional intelligence in senior level managers. *Leadership and Organization Development Journal*, *23*(2), 68–78. doi:10.1108/01437730210419198

Gardner, W., Cogliser, C., Davis, K., & Dickens, M. (2011). Authentic leadership: A review of the literature and research agenda. *The Leadership Quarterly*, *22*(6), 1120–1145. doi:10.1016/j.leaqua.2011.09.007

Garland, E., Fredrickson, B., Kring, A., Johnson, D., Meyer, P., & Penn, D. (2010). Upward spirals of positive emotions counter downward spirals of negativity: Insights from the broaden-and-build theory and affective neuroscience on the treatment of emotion dysfunctions and deficits in psychopathology. *Clinical Psychology Review*, *30*(7), 849–864. doi:10.1016/j.cpr.2010.03.002 PMID:20363063

Gilstrap, C., & Morris, A. (2015). Role fluidity and emergent followership: How nonprofit leaders communicate their relationship as/to followers. *Journal of Nonprofit Education and Leadership*, *5*(3), 153–173.

Goffee, R., & Jones, G. (2006). *Why should anyone be led by you: What it takes to be an authentic leader*. Boston, MA: Harvard Business School Press.

Grant, M., & Booth, A. (2009). A typology of reviews: An analysis of 14 review types and associated methodologies. *Health Information and Libraries Journal*, *26*(2), 91–108. doi:10.1111/j.1471-1842.2009.00848.x PMID:19490148

Hambley, L., O'Neill, T., & Kline, T. (2007). Virtual team leadership: The effects of leadership style and communication medium on team interaction styles and outcomes. *Organizational Behavior and Human Decision Processes*, *103*(1), 1–20. doi:10.1016/j.obhdp.2006.09.004

Hatch, M. (1993). The dynamics of organizational culture. *Academy of Management Review*, *18*(4), 657–693. doi:10.5465/amr.1993.9402210154

Helfat, C., & Martin, J. (2015). Dynamic managerial capabilities: a perspective on the relationship between managers, creativity, and innovation in organizations. In J. Zhou (Ed.), *The Oxford handbook of creativity, innovation, and entrepreneurship*. Oxford, UK: Oxford University Press.

Hoerl, R., & Gardner, M. (2010). Lean Six Sigma, creativity, and innovation. *International Journal of Lean Six Sigma*, *1*(1), 30–38. doi:10.1108/20401461011033149

Joo, B., Lim, D., & Kim, S. (2016). Enhancing work engagement: The roles of psychological capital, authentic leadership, and work empowerment. *Leadership and Organization Development Journal*, *37*(8), 1117–1134. doi:10.1108/LODJ-01-2015-0005

Kane, G., Palmer, D., Phillips, A., Kiron, D., & Buckley, N. (2016). Aligning the organization for Its digital future. *MIT Sloan Management Review*, *58*(1), n.p.

Kellerman, B. (2008). *Followership: How followers are creating change and changing leaders*. Boston, MA: Harvard Business School Press.

Kelley, R. (2008). Rethinking followership. In R. Riggio, I. Chaleff, & J. Lipman-Blumen (Eds.), *The art of followership: How great followers create great leaders and organizations* (pp. 5–16). San Francisco, CA: Jossey-Bass.

Khan, S. (2016). Leadership in the digital age: A study on the effects of digitalization on top management leadership. *Masters Dissertation, Stockholm Business School.*

Leimeister, J., Österle, H., & Alter, S. (2014). Digital services for consumers. *Electronic Markets*, *24*(4), 255–258. doi:10.100712525-014-0174-6

LinkedIn. (2018). *2018 workplace learning report: The rise and responsibility of talent development in the new labor market*. Carpinteria, CA: LinkedIn Learning.

Liu, D., Liao, H., & Loi, R. (2012). The dark side of leadership: A three-level investigation of the cascading effect of abusive supervision on employee creativity. *Academy of Management Journal*, *55*(5), 1187–1212. doi:10.5465/amj.2010.0400

Luthans, F. (2002). The need for and meaning of positive organizational behavior. *Journal of Organizational Behavior, 23*(6), 695–706. doi:10.1002/job.165

Luthans, F., & Avolio, B. (2003). Authentic leadership development. In K. Cameron, J. Dutton, & R. Quinn (Eds.), *Positive organisational scholarship: Foundations of a new discipline* (pp. 241–258). San Francisco, CA: Berrett Koehler Publishers.

Luu, T. (2019). *Linking authentic leadership to salespeople's service performance: The roles of job crafting and human resource flexibility. Industrial Marketing Management*. Advanced Online Publication.

Martin, A., Jones, E., & Callan, V. (2005). The role of psychological climate in facilitating employee adjustment during organizational change. *European Journal of Work and Organizational Psychology, 14*(3), 263–289. doi:10.1080/13594320500141228

Martin, A., Jones, E., & Callan, V. (2006). Status differences in employee adjustment during organizational change. *Journal of Managerial Psychology, 21*(2), 145–162. doi:10.1108/02683940610650758

McKinsey Global Institute. (2018). *Skill shift. Automation and the future of the workforce*, (accessed 12 June 2019) https://www.mckinsey.com/featured-insights/future-of-work/skill-shift-automation-and-the-future-of-the-workforce

Milić, B., Grubić-Nešić, L., Kuzmanović, B., & Delić, M. (2017). The influence of authentic leadership on the learning organization at the organizational level: The mediating role of employees' affective commitment. *Journal for East European Management Studies, 22*(1), 9–38. doi:10.5771/0949-6181-2017-1-9

Mulgan, G. (2006). The process of social innovation. *Innovations: Technology, Governance, Globalization, 1*(2), 145–162. doi:10.1162/itgg.2006.1.2.145

Murawski, M., & Bick, M. (2017). Digital competences of the workforce – a research topic? *Business Process Management Journal, 23*(3), 721–734. doi:10.1108/BPMJ-06-2016-0126

Neumeier, M. (2014). *The 46 rules of genius: an innovator's guide to creativity*. San Francisco, CA: New Riders.

Organisation for Economic Co-operation and Development. (2001). Understanding the digital divide. Paris: Directorate for Science, Technology and Industry, OECD.

Özaralli, N. (2003). Effects of transformational leadership on empowerment and team effectiveness. *Leadership and Organization Development Journal, 24*(6), 335–344. doi:10.1108/01437730310494301

Podsakoff, P., MacKenzie, S., Lee, J., & Podsakoff, N. (2003). Common method biases in behavioral research: A critical review of the literature and recommended remedies. *The Journal of Applied Psychology, 88*(5), 879–903. doi:10.1037/0021-9010.88.5.879 PMID:14516251

Powell, W., Oberg, A., Korff, V., Oelberger, C., & Kloos, K. (2017). Institutional analysis in a digital era: Mechanisms and methods to understand emerging fields. In G. Krucken, C. Mazza, R. Meyer, & P. Walgenbach (Eds.), *New themes in institutional analysis: Topics and issues from European research*. Cheltenham, UK: Edward Elgar. doi:10.4337/9781784716875.00016

Pricewaterhouse Cooper. (2017). *Workforce of the future. The competing forces shaping 2030.* (accessed 12 June 2019) https://www.pwc.com/gx/en/services/people-organisation/workforce-of-the-future/workforce-of-the-future-the-competing-forces-shaping-2030-pwc.pdf

Rego, P., Lopes, M., & Simpson, A. (2017). The authentic-Machiavellian leadership grid: A typology of leadership styles. *The Journal of Leadership Studies*, *11*(2), 48–51. doi:10.1002/jls.21524

Ribeiro, N., Duarte, A., Filipe, R., & Torres de Oliveira, R. (2019). How authentic leadership promotes individual creativity: The mediating role of affective commitment. *Journal of Leadership & Organizational Studies*; Advanced Online Publication. doi:10.1177/1548051819842796

Ridgeway, C. (2001). Gender, status, and leadership. *The Journal of Social Issues*, *57*(4), 637–655. doi:10.1111/0022-4537.00233

Schein, E. (1985). *Organizational culture and leadership*. San Francisco, CA: Jossey-Bass.

Seligman, M., & Csikszentmihalyi, M. (2014). Positive psychology: An introduction. In M. Csikszentmihalyi (Ed.), *Flow and the foundations of positive psychology* (pp. 279–298). Dordrecht, The Netherlands: Springer.

Shepherd, J. (2004). What is the digital era? In G. Doukidis, N. Mylonopoulos, & N. Pouloudi (Eds.), *Social and economic transformation in the digital era* (pp. 1–18). IGI Global. doi:10.4018/978-1-59140-158-2.ch001

Shirey, M., White-Williams, C., & Hites, L. (2019). Integration of Authentic Leadership Lens for Building High Performing Interprofessional Collaborative Practice Teams. *Nursing Administration Quarterly*, *43*(2), 101–112. doi:10.1097/NAQ.0000000000000339 PMID:30839447

Somech, A., & Drach-Zahavy, A. (2013). Translating team creativity to innovation implementation: The role of team composition and climate for innovation. *Journal of Management*, *39*(3), 684–708. doi:10.1177/0149206310394187

Srivastava, A., Bartol, K., & Locke, E. (2006). Empowering leadership in management teams: Effects on knowledge sharing, efficacy, and performance. *Academy of Management Journal*, *49*(6), 1239–1251. doi:10.5465/amj.2006.23478718

Ugboro, I., & Obeng, K. (2000). Top management leadership, employee empowerment, job satisfaction, and customer satisfaction in TQM organizations: An empirical study. *Journal of Quality Management*, *5*(2), 247–272. doi:10.1016/S1084-8568(01)00023-2

van Laar, E., van Deursen, A., van Dijk, J., & de Haan, J. (2017). The relation between 21st-century skills and digital skills: A systematic literature review. *Computers in Human Behavior*, *72*, 577–588. doi:10.1016/j.chb.2017.03.010

Van Vugt, M. (2006). Evolutionary origins of leadership and followership. *Personality and Social Psychology Review*, *10*(4), 354–371. doi:10.120715327957pspr1004_5 PMID:17201593

Vitello-Cicciu, J. (2019). Am I an Authentic Nursing Leader for Healthy Workplace Environments? *Nurse Leader*, *17*(3), 201–206. doi:10.1016/j.mnl.2019.03.011

Williams, K., Chatterjee, S., & Rossi, M. (2008). Design of emerging digital services: A taxonomy. *European Journal of Information Systems, 17*(5), 505–517. doi:10.1057/ejis.2008.38

Wong, C., & Laschinger, H. (2013). Authentic leadership, performance, and job satisfaction: The mediating role of empowerment. *Journal of Advanced Nursing, 69*(4), 947–959. doi:10.1111/j.1365-2648.2012.06089.x PMID:22764828

World Economic Forum. (2016). *The future of jobs. Employment, skills, and workforce strategy for the fourth industrial revolution*, (accessed 12 June 2019) https://www.weforum.org/reports/the-future-of-jobs

ADDITIONAL READING

Butler-Henderson, K. (2017). *Health Information Management 2025: What is required to create a sustainable profession in the face of digital transformation?* Launceston, Australia: University of Tasmania.

Crawford, J., Dawkins, S., Martin, A., & Lewis, G. (2018). Conceptualising authentic followers and developing a future research agenda. In D. Cotter-Lockard (Ed.), *Authentic Leadership and Followership: International Perspectives* (pp. 271–293). Cham, Switzerland: Palgrave Macmillan. doi:10.1007/978-3-319-65307-5_11

Crawford, J., Dawkins, S., Martin, A., & Lewis, G. (2019). Putting the leader back into authentic leadership: Reconceptualising and rethinking leaders. *Australian Journal of Management*; Advanced Online Publication. doi:10.1177/0312896219836460

Liu, D., Liao, H., & Loi, R. (2012). The dark side of leadership: A three-level investigation of the cascading effect of abusive supervision on employee creativity. *Academy of Management Journal, 55*(5), 1187–1212. doi:10.5465/amj.2010.0400

Pricewaterhouse Cooper. (2017) *Workforce of the future. The competing forces shaping 2030.* (accessed 12 June 2019) https://www.pwc.com/gx/en/services/people-organisation/workforce-of-the-future/workforce-of-the-future-the-competing-forces-shaping-2030-pwc.pdf

World Economic Forum. (2016) *The future of jobs. Employment, skills and workforce strategy for the fourth industrial revolution*, (accessed 12 June 2019) https://www.weforum.org/reports/the-future-of-jobs

KEY TERMS AND DEFINITIONS

Authentic Follower: An individual who, through their capacity for authenticity and positive organisational engagement, are self-managing and follow leaders with whom they share values (Crawford et al., 2018, p. 274).

Authentic Leader: A leader who “influences and motivates followers to achieve goals through their sincerity and positive moral perspective, enabled through heightened awareness and balanced processing” (Crawford et al., 2019, n.p.).

Digital Era: "Characterized by technology which increases the speed and breadth of knowledge turnover within the economy and society" (Shepherd, 2004, p. 1).

Digital Innovation: The creation or application of technology to enhance existing practices and solve emergent problems.

Digitally Empowered Worker: A digital worker who is aware, creative, agile, and willing to learn.

Digital Productivity: The digital output value of workers.

Digital Services: An activity or benefit that one party can give to another, that is, provided through a digital transaction" (Williams, Chatterjee, & Rossi, 2008, p. 507).

MOOC: Massive open online courses.

STEMM: Science, technology, engineering, mathematics, and medicine.

Chapter 7
Digitalisation of Mandatory Social Expenditure Targets in India:
Cost or Value Addition?

Ajit Dayanandan
University of Alaska, Anchorage, USA

Sudershan Kuntluru
Indian Institute of Management, Kozhikode, India

ABSTRACT

The chapter examines the impact of mandatory CSR expenditure targets of firms on financial performance in India. Extant studies have shown using event study methodology that announcement effect of compulsory CSR expenditure targets on abnormal returns of these firms is negative. Further, it is examined on how financial performances of these firms which met or exceeded the targets (Socially responsible firms) or were not able to meet the targets (socially irresponsible firms) differ. Based on the data for 1,460 firm years for the period 2015 to 2018 in the Indian context, the empirical analysis shows that their financial performance was negatively impacted because of these mandated CSR targets. Available evidence shows that CSR expenditure that spurs digitalisation and technological innovation helps the poorest and most vulnerable which enables them to lead a healthy, productive life.

INTRODUCTION

Corporate Social Responsibility (CSR) is widely discussed among academics, firms, regulators and other market participants (Maragolis et al. 2009; Barnea & Rubin, 2010; Kitzmueller and Shimshack, 2012, Ali et al. 2017). By engaging in CSR activities, firms can not only generate favorable stakeholder attitudes but also better support behaviors like the purchase of its product, seeking employment, investing in the company. In the long run, it enables firms to build corporate image, strengthen stakeholder-company

DOI: 10.4018/978-1-7998-2799-3.ch007

relationships, and enhance stakeholders' advocacy behaviors. The increasing power of activist organizations and the media has made CSR strategies increasingly relevant and strategies of "doing well by doing good" more relevant in the modern corporate world (Economist, 2005). Firms engage in "good" corporate conduct because it provides benefits to the society, although they have operational and disclosure costs (Benabou and Tirole, 2010). The incorporation of environmental, social and governance principles into corporate decisions has shifted the paradigm from conventional view that the firms' purpose was no longer to serve their owners alone, but also customers, staff, suppliers, communities and other stakeholders (Economist, 2019). There is also impressive body of research which examined the relationship between CSR investments and a firm's performance and these have concluded that the relationship is inconclusive (Orlitzky et al. 2003; Mackey et al. 2007).

But in the context of developing economies, it is increasingly argued that CSR should embrace not only corporate conduct, social, environmental and human right issues but also the role of business in relation to poverty reduction (Prieto-Carrón et al. 2006). But many firms in emerging market economies firms are reluctant to undertake CSR expenditures on a voluntary basis. India had taken legislative measures to mandate CSR expenditures. Under mandated CSR expenditure provisions in India, larger Indian firms are mandated to spend annually 2 per cent of their 3-year average annual net profits on CSR activities since 2014. In the delivery of CSR outcomes, the corporate sector in India had played an active role in digitally empowering the "have-nots" by addressing economic and social divide, tackle social injustice, climate change, inequality etc. The study examines whether the meeting mandatory CSR targets in India had positively/adversely affected the profitability of firms.

INDIA: BACKGROUND

In India, CSR conduct of firms has played an important role in the digitalisation of the economy- the corporate funding of social infrastructure like schools, health, community infrastructure and providing computers to rural school children, connecting communities through internet, devising mobile and payment mechanisms has fostered digitalisation of the Indian economy in a substantial way. In India, the digital industry has played an important role in the economy. India has the world second internet subscribers - 560 million in 2018 (Kaka et al. 2019). The industry rose from $2.5 billion in the late 1990s to more than $110 billion in recent times (McKinsey, 2016). The industry created more than 50 per cent of all organized sector jobs in the last five years. India, is the fastest-growing wireless (mobile phone) market in the world. About 15 per cent of India's 1.2 billion people use mobile- phone services, which are offered at rates as low as 2 U.S. cents a minute. Indian operators spend between 60 per cent and 65 per cent of their network expansion costs on so-called passive infrastructure. With the sharing of passive infrastructure, mobile services is extended to areas that are unviable in the past, especially rural towns and villages. In September 2019, the Government widened the scope of CSR activities by including funding towards research across various fields such as science, technology, medicine. This is expected to attract more funds on research and development including the push towards digitalisation in the country.

Many critics argue that investments in digitalisation should not be a priority for a developing country given the more basic needs in areas such as nutrition, health and education (Gates, 2000). But digitalisation of Indian economy especially digital have-nots (poor and rural households) has derived tremendous digital dividends either in terms of connectivity, improving their livelihood by enhancing the income of households engaged in agriculture, forestry, fisheries etc. (Jensen, 2007).

The Aadhaar card (meaning authentication) introduced in India is the world biggest and sophisticated ID programme; it is a 12-digit unique identity number that can be obtained voluntarily by residents of India, based on their biometric and demographic data. The data is collected by the Unique Identification Authority of India (UIDAI), a statutory authority established in January 2009 by the government of India. The Aadhaar card ID system is used by the government for disbursement of cooking gas subsidies, house allotments, school scholarships, admission into remand and welfare houses, passports, e-lockers (Digilocker), for archiving documents, bank accounts, tax payments etc.

In the delivery of CSR outcomes, the corporate sector plays an active role in digitally empowering the "have-nots" at the bottom-of-the pyramid of the Indian society (Prahlad, 2005). For example, Google Pay is setting up link to and between 12 million kirana (family-run) retail stores in India. The idea that firms through their digitalisation strategies can facilitate a sense of purpose in the society by addressing economic and social divide, tackle social injustice, climate change and inequality and benefit from it is sweeping currently the mainstream narrative (Economist, 2019). In fact, study by Banerjee et al. (2007) found that in randomised trials, computer assisted learning programmes in India (both in urban and rural areas) had a positive impact on children's academic achievement (increased academic achievement by 0.14 standard deviation in the first year and 0.28 standard deviation in the second year).

In India spending on CSR has gone up from Rs.10,066 crore in 2014-15 to Rs.13,327 crore in 2017-18 (Prasad, 2019). CSR spending between 2014-15 and 2017-18 was the highest in education, health, fight against poverty and malnutrition, access to clean drinking water, livelihood and for the differently-abled. But paradoxically, most of the CSR expenditure by firms in India went to states like Maharashtra, Karnataka, Andhra Pradesh, Gujarat, Tamil Nadu and Delhi (around 40 per cent) of the Rs. 52,208 crore that companies spent across India on CSR activity during 2014-15 to 2017-18. On the other hand, Jharkhand, Bihar, Chattisgarh, Madhya Pradesh, and Uttar Pradesh, which together have more than half of 117 most backward districts in the country, received only 9 per cent of the total CSR spending during the period (Prasad, 2019). In the Indian context, existing studies which examine CSR-firm value relationship report mixed results (Manchiraju and Rajagopal, 2015; Sarkar and Sarkar, 2015; Mukherjee and Bird, 2016; Dharmapala and Khanna, 2016). The main aim of the present study is to examine link between mandated CSR expenditure in India and its impact of on their financial performance.

But many firms in emerging market economies firms are reticent to undertake CSR expenditures on a voluntary basis. Based on the concerns of economic growth and uneven distribution of income, many governments and regulators in many countries (like India, Indonesia and Philippines) had taken legislative measures to mandate CSR expenditures. The underlying challenges in emerging market economies are to link the interests of shareholders with the wider goal of the society given the fact that these countries have large segments of the population which is at the "bottom of the pyramid". Indonesia was the first country which mandated in 2007 that firms operating in natural resources industry create an "obligatory funding" for implementing CSR (Waagstein, 2011). Philippines legislative arm (Congress) in 2013 while approving the CSR Act argued that "…many corporations and other business organizations have very little care for the welfare of the society, community where they operate and the natural environment around them" (p.1). Philippines CSR Act 2013 only calls for CSR spending for which they are entitled to "full deductions" for such expenses. India was the first country which passed legislation mandating CSR spending mandatory for its large companies with its amendment of Indian Companies Act 2013. Under mandated CSR expenditure provisions in India, larger Indian firms (which any of the three criterion like net worth, sales and net profit) are mandated to spend annually 2 per cent of their 3-year average annual net profits on CSR activities and disclose such expenditures in their financial statements and a separate individual CSR report (Dhanesh, 2015; Subramaniam et al. 2015; Bansal et al. 2018). Nearly 59 percent of Indian firms listed in the National Stock Exchange in India come under the mandated CSR expenditure targets (Manchiraju and Rajgopal, 2017).

The mandated CSR expenditure targets in India is best described as "comply or explain" model where if a company is not in a position to spend the prescribed amount on CSR, the board is required to disclose it and give reasons for its non-compliance. The mandatory CSR regime has created "quagmire" situations for Indian firms. The issue has become a high profile public issue with Mr. Ratan Tata owner of the biggest Indian conglomerate comparing mandatory CSR to "another tax on business" (Economist, 2017). Economist (2017) called mandatory CSR in India "coerced do-goodery," and critics of mandated CSR expenditure targets in India argue that these mandated CSR expenditures represent corporate costs which undermine their competitive positions in the industry/country. Critics also point out that that making CSR activities mandatory is essentially an exercise in outsourcing government social responsibility to the private sector to cover up for the government failure to address social problems. The extant literature has shown that CSR spending can be costly, and also they can compete for firms' limited financial resources and the empirical findings regarding the impact of CSR on firms' financials are mixed (Margolis et al. 2009). In the Indian context, the initial studies show that it varied by size- for large companies, making CSR spending mandatory did not elicit the forecasted increase in CSR expenditure while non-mandatory companies reduced their level of CSR expenditures (Mukherjee and Bird 2016). The study by Manchiraju and Rajgopal (2017) shows the mandatory CSR expenditure law in 2013 caused a significant drop in the stock price of firms forced to spend money on CSR. The availability of data of meeting mandatory CSR targets in India since 2014 provides an opportunity to examine whether meeting compulsory CSR targets positively/adversely affected the profitability of firms. This enables us to dig deep into the behavior of firms in terms of CSR expenditure and its impact of on the financial performance parameters.

Among the emerging market economies, India represents an interesting case study. It is one of the fastest growing economies and fifth largest economy of the world (with GDP of around $ 2 trillion) which has attracted substantial global portfolio allocation. The philanthropic orientation of Indian firms has its roots in history; most of the prominent family-owned enterprises undertook community development expenditures as part of their corporate conduct. Secondly, the fruits of rapid growth are unevenly distributed - it is home to the largest number of people living in absolute poverty and the gap between have and have-nots has increased. Thirdly, India's corporate institutional structure has peculiar features. In contrast to dominant institutional ownership in the United States, India is dominated by concentrated ownership from family-owned enterprises (Khanna and Palepu, 2000,2005) and promoter driven firms (like Tatas, Infosys etc.) who dominate the market. Most of the family owned and promoter driven firms in India dominate minority shareholders and occupy a lead role in decision making and often influence the decisions of the board. Most of the firms in India is 'family centered' style of management' and has voluntarily spent considerable amount of money on CSR activities (mostly on community development) while others have done so only when mandated by regulators. Increasingly, firms in India use CSR not only as a tool of philanthropy but also to boost their bottom line. For e.g., Godrej, a consumer group, offers three-month training courses for beauticians around the country hoping to boost the demand for cosmetic products. Similarly, Ashok Leyland, a truck maker, provides free driving lessons helping to fill the shortage of truck drivers (Economist, 2017). Fourthly, with successful implementation of Aaadhar card as an authentication, its use in banking, government disbursement of loans and grants, railway/air bookings, taxes etc., the country has been adopted technology and digitalisation at a rapid pace. The Prime Ministers Digital India, Startup India, Skill India are some of the technological initiatives in this area. In the creation of digital infrastructure especially in rural areas, the corporate sector has played a crucial role in funding digital India. Fifthly, India has become the first country to legally mandate expenditures (2 per cent of net income) on CSR with effect from 2014. It is further examined in the Indian context the relationship between mandatory CSR expenditure targets and performance subsequent to the mandatory disclosure using recent data (last 4 years i.e., 2015-2018) of 365 firms (which are mandated to CSR expenditure targets) and which account for 88 per cent of market capitalization of firms in India.

RELATED LITERATURE REVIEW

There is an impressive body of research which examined the relationship CSR and its relationship with firm size/age, profitability, corporate governance mechanisms, earnings quality, earnings management (Mackey et al. 2007).The debate on this topic was sparked when Friedman (1970) declared in his well-known New York Times essay that the sole responsibility of the firm is to increase its profits. The "shareholder expense " view of Friedman (1970) is based on the argument that CSR activities is a manifestation of "moral hazard" towards the shareholder (owner) and represents a "donation" from shareholders to stakeholders. On the other hand, the "stakeholder value maximization" view have argued that firms have a duty to society that goes well beyond simply maximization of the wealth of equity holders and firms have a social role along with their commercial interests" (Freeman, 1984). In the modern world, firms are responsible for the interests of all stakeholders. A way to resolve this conflict is to ensure that socially responsible behavior improve the firm's present value of cash flows (Mackey et al. 2007). Drucker (1984) suggested that firms should ensure that social responsibilities also become business opportunities. Such strategically motivated CSR activities are referred to in the literature as "doing well is doing good".

There is a growing literature called "sustainable finance" defined as the incorporation of environmental, social and governance (ESG) principles into business decisions. ESG can have substantial impact on firms' performance, risks and stability of the financial system (IMF, 2019). Exposing a firm to environmental risks can lead to large losses and bankruptcies for firms. Examples of such risk include liabilities related to asbestos, toxic spills in mining, chemicals and oil & gas. Integration of ESG factors into business models and strategic strategies can mitigate these risks. Social risks in the form of inequality could contribute to social, economic and financial instability (Rajan, 2010). Governance failures in the corporate sector especially in the banking and financial sector has contributed and accentuated financial crisis like Asian Financial Crisis, 2008 Financial Crisis in United States. ESG factors are now part of the "core" of long-term corporate strategy of generating revenue while also benefiting society.

Studies have also found that CSR activities provide a number of benefits to a firm's stakeholders including increase employee morale, better firm reputation and more harmonious growth (Servaes and Tamyo, 2013). Dhaliwal et al. (2011) argue that voluntary disclosure of CSR activities attracts institutional investors and analysts and reduces the firm's cost of capital. Sacconi and Antoni (2011) suggests that a firm's CSR activities are a good proxy for its social capital and that a firm's CSR activities general social capital and trust. Based on this argument, the study by Lins et al. (2017) found that for US firms, firms with high CSR ratings outperform firms with low CSR ratings during the crisis by at least four percentage points, after controlling for firm and risk characteristics. Similarly, studies have also found firms with good CSR record have less firm-level tail risk from carbon emissions (Ilhan, Sautner & Vilkov, 2019).

The empirical evidence regarding the relationship between CSR and financial performance is divergent. There are evidence which show positive relationship between CSR and financial performance (De Velde et al. 2005; Orlitzky et al. 2003; Gregory et al. 2014). On the other hand, some studies find significant negative relationship between CSR and financial performance (Griffin and Mahon, 1997; Wright and Ferris, 1997; Brammer et al. 2006). There are also studies which show no significant relationship (Mc-Williams and Siegal, 2000; Moore, 2001; Soana, 2011). Given the divergent results, there is a need for further research in this area especially in the context of emerging market economies where there is an attempt to mandate CSR spending by government/regulators. In theory, the application of mandatory CSR should act as a complement to voluntary CSR. However there is an immense risk of such mandatory CSR would make them defensive and discourage them from embracing that responsibility and end of the process (Waagstein, 2011).

In India, most of the studies relate to nature of CSR reporting and its measurement (Singh and Ahuja, 1983; Vasal, 1995; Murthy and Abeysekera, 2008; Kansal et al. 2014; Prasad et al. 2017). Most of these studies examine CSR disclosures, patterns and trends in India. Latterman et al. (2009) undertook a comparative study of 68 largest multinational firms in China and India from the Forbes list and found that Indian firms communicate more CSR primarily due to more "rule-based" as opposed to "relation-based" governance environment. A study by Mishra et al. (2010) is perhaps the most comprehensive in the Indian context. Using a sample of firms using perceptual data on CSR and non-financial data from CEO's of Indian firms for six stakeholder groups, supplemented by financial data of these firms, the study finds that stock-listed firms show responsible business practices compared with non-listed firms. The study by Kansal et al. (2014) finds in the Indian context that corporate size and industry category are found to correlate with corporate social disclosures. Prasad et al. (2017) find that CSR disclosures during 2011-12 and 2014-15 have increased over time and the quality of disclosure is largely descriptive. The study also finds that CSR disclosure varies positively with industry, size, age and foreign customers and negatively with leverage. But the recent study by Manchiraju and Rajgopal (2017) and Mukherjee et al. (2018) examined the effects of mandatory CSR expenditure in India. Manchiraju and Rajgopal (2017) show the mandatory CSR expenditure of affected (mandated) compared with unaffected firms saw a significant drop in abnormal returns for various events associated with the implementation of mandatory CSR in India. Mukherjee et al. (2018), on the other hand, looks at the behavior of firms' pre and post-mandatory CSR in India and found that mandatory CSR legislation affected adversely profitability of firms. Our study is different from Manchiraju and Rajgopal (2017) and Mukherjee and Bird (2018) in the sense that it is examined the impacts of CSR mandate on CSR-mandated firms who met the CSR threshold criteria (called CSR firms) and those who did not meet the CSR targets (non-CSR firms) on their profitability's using data for 2015 to 2018. It is posited that mandatory CSR expenditure targets lead to a change in firm behavior and would lead to a decrease in firm financial performance. The intuition behind this argument is that if firms are mandated to meet certain CSR targets, those firms who are spending more than this target is likely to scale back their excess CSR expenditure, and those who are below these CSR expenditure targets in the absence of any tax benefit might opt for "explain" mode of behavior. Moreover, if firms believe that CSR activities benefited them, they would have been in place before the mandate.

DATABASE AND RESEARCH METHODOLOGY ADOPTED FOR THIS STUDY

Database

The study is based on the Centre for Monitoring Indian Economy's Prowess database. Prowess database has information on CSR targets for each listed company in India since 2014 (those firms with a net worth of Rs. 4,500 crores or more or turnover of Rs.1,000 cores or more or net profit of Rs. 5 cores of more during any financial year). In other words, the dataset covers firms in India for which CSR targets are mandated (the treatment group) and this group consist of firms which met or exceeded the targets (Socially responsible firms) or were not able to meet the targets (socially irresponsible firms). The related financial data are also sourced from Prowess database.

We started the data collection from Nifty 500 companies. The study was successful in collecting 365 out of 500 companies for which CSR targets and compliance data for the period 2015 to 2018 were available. For the estimation period data relates to 1,460 firm years. These 365 companies account for 80 to 85 per cent of the market capitalization of NSE Nifty companies for the period 2015 to 2018 and hence covers slightly over four-fifth coverage of companies in the sample could be considered well-represented of the overall population. Most of the firms in the sample are from manufacturing sector (around 50 per cent), followed by financial sector (14 per cent), information and communication (8.5 per cent) (See Table 1 and 2).

Methodology

In the literature especially in the US context, there are a number of measures of CSR in the literature include CSR ratings (e.g., Kinder, Lydenberg, Domini Research and Analytics ratings [KLD]); environmental responsibility (e.g., reduction of harmful emissions); and the existence of a code of ethics or measures of charitable giving. Our study uses CSR spending information identified by the KLD as a measure of a firm's CSR engagement. The Companies Act 2013 mandates that 2 per cent of average net profits from the past 3 years be spent on CSR activities in the areas of community relations, environment, energy, and product.

Table 1. List of variables used in the study

Variables		Description
Total Assets	A	Total assets
Age	AGE	Age of firms in number of years
Revenue	REV	Revenue (Sales) in Rupees Millions
Net Income	NI	Net Income in Rupees Millions
Market Capitalisation	MC	Total market Capitalization at the end of each financial year
Return on Assets	ROA	Net Income/Total Assets
Return on Equity	ROE	Net income/Total Equity
Leverage	LEV	(Total Liabilities- Total Equity)/Total Assets
CSR Firms	CSRF	Firms meeting and exceeding CSR regulatory targets are assigned a value of 1 and 0 otherwise.

Table 2. Industry-wise distribution of firms used in the study

Industry Code	Industry Name	Percentage Distribution
1	Agriculture, forestry and fishing	5.75
2	Mining and quarrying	0.82
3	Manufacturing	50.14
5	Water supply; sewerage, waste management and remediation activities	2.47
6	Construction	7.40
7	Wholesale and retail trade	4.66
8	Transportation and storage	3.56
9	Accommodation and Food service activities	0.82
10	Information and communication	8.49
11	Financial and insurance activities	14.25
13	Professional, scientific and technical activities	0.55
14	Administrative and support service activities	0.55
17 & 18	Others	0.55
	Total (N=365)	100

The present study will use a CSR spending ratio, which is the amount spent by a firm on CSR activities divided by the amount required to be spent by a firm per the Companies Act 2013. A ratio of 1 or greater than 1 implies that the firm spent adequate amount or more than required by law and a ratio less than 1 implies that the firm spent less than required. Socially responsible firms are defined in terms of the ratio equal to 1 and or greater than 1. Socially irresponsible firms are defined in terms of the ratio less than 1. Nearly 56.7 per cent of the firms in the sample was classified as socially responsible firms and nearly 43.3 per cent of the firms were classified as socially irresponsible firms. Logit and probit models are used to estimate the impact CSR on performance of firms, viz.,

$$CSRF_{it} = \beta_0 + \beta_1 SIZE_{it} + \beta_2 LEV_{it} + \beta_3 Age_{it} + \beta_4 Performance_{it} + \varepsilon_{it} \tag{1}$$

$$\beta_1 > 0, \beta_2 \langle 0, \beta_3 \rangle 0, \beta_4 < 0.$$

Where CSRF refers to CSR Firms (value of 1 for socially responsible firms, 0 for socially irresponsible firms), SIZE (proxied by market capitalization), LEV is leverage (Total liabilities to total assets), Performance is measured by ROA is return on assets (Net income to total Assets) and ROE (Net income to Total Equity). In equation (1), the control variables are Size, Leverage, Age and the independent variable is ROA/ROE.

Table 3. Descriptive statistics of variables used in the study: 2015-2018.

Variables	Mean	Median	Standard Deviation	Co-efficient of Variation	Minimum	Maximum
CSR Firms (CSRF)	0.5671	1.0000	0.4956	0.8740	0	1
Age	42	34	23.9690	0.5698	3	123
Total Assets (Rs. Million)	366289	45814	1663718	4.5421	2236	345
Revenue (Rs. Million)	97348	23073	351380	3.6095	3235	5130167
Market Capitalisation (Rs. Million)	255335	72605	559535	2.1914	3121	5591596
Leverage	0.4987	0.4935	0.2264	0.4579	0.0025	1.0117
Return on Assets (ROA)	0.1553	0.1303	0.1004	0.6473	-0.2149	0.8223
Return on Equity (ROE)	0.1556	14.7000	0.1622	1.0428	-0.1372	1.3136

EMPIRICAL RESULTS

This section presents the empirical results. Table 3 presents the descriptive statistics of variables used in the investigation. The mean value of the CSR firms is 0.5671 which means that nearly 57 per cent of the sample firms are classified as CSR firms and the remaining 43 per cent are Non-CSR firms. The average age of the firms is 42 years with a median age of 34 years. The average market capitalization (proxy for size) is around Rs.255,335 million with the median substantially lower (Rs. 72,605 million). The average leverage of firms in the sample is 0.4987 with the median (0.4935) slightly lower which shows the leverage of firms is not substantial. The average return on assets (ROA) is around 0.1553 and the average return on equity is around 0.1556.

Table 4 presents the correlation matrix of variables used in the study for the period 2015-2018. The correlation coefficient among the variables is generally low except among performance variables (ROA, ROE), size variables (market capitalization, total assets and revenue). The correlation coefficient between CSR firms (CSRF) and performance variables (ROA and ROE) is generally negative. There is positive correlation between CSR firms (CSRF) and size variables (market capitalization, total assets and revenue). There is also a positive correlation between CSR firms and age of firms. There is negative relation between CSR firms and leverage of firms. Leverage is also negatively correlated with performance variables.

Table 5 reports the results of logit and probit regressions of equation (1) for financial performance (profitability) variables ROA and ROE. The age of the firm is positively related to CSR performance variable (CSR Firms) and is statistically significant (at 1 per cent level) in both logit and probit versions of the model. The ROA version of the equation (1) shows that CSR performance variable is negatively related to financial performance both in logit and probit models. Similarly, in the ROE version of the model also it is inversely related with CSR firms and statistically significant at 1 per cent level. The negative relation between CSR firms and profitability variables (ROA and ROE) shows that mandated CSR expenditures in India adversely affects profitability of firms in India. The size of the firms as proxied by market capitalization is positively related to CSR firms. Leverage variable is not statistically significant.

Table 4. Correlation matrix of variables used in the study: 2015-2018.

Variables	CSR	Total Assets	Sales	Market Capitalization	Leverage	Return on Assets (ROA)	Return on Equity (ROE)	Age
CSR Firms	1.0000	0.0223	0.0840	0.0242	-0.0063	-0.0760	-0.0840	0.0801
Total Assets	0.0223	1.0000	0.1441	0.4322	0.2535	-0.1292	-0.0878	0.0479
Sales	0.0840	0.1441	1.0000	0.4391	0.0407	-0.0044	0.0294	0.0857
Market Capitalisation	0.0242	0.4322	0.5440	1.0000	-0.0132	0.1441	0.1443	0.0740
Leverage	-0.0063	0.2535	0.0407	0.4391	1.0000	-0.3910	-0.1718	0.0380
Return on Assets (ROA)	-0.0760	-0.1292	-0.0044	0.1441	-0.3910	1.0000	0.7803	-0.0441
Return on Equity (ROE)	-0.0803	-0.0866	0.0294	0.1443	-0.1718	0.7803	1.0000	-0.0259
Age	0.0801	0.0479	0.0857	0.0740	0.0380	-0.0441	-0.0259	1.0000

The fit of the equations (logit and probit) are good given the nature of the data and LR Chi Squared statistics which is a joint test of the significance of independent variable is also statistically significant.

Table 5. Determinants of CSR spending ratio of firms in India: 2015-2018.

Variables	Hypothesized Signs	Dependent Variable			
		CSR Firms			
		Logit	Probit	Logit	Probit
Constant	+/-	-0.7097 (0.51118)	-0.4529 (0.3177)	-0.9909 (0.5108)*	0.6221 (0.3176)*
Age of the Firm	+	0.0046 (0.0023)**	0.0028 (0.0014)*	0.0047 (0.0023)**	0.0029 (0.0014)**
ROA	+/-	-2.5455 (-0.6143)***	-1.5754 (0.3743)***		
ROE	+/-			-0.0136 (0.0036)***	-0.0085 (0.0022)***
Log(Market Capitalization)	+	0.1041 (0.0424)***	0.0658 (0.0263)**	0.0999 (0.0424)**	0.0627 (0.0262)**
Leverage	-	0.0738 (0.2731)	0.0505 (0.1688)	0.3172 (0.2623)	0.2000 (0.1613)
Industry Dummies		Yes	Yes	Yes	Yes
Diagnostics					
N		1,460	1,460	1,460	1,460
Pseudo R^2		0.0253	0.0254	0.0238	0.0239
LR Chi^2		50.67***	50.00***	47.64***	47.81***
No of cases correctly predicted(%)		59.8	59.7	59.6	59.6

Note:
1. Figures in brackets are standard errors.
2. ***, **, * indicate statistical significance at 1%,5% and 10% respectively.

SOLUTIONS AND RECOMMENDATIONS

The study finds that mandated CSR expenditure targets in India had a negative impact on firms' profitability. This could pose a challenge on sustainability of CSR programmes. The mandatory nature of CSR dispensation in India needs a relook. A solution to this challenge could be to allow "tax deduction" on CSR expenditures so that it does not become another "tax on the industry"

Increased CSR interface especially through the digitalisation of the transactions can foster trust and social capital especially among the poor and enhance market participation.

For CSR to have a profound impact on society, promotion of trust among market participants is essential. In order to enhance trust, build social capital, firms could route their CSR expenditures through non-governmental agencies which have a good record of serving the poor and sections of the population at the bottom of the pyramid.

Digitilisation and competitive spirit underlying it has made some firms more powerful in the corporate domain especially in the social media platform. In order to prevent regulation crippling their growth and power, these firms need to forge a close link with society to maintain trust and legitimacy.

Firms should be encouraged to promote "social entrepreneurship" like microfinance which is popular with the customers, investors and employees. This would mean that the purpose of the firms would be set by owners and not executives of firms.

Responsible conduct of firms vis-a-vis their employees should be encouraged. Although there is considerable improvement in firms in matters of providing better workplace environment, there are wide disparities among different levels of employees and many needs to be done to improve the plight of those at the lower end of the organisational hierarchy.

There is also a need to ensure that ethics and corporate governance become an essential part of corporate culture.

FUTURE RESEARCH DIRECTIONS

So far research on impact of CSR on firms and society has been confined to only macro level like impact of mandated CSR target announcements on stock prices of firms, impact on profitability, earnings management etc. There is a need to more micro level studies regarding the impact on various sections of the society. Case studies on various types of CSR expenditure on various stakeholders is another area which require attention. Moreover, there is also a need for research on welfare implications of various CSR initiatives (like Digitalisation).

CONCLUSION

The aim of this paper is to study the impacts of mandatory CSR expenditure targets of India (amendment to Section 135 of Indian Companies Act, 2013) which requires firms which meet certain thresholds (like turnover, net worth, profit) to spend 2 per cent of their profits averaged over the past three years on CSR expenditure. These targets are in the nature of "comply or explain" model. Extant studies have shown using event study methodology that announcement effect of compulsory CSR expenditure targets on abnormal returns of these firms is negative. Further, it is pursued by examining on how financial performances of these firms which met or exceeded the targets (Socially responsible firms) or were not able to meet the targets (socially irresponsible firms) differ. Based on the data for 1,460 firm years for the period 2015 to 2018 in the Indian context, our empirical analysis shows that their financial performance was negatively impacted because of these mandated CSR targets. The study shows that in the absence of any tax benefit for mandated CSR expenditures, these CSR expenditures assume the role of "donation" from shareholders to stakeholders without any "financial good". But CSR still has a huge opportunity to deliver a truly unique model for India.

The example of Google Pay setting up connectivity to 12 million kirana (family run) retail stores in India is a classic example of "win-win" strategy for both Google and kirana retail store owners. The availability of internet and connectivity among the kirana retail owners could improve information available to these segments of the population thereby matching their supplies with demand and generate better prices for their products/services. This in turn would improve the living standards of these segments. At the same time, Google can generate substantial advertisement revenue by placing advertisements on new clients (eyeballs).

An important caveat of the study is that our analysis does not analyse the social-welfare implications of the mandatory CSR rule. However, the extant literature on the subject shows that CSR expenditure that spur digitalisation and technological innovation helps the poorest and most vulnerable which enable them to lead a health, productive life, improve their livelihood and bring long-term prosperity to the communities. Firms are catalysts of transformation in the society in addition to maintaining the financial health of the company.

REFERENCES

Ali, W., Frynas, J. G., & Mahmood, Z. (2017). Determinants of corporate social responsibility (CSR) disclosure in developed and developing countries: A literature review. *Corporate Social Responsibility and Environmental Management*, *24*(4), 273–294. doi:10.1002/csr.1410

Banerjee, A. V., Cole, S., Duflo, E., & Linden, L. (2007). Remedying Education: Evidence from two randomized experiments in India'. *The Quarterly Journal of Economics*, *122*(3), 1235–1264. doi:10.1162/qjec.122.3.1235

Bansal, S., Khanna, M., & Jain, S. (2018). Corporate social responsibility rules in India: An Assessment. *Economic and Political Weekly*, *53*, 44–51.

Barnea, A., & Rubin, A. (2010). Corporate social responsibility as a conflict between shareholders. *Journal of Business Ethics*, *97*(1), 71–86. doi:10.100710551-010-0496-z

Benabou, R., & Tirole, J. (2010). Individual corporate social responsibility. *Economica*, *77*, 1–19. doi:10.1111/j.1468-0335.2009.00843.x

Brammer, S., Brooks, S., & Pavelin, S. (2006). Corporate social performance and stock returns: UK Evidence from disaggregate measures. *Financial Management*, *35*(3), 97–116. doi:10.1111/j.1755-053X.2006.tb00149.x

De Velde, E. V., Vermier, W., & Corten, F. (2005). Finance and accounting: Corporate social responsibility and financial performance. *Corporate Governance*, *5*(3), 129–137. doi:10.1108/14720700510604760

Dhaliwal, D. S., Li, O. S., Tsang, A., & Yang, Y. G. (2011). Voluntary nonfinancial disclosure and cost of equity capital: The initiation of corporate social responsibility reporting. *The Accounting Review*, *86*(1), 59–100. doi:10.2308/accr.00000005

Dhanesh, G. S. (2015). Why corporate social responsibility? An analysis of drivers of CSR in India. *Management Communication Quarterly*, *29*(1), 114–129. doi:10.1177/0893318914545496

Dharmapala, D., & Khanna, V. (2016). The impact of mandated corporate social responsibility: Evidence from India's companies act of 2013, *Coase-Sandor Working Paper Series in Law and Economics*.

Drucker, P. E. (1984). The new meaning of corporate social responsibility. *California Management Review*, *26*(2), 55–63. doi:10.2307/41165066

Economist. (2005). *The good company-Special Report*. Retrieved from https://www.economist.com/special-report/2005/01/22/the-good-company

Economist. (2017). *Donate, or else: Indian firms make the best of coerced do-goodery*. Retrieved from https://www.economist.com/business/2017/11/16/indian-firms-make-the-best-of-coerced-do-goodery

Economist. (2019). *I'm from a company, and I'm here to help*, August 24, 14-16.

Freeman, R. E. (1984). *Strategic management: A stakeholder approach*. Boston, MA: Pittman-Ballinger.

Friedman, M. (1970). The social responsibility of business is to increase its profits. *New York Times Magazine*, *32*, 122-126.

Gates, B. (2000). Digital dividends. www.microsoft.com/billgates/speeches/2000/10-18digitaldividends

Gregory, A., Tharyna, R., & Whittaker, J. (2014). Corporate social responsibility and firm value: Disaggregating the effects on cash flow, risk and growth. *Journal of Business Ethics*, *124*(4), 633–657. doi:10.100710551-013-1898-5

Griffin, J. J., & Mahon, J. F. (1997). The corporate social performance and corporate financial performance debate: Twenty five years of incomparable research. *Business & Society*, *36*(1), 5–31. doi:10.1177/000765039703600102

Ilhan, E., Sautner, Z., & Vilkov, G. (2019). Carbon Tail Risk. *Available at SSRN 3204420*.

International Monetary Fund. (2019). Sustainable finance, Chapter 6. *World Economic Outlook*, (October): 81–92.

Jensen, R. (2007). The Digital provide: Information (Technology), market performance and welfare in the south Indian fisheries sector. *The Quarterly Journal of Economics*, *122*(3), 879–924. doi:10.1162/qjec.122.3.879

Jones, T. (1995). Instrumental stakeholder theory: A synthesis of ethics and economics. *Academy of Management Review*, *20*(2), 404–437. doi:10.5465/amr.1995.9507312924

Kaka, N., Madgavkar, A., Kshirsagar, A., Gupta, R., Manyika, J., Bahl, K & Gupta, S. (2019). *Digital India: Technology to transform a connected nation*, McKinsey Global Institute, March.

Kansal, M., Joshi, M., & Batra, G. S. (2014). Determinants of corporate social responsibility disclosures: Evidence from India. *Advances in Accounting, incorporating. Advances in International Accounting, 30*(1), 217–229. doi:10.1016/j.adiac.2014.03.009

Khanna, T., & Palepu, K. (2000). Is Group Affiliation Profitable in Emerging Markets? An analysis of Diversified Indian Business Groups. *The Journal of Finance, 55*(2), 867–891. doi:10.1111/0022-1082.00229

Khanna, T., & Palepu, K. G. (2005). *The Evolution of Concentrated Ownership in India: Broad Patterns and a History of the Indian Software Industry. A History of Corporate Governance Around the World: Family Business Groups to Professional Managers* (R. K. Morck, Ed.). University of Chicago Press.

Kitzmuller, M., & Shimshack, J. (2012). Economic perspectives on corporate social responsibility. *Journal of Economic Literature, 50*(1), 51–84. doi:10.1257/jel.50.1.51

Latterman, C., Fetscherin, M., Alon, I., & Li, S. (2009). CSR communication intensity in Chinese and Indian multinational companies. *Corporate Governance, 17*(4), 426–442. doi:10.1111/j.1467-8683.2009.00758.x

Lins, K. V., Servaes, H., & Tamayo, A. (2017). Social capital, trust and firm performance: The value of corporate social responsibility during the financial crisis. *The Journal of Finance, 72*(4), 1785–1823. doi:10.1111/jofi.12505

Mackey, A., Mackey, T. B., & Barney, J. B. (2007). Corporate social responsibility and firm performance: Investor preferences and corporate strategies. *Academy of Management Review, 32*(3), 817–835. doi:10.5465/amr.2007.25275676

Manchiraju, H., & Rajgopal, S. (2017). Does corporate social responsibility (CSR) create shareholder value? Evidence from the Indian Companies Act 2013. *Journal of Accounting Research, 55*(5), 1257–1300. doi:10.1111/1475-679X.12174

Margolis, J. D., Effenbein, H. A., & Walsh, J. P. (2009). *Does it pay to be good and does it matter? A meta-analysis of the relationship between corporate social and financial performance.* Working Paper, Cambridge, MA: Harvard University.

McKinsey. (2016). *Indian's economy: Why the time for growth is now*, September, 1-11.

McWilliams, A., & Siegel, D. (2000). Research notes and communications- Corporate social responsibility and financial performance: Correlation or misspecification. *Strategic Management Journal, 21*(5), 603–609. doi:10.1002/(SICI)1097-0266(200005)21:5<603::AID-SMJ101>3.0.CO;2-3

Mishra, S., & Suar, D. (2010). Does corporate social responsibility influence firm performance of Indian companies? *Journal of Business Ethics, 95*(4), 571–601. doi:10.100710551-010-0441-1

Moore, G. M. (2001). Corporate social performance: An investigation in the UK supermarket industry. *Journal of Business Ethics, 34*(3/4), 299–315. doi:10.1023/A:1012537016969

Mukherjee, A., & Bird, R. (2016). Analysis of mandatory CSR expenditure in India: A survey. *International Journal of Corporate Governance, 7*(1), 32–59. doi:10.1504/IJCG.2016.077982

Mukherjee, A., Bird, R., & Duppati, G. (2018). Mandatory corporate social responsibility: The Indian experience. *Journal of Contemporary Accounting & Economics*, *14*(3), 254–265. doi:10.1016/j.jcae.2018.06.002

Murthy, V., & Abeysekera, I. (2008). Corporate social reporting practices of top Indian software firms. *The Australian Accounting Business & Finance Journal*, *2*(1), 36–59. doi:10.14453/aabfj.v2i1.4

Orlitzky, M., Schmidt, F. L., & Rynes, S. L. (2003). Corporate social and financial performance: A meta-analysis. *Organization Studies*, *24*(3), 403–441. doi:10.1177/0170840603024003910

Prahlad, C. K. (2005). *The fortune at the bottom of the pyramid: eradicating poverty through profits*. New Jersey: Wharton School.

Prasad, G. C. (2019). Who gains the most from corporate charity, LiveMint, August 23, 2019. https://www.livemint.com/companies/news/who-gains-the-most-from-corporate-charity-1566536241870.html. (Accessed on November 22, 2019).

Prasad, M., Mishra, T., & Kalro, A. D. (2017). Environmental disclosure by Indian companies: An empirical study. *Environment, Development, and Sustainability*, *19*(5), 1999–2022. doi:10.100710668-016-9840-5

Prieto-Carrón, M., Lund-Thomsen, P., Chan, A., Muro, A., & Bhushan, C. (2006). Critical Perspectives on CSR and development: What we know, what we don't know and what we need to know. *International Affairs*, *82*(5), 977–987. doi:10.1111/j.1468-2346.2006.00581.x

Rajan, R. (2010). How inequality fueled the crisis, Project Syndicate, July 9.

Sacconi, L., & Antoni, G. D. (2011). *Social capital, corporate responsibility, economic behavior and performance*. New York, NY: Palgrave MacMillan. doi:10.1057/9780230306189

Sarkar, J., & Sarkar, S. (2015). Corporate social responsibility in India- An effort to bridge the welfare gap. *Review of Market Integration*, *7*(1), 1–36. doi:10.1177/0974929215593876

Servaes, H., & Tamayo, A. (2013). The impact of corporate social responsibility on firm value: The role of customer awareness. *Management Science*, *59*(5), 1045–1061. doi:10.1287/mnsc.1120.1630

Singh, D., & Ahuja, J. (1983). Corporate social reporting in India. *The International Journal of Accounting*, *18*, 151–169.

Soana, M. G. (2011). The relationship between corporate social performance and corporate financial performance in the banking sector. *Journal of Business Ethics*, *104*(1), 133–148. doi:10.100710551-011-0894-x

Subramaniam, R. K., Samuel, S. D., & Mathenthiran, S. (2015). Liquidity implications of corporate social responsibility disclosures: Malaysian evidence. *Journal of International Accounting Research*, *15*(1), 133–153. doi:10.2308/jiar-51204

Vasal, V. (1995). A spatio-temporal analysis on social reporting by companies-Evidence from Indian public sector. *Journal of Accounting and Finance*, *2*, 256–269.

Waagstein, P. R. (2011). The mandatory corporate social responsibility in Indonesia: Problems and implications. *Journal of Business Ethics*, *98*(3), 455–466. doi:10.100710551-010-0587-x

Wright, P., & Ferris, S. (1997). Agency Conflict and corporate strategy: The effect of disinvestment on corporate value. *Strategic Management Journal*, *18*(1), 77–83. doi:10.1002/(SICI)1097-0266(199701)18:1<77::AID-SMJ810>3.0.CO;2-R

Chapter 8
Managerial Innovation for Digital Healthcare Transformation

Dina Ziadlou
Colorado Technical University, USA

ABSTRACT

Digital technologies adoption has modernized the performance of healthcare organizations to a superior level. Nowadays, healthcare organizations in the digital transformation journey have encountered the challenges to adopt and adapt their strategies to achieve effective, desired outcomes. In the digital era, management innovation plays a critical and significant role in removing barriers that hinder progress. This chapter has focused on one of the components of organization innovation, which is managerial innovation. The healthcare service modernization based on digital transformation needs a higher level of managerial innovation. The author has elaborated five critical factors including change, digitalization, leadership, healthcare stakeholders, and globalization in managerial innovation. These factors lead the organizations to be capable of achieving sustainable effectiveness and making a value-creating system through managerial innovation.

INTRODUCTION

The paradigm for contemporary healthcare adopting modern digital technologies has changed the role of value-based performance of healthcare services, healthcare deliveries, disease management, patient experience, and population health. Today healthcare leaders and practitioners are adopting modern digital techniques to solve severe problems. The data from the statistical reports have revealed that the digital online market is dramatically expanding and fast penetrating the medical field and digital technologies applications are increasing productivity and dynamics of improving healthcare; from improved quality, accessibility, communication to reduce cost, recuperating performance, and developing decision support (Schwarzmüller, Brosi, Duman, & Welpe, 2018; Kane, 2015; Bowersox, Closs, & Drayer, 2005). Despite all advancements, healthcare organizations are still in the primary stage of transforming into full

DOI: 10.4018/978-1-7998-2799-3.ch008

smart healthcare. In the transformation journey, many hospitals have challenges to maximize the use of digital services (Hadeel & Sandhu 2021; Sandhu 2020; Venkatesh & Davis 2000) due to the high cost of technologies, employees' resistance to change, complex processes, lack of leadership support, and disparities in healthcare deliveries(Buvat et al., 2018; Faddis, 2018; Agarwal, DesRoches, & Jha, 2010). The transition from the traditional model of hospitals to modern medical centers is a sophisticated and complex journey that must be structured in different stages of transformation.

An important factor is the capabilities of organizations is the ability of improvement, change, renewal, and modification of the structure, process, product, services, and operation that mean the ability of innovation. Innovation is an ambarella concept for the generation (creating), diffusion (connecting generation to adoption), and adoption (using) innovations in different levels (Kumar, 2013; Zennouche, Zhang, & Wang, 2013; Rogers 1995; Davis 1989). In organizational level, how the organization manages and applies innovation in the process of change, how it creates a new idea, service, product, and good, or, creates change in an existing service, product, good, how to adopt new management models in tangible and intangible resources, and how it distinct between generating innovative and adopting innovation have been answered (Damanpour, 2017). The organization innovation has three angles: (1) environmental innovation, which is the change, addition or improvement in external or internal context of the organization, (2) managerial innovation referred to human capital, leadership, performance, and (3) organizational innovation related to structure and culture of the organization shown in figure 1 (Damanpour, 2017). However, in this chapter, the author has explored the organizational innovation from the perspective of managerial innovation to elaborate on the important factors involving in innovation development.

RESEARCH METHODOLOGY

The research methodology adopted for this study is exploratory research to identify a managerial innovation solution in healthcare organizations. The researcher has explored the intertwined relationship of technology, leadership, employees, change, and globalization in modern healthcare by conducting an in-depth literature review. The goal of the study is to understand organizational adaptation in digital transformation to develop a strategy for improving managerial innovation in digital healthcare transformation.

Figure 1. Organization innovation

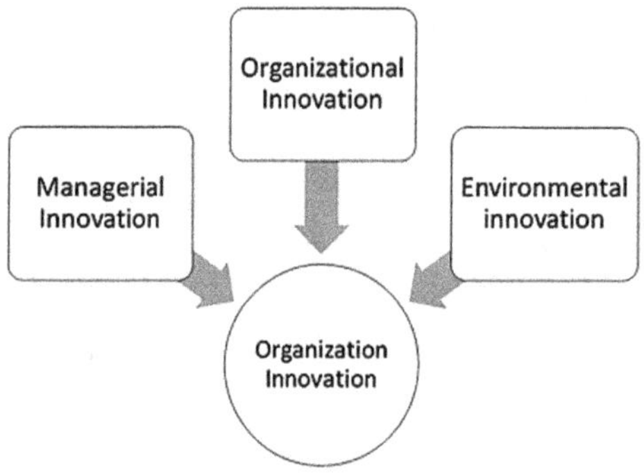

Figure 2. The contributing factors in managerial innovation

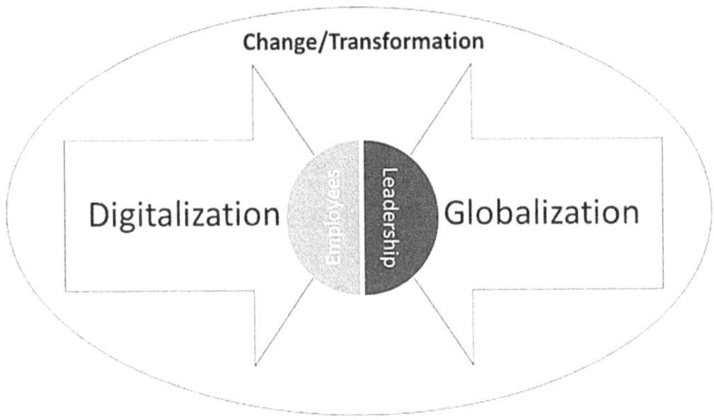

BACKGROUND

Digital transformation phenomena in healthcare have created favorable and unfavorable consequences for healthcare organizations. Many hospitals are endeavoring to establish digital health technologies to stay competitive in the digital arena and also deal with the consequences of technologies. However, managerial innovation influenced by change management is an essential point in the adoption of innovation is to ensure that all individuals accept the innovation and use it in their activities to fill out the gap between knowledge and practice in healthcare (Damanpour, 2017; Rogers 1995). This research has advocated research on factors affecting managerial innovation, including change, digitalization, leadership, people, and globalization, as shown in figure 2.

MAIN FOCUS OF THIS CHAPTER

The focus of this chapter is to explore the intertwined relationship of significant factors impacting managerial innovation in healthcare organizations, including (1) change, (2) digitalization, (3) leadership, (4) individuals, and (5) globalization. Therefore, this chapter has been allocated exclusively to delineate these critical components to improve the healthcare continuum for value creation.

Exploring the Important Role of Change

In classic Organization Development (OD), the organizations are living- system, in contrast, in modern OD, the organizations are meaning-making systems; the classic OD emphasizes the change in behavior and action of people, whereas modern OD focuses on the shift in mind-sets and thoughts of people (Watkins, Mohr, & Kelly, 2011). Moreover, two arguable concepts of change management and change leadership are used interchangeably in the management context. However, in digital transformation, the distinction between these two concepts boosts a better understanding of change aspects and constructing organizational development. Change management is defined as the set of processes and activities for preparing, planning, handling, and evaluating the status quo to shift from the current situation to the desired condition (Shafran, 2017; Kotter, 2012; Atchison, 1999).

On the other hand, change leadership has referred the art of leadership to create a shared vision, inspire and motivate people, and empower the followers for transformation (Arbab Kash et al., 2014; Allman, 2009; Gill, 2003; Kotter, 1995). Although change management and change leadership differ in concept, they complete each other. According to Kotter (2012), change management focuses on processes to make the organization profitable and efficient; whereas, the change leadership focuses on the human capacities and their capabilities to nurture them as the change agents. Gardner (2009) stated that the root cause of most of the failures in organizational development is people, not systems. The most significant constraint for organizational transformation is related to human behavior, action, and relation derived from employees' resistance, poor leadership, and cultural issues, not the complexity of technologies. Therefore, the bottleneck of creating effective transformation is to deeply contemplate the nature of change, future circumstance, and black swan in the digital era.

According to Henden (2019), Black Swan is, "Unpredictable event that is beyond what is normally expected of a situation and has potentially severe consequences" [para.3]. The black swan can happen in any business with a complex system, and it can also impact society. For instance, digital technology in its nature has been entirely one of the sources of a black swan while it has influenced fundamentally not only companies but also human behavior and lifestyle. Moreover, non-linear changes, failure in risk management, decentralized decision making, full reliance on technology, and high self-confidence in using the technology are the factors that can increase the likelihood of black sawn in the long-term (Davis, 2014). The managerial innovation must identify the black sawn to minimize the risks during the digital transformation journey. Taleb (April 2007) stated,

There are two possible ways to approach phenomena. The first is to rule out the extraordinary and focus on the "normal." The examiner leaves aside "outliers" and studies ordinary cases. The second approach is to consider that in order to understand a phenomenon, one needs first to recognize the extremes-particularly if, like the Black Swan, they carry an extraordinary cumulative effect.

Another approach of the managerial innovation to adopt changes must be to enable individuals to learn from their experience to innovate new solutions (Stary, 2014). In doing so, the organization must inject the dynamic of learning to transform the mindset of employees. In digital transformation, the learning approach has a transformative approach to encourage members to rethink their assumptions and learn how to think differently from a new perspective to create an effective desired future (Odor, 2019). The dynamic of learning consists of single-loop learning, double-loop learning, and triple-loop learning showed in figure 3.

In single-loop learning, individuals perform policies and procedures based on determined, rigid orders, as the fixed rules in organizations, that they must follow it to mitigate or improve malfunctions and failures in the systems to reach out the answer of this question: "Are we doing things, right?" (Bleich, 2014; Organizational Learning, 2014). In this model, individuals improve their behavior to avoid redundancy and repeat errors (Argyris, 1977). In double-loop learning, individuals step out beyond the box of operation, and they learn knowledge processing and how one solution can create a better outcome rather than the other solutions (Stary, 2014). Individuals rethink their assumptions to understand the situation deeply for creating an alternative solution to answer this question, "Are we doing the right things?" (Bleich, 2014; Organizational Learning, 2014). In this type of learning, innovation, and critical thinking have the roles in achieving the organization's goals, and people intensively enhance their self-awareness, honesty, and responsibilities (Argyris, 1977). Ultimately, triple-loop learning is about

Figure 3. Loop-Learning

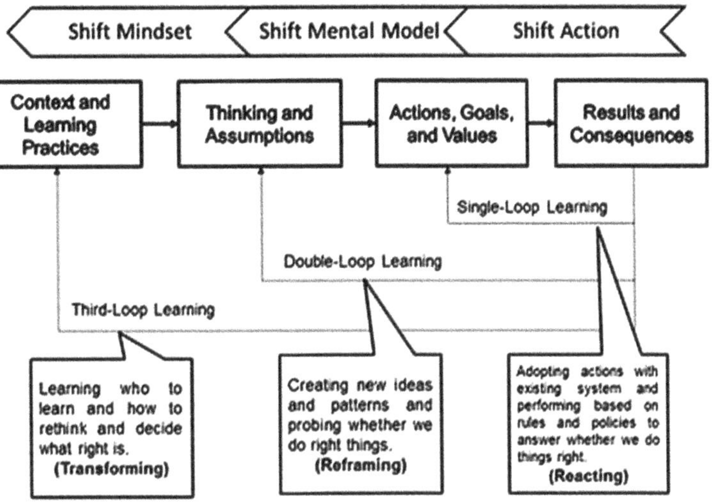

learning (Romme & Witteloostuijn, 1999). It stimulates the mindset of individuals to rethink learning strategies via activation of the iterative learning process to answer this question: "How we decide what is right?" (Bleich, 2014; Organizational Learning, 2014). In this journey, individuals would learn that all previous actions have affected the present, and the present actions are building the future.

Exploring the Important Role of Digitalization

The digital technology revolution is the root cause of changes in many modern businesses that create digital transformation (DT). DT refers to the integration of digital technologies in all dimensions of a business; the digital transformation flow is revolutionizing the business context from infrastructure, process, technology, financial, and operational services to cultural, behavioral, and climate of the organization (Roth & Keller, 2019; Schwarzmüller et al., 2018). The digitalization via the Artificial Intelligence (AI), Big Data, Cloud Computing, and the Internet of Things (IoT) has shaped the digital medical systems, such as robotic surgery, wearable sensors, real-time location system (RTLS), Clinical Decision Support Systems (DSS), Electronic Medical Records (EMR), telehealth, m-Health, and virtual nursing in order to improve quality, develop access, and reduce cost of healthcare (McCullough, Casey, Moscovice, & Prasad, 2010; Bauer, 2002; Coile Jr, 2000). For instance, the hospitals establish paperless systems such as PACS (Picture Archiving and Communication Systems), as one of the revolutions of digital services, to (1) minimize papers waste; (2) improve access of radiologists' network nationwide; (3) develop telecommunication systems (4) facilitate communication among patients, providers, nurses, and staff; and (5) improve automated systems to provide patient privacy and medical data security (Wang & Alexander, 2013).

On the other hand, the healthcare organizations are cumbersome and complicated environments; establishing technologies, managing stakeholders, accommodating financial resources for creating a balance between quality, cost, and access are sophisticated processes that need the innovative approach. Therefore, the role of innovation is going to be highlighted in the managerial aspects of the organizations. Innovation is a systematic process from generating an idea to implementation and utilization of the concept in reality toward increasing the effectiveness and efficiency of a system/organization.

Kumar (2013) stated, "Innovations conceived by carefully integrating design processes with business and technology have a better chance of achieving high user value and economic value, leading to greater adoption and market leadership" (p. 7). The transformation is manifested in shifting from traditional models to modern and post-modern organizations, from analog systems to digital world, from work-oriented employees to work-life balance-oriented workforce, from organizational hierarchy to holacracy organizations, and from linear thinking to nonlinear thinking. Agarwal, DesRoches, and Jha (2010) stated, "There is a substantial consensus that the digital transformation of healthcare through broad and deep use of health information technology (HIT) across the health-care ecosystem, in conjunction with other complementary changes, can reduce costs and improve quality" (p. 796).

The promise of IoT, AI, telemedicine, wearable sensors, remote patient monitoring, and all digital health technologies are real. Notwithstanding, to keep up with the digital revolution, the hospitals must either launch comprehensive digital health strategies or reinvigorate the existing digital health strategies. The new practices based on new digital technologies and new knowledge management models to utilize IT services are thoughtful factors in managerial innovation (Damanpour, 2017). Dess, McNamara, Eisner, & Hyun (2019) describe, "strategic management of an organization entitled three ongoing processes: analyses, decisions, and actions" (p. 6).

The leaders and senior managers need analytics integration skills to produce actionable solutions through forecasting the future of health technologies to align the digital health strategies continuously. However, despite developing sustainable strategies to promote digital technologies such as Electronic Health Record (EHR) and telehealth as efficient and practical ideas in healthcare organizations, these technologies have created challenges in many medical centers due to the lack of a trade-off between present technologies and upcoming digital systems (Kane, 2015). Also, given the reports, these technologies caused nurses' and physicians' burnout and harm interaction of physicians and patients in their operations (Ommaya et al., 2018; Morse, 2018; Kuza, 2017; Mary Norine & Rumsfeld, 2017).

Given the reports, between 35% to 75% of projects change are failed, and most of HIT projects' failures are derived from lack of well-developed planning phase, 50% of failures are due to managerial issues, and 20% of failures come from a lack of effective communication (Scheer, 2016; Tait, 2013).

The organizations can adopt innovation for maintaining changes in technology, performance, and behavior of employees. They can utilize innovation for improving productivity, market share, and profitability. Moreover, they may adopt innovation for strategic entrepreneurship (Kumar, 2013). In all terms, the Research and Development (R&D) division, design unit, product development unit, strategic planning unit, or innovation center present the ideas, and the departments such as management, marketing, human resource, and manufacturing adopt the innovations. However, organizational innovation refers to managerial and non-technological innovation. Indeed, organizational innovation influenced by change management, as a subprocess for organizational change (W. Wade, 2012). The essential point in the adoption of organizational innovation is to ensure that all individuals accept the innovation and use it in their activities (Hadeel & Sandhu 2021; Venkatesh & Davis 2000; Damanpour, 2017; Rogers 1995). In this regard, Reifsteck and Reese (2016) recommend the hospitals to establish a committee for oversight of digital health technologies as well as inviting community advisors with digital technology and business strategy experience to work closely with the committee before developing technologies.

Exploring the Important Role of Leadership

The leadership and management are the agents of digital transformation; the organizations need both management and leadership to stay alive and make progress. However, there is a difference between these two concepts. According to Nahavandi (2015), Mintzberg (2009), and Gill (2003) the leadership is an art of human resource empowerment, inspiration, and motivation toward accomplishing goals, values, integrity, trust, and dignity, and the management refers to maintain the existing structure through delegating, communicating, controlling, and monitoring the subordinates to perform their assigned tasks. The leaders endeavor to lead organizations to attain goals, whereas managers. Indeed, leadership is relationship-oriented to create shared value and influencing followers to achieve the desired future, and it is about people's leadership; on the other hand, management is task-oriented, and it is about processes improvement and task management. Nahavandi (2015) stated that leadership focuses on the future to create change based on shared values and emotional connection, but management focuses on the present and to maintain policies and procedures and stabilize the status quo.

The postmodern and modern organizations are endeavoring to inject the leadership skills not only in top leaders but also among directors and managers to combine task-oriented activities and technical skills with relationship-oriented skills. The managers who motivate the employees to accomplish goals, discover talents, and create harmony between employees' performance and organizational goals are the managers who are stepping toward leadership skills, the managers filling the gap between employees and leadership, and the managers who can perceive the emergence of change in the system and facilitate the process of cultivating the change agents in the system (Bennis, Sample, & Asghar, 2015; Nahavandi, 2015).

According to Lipinski (2013), "Organization Development (OD) is the process of creating a model of influence based on the vision and goals of an organization that, when implemented, attempt to obtain a successful outcome for that organization" (p. 64). For this aim, three leadership skills require to be added in future leaders' skills.

Four Leadership Skills in Digital Transformation

One of the most important factors in digital transformation is leadership. The leaders create a vision and motivate followers to achieve the vision. The leadership style and the way that leaders communicate the goals and inspire the individuals to have a critical role to build the atmosphere of the organization and create dynamic and productivity (Robbins & Judge, 2017). Indeed, the leadership effectiveness to maintain internal stability and sustain external adaptability does have an impact on the transformation journey (Nahavandi, 2015).

In this journey, nurturing three skills in future leaders help to facilitate the digital transformation, including systems thinking, global thinking, and innovative thinking in order to create a value-based system deriving from a shared vision that works for prominent goals in human life: people, prosperity, partnership, peace, and planet (Fleming, Wise, Hansen, & Sams, 2017). The system thinking provides a holistic view that stimulates interrelationship between all parts of a whole system to create meaningful solutions (Whitehead, Scherer, & Smith, 2015). Innovative thinking enlightens the ways of new thinking to transform our perceptions (Lin, Wang, & Tsai, 2013). Moreover, global thinking illustrates the horizon of healthcare and foster understanding of individuals to perceive the interconnectedness of parts and a whole system (Briscoe, 2016). Ultimately, scenario thinking empowers the organizations in decision making and seeing alternative futures not easily estimated or predicted by trends.

1. Systems Thinking

The review of literature in a leadership role in organizational changes toward transformation from traditional style to modern style revealed that the system is shifting from leadership-centric role to human values development (Arbab Kash, Spaulding, Johnson, & Gamm, 2014; Van Fleet & Peterson, 2016; Ziadlou, 2019, September). Digital transformation seeks to create integration and meaningful use not only in a technology setting but also in human culture and their mindset setting and the way how they implement their assumptions and values. Bollington (2015) stated, "there is different between being digital and thinking digital" (p. 6). Therefore, while healthcare system is a social system connecting patients, families, healthcare providers, communities, a system thinking agent can reform the interpretation of individuals from digital services in a common frame, reconcile disparities and conflicts of views, and articulate the impact of behavior, action, and decision of care teams on individuals, society, and world (Bleich, 2014a, 2014b).

Through the systems thinking development in healthcare, healthcare can promote relationship, enjoyment, contribution, and joy of life and work (Bleich, 2014b). Moreover, Domegan et al. (2016) stated that systems thinking is a co-creating solution for advocating change. Thus, it must be actively one of the top skills in management attributes. The systems thinking engages people in contemplating resolutions rather than starting actions and reactions. Manuele (2019) stated, "To achieve a broad adoption of systems/macro thinking in operational risk management, a shift must occur from a focus on unsafe acts of employees as being the principal causal factors for incidents and illnesses to a focus on the work systems and work methods as the principal sources for causal factors" (p. 37).

The systems thinking provides the opportunity of viewing the organization as a whole system. The whole system navigates mindsets to understand how systems and their parts are related to one another and how to work together within the whole toward attaining a common goal (Bilodeau, 2019). While the organizations are the systems, including a set of elements (tangible and intangible elements) with interrelationships that keep the organization stable and dynamic, the systems thinking must be one of the skills of digital leaders.

2. Innovative Thinking

Innovation is not only inventing services, product, or process; it is about creating value, improving triple-loop learning, and engaging people to attain higher goals; the goals for improving the life and health of all people around the world. Mehta (2019) states, "We need lots of innovative thinking while dealing with the volume of patients. There can be innovating thinking in front of technology, processes, and management".

HIMSS-Elsevier Digital Healthcare (2017) recognized the Royal Children Hospital in Melbourne of Australia for reducing readmissions and cost by utilizing electronic medical records and Fudan University Huashan Hospital in China for hybrid cloud environment for individualized medicine as winners for innovative thinking of best usage of health information technology to improve healthcare outcomes. Azemi (2018) addressed innovation development in five steps: "idea finding, idea shaping, idea defining, idea refining, and idea communicating" [para.3] and revealed that systems thinking influences innovation. However, Azemi did not proof this assertion through a quantitative or qualitative study.

3. Global Thinking

In system thinking, the researcher discussed that through a whole-system perspective, the organizations could improve the relationship, develop strategic initiatives, and redesign the organization through systems thinking. However, this era is more interconnected and interdependent. It needs thinking beyond a whole organization. It requires members to reform their thoughts and their actions not only at the organizational level with considering an organization as a whole but also; they must view themselves as a part of a global system that needs to care about all people around the globe.

According to Daily News (2019), Sobowale and Ortigara (2017), MENA Report (2017), Nanaimo News Bulletin (2017), the need for global thinking is one of the top priorities and digital technologies, connectivity, and innovation are the opportunity to save the planet. Therefore, global thinking must be a skill for digital leaders to shift the mindset toward a holistic worldview. Leverage of internet and digital technologies in all businesses, as well as human society, results in globalization; removing distance, time zone, and language among people around the globe has been one of the consequences of globalization.

Briscoe (2016) stated, "A global system's lens gives us the ability to hone our understanding of parts and their interconnectedness; then to look for leverage points for change and to make decisions based on the interdependency of all part" (p. 5).

According to the Cambridge International Examination (2016), global competence is the capability of understanding the global challenges through disciplinary and interdisciplinary study, communicate ideas, recognize perspectives, and take action for improvement. The characteristics of global thinkers are (1) contribute to global culture, (2) support development of global skills, (3) be flexible in individual and group global thinker development, (4) make visible global thinker, and (5) act as micro-intervention. Long-lasting transformation needs a sense of ownership, and a leader with the ability of global thinking can cultivate motivation and transform the mindset of individuals to take responsibility for improving the world and for achieving SDGs, particularly, SDG3.

4. Scenario Thinking

Scenario planning is essential for the businesses because whereas standard project planning tends to focus on decisions, policies, and processes, scenarios focus on internal and external driving forces, namely, changes in social, technical, economic, legal, financial, environmental, and demographic change (Harris, 2014). It can evaluate the robustness of current strategic plans, illustrate new opportunities on the horizon, increase innovation, reduce risks and uncertainties (Blyth, 2005; Bowman, 2016). "It assists them with thinking, in advance, about the many ways the future may unfold and how they can be responsive, resilient, and effective, as the future becomes [a] reality [para.1]." Scenario planning identifies current reality, forecasts future needs, and determines internal and external driving forces to produce potential future (Sargent, 2011). Therefore, in healthcare organizations, through scenario thinking, the leaders can stimulate the mindset of change, develop contingency plans, and prepare for a timely response. Boisot and MacMillan (2004) classified thinking about the future in three modes: probable future (predictionary mind), possible future (possibilities thinking), and preferable future (visionary mind). While managerial innovation is more laborious than technological innovation, these three modes of thinking can underpin managerial innovation to create innovation in management process, operation, knowledge development, and employees support.

Exploring the Important Role of Individuals

Comparatively speaking, the shift from traditional hospitals to modern healthcare organization was the revolution of the 21st century. The digital transformation has developed employees-leaders connections via Big Data. It has created decision making support via artificial intelligence. It has expanded accessibility and convenience to care services through telemedicine. It has engaged patients with their health via mobile health. Moreover, it has increased the efficiency of performance via Electronic Medical Records. However, the role of human capital in embracing change and implementing transformation is the critical part. The human capital is defined as "individual capabilities, knowledge, skills, and experience of the company's employees and managers" (Dess et al., 2019, p. 106).

Three significant factors impacted on leading employees in digital transformation: (1) engagement, (2) empowerment, and (3) culture development (Walsh & Rumsfeld, 2017; Weston, 2017; Adeyemi, 2017; Herscovitch, 2004; Kelly, 2016; Shuck, Twyford, Reio, & Shuck, 2014; Thomas, 2014; Weinstein et al., 2008).

Individuals' Engagement

In modern and postmodern organizations, the attention of leaders must be on human-related factors in order to influence and transform the mindset and behavior of employees (Marshak & Grant, 2008). Nowadays, human capital is the core value of all organizations, and the cooperation of human resources delineates the success of the organization for achieving the desired future.

The employees' engagement is a process of promoting emotional and behavioral energy to attain organizational outcomes. In order to engage the employees in the change process, the leaders of the hospitals must contemplate how to empower and engage different types of stakeholders to create a healthy environment and develop a collective level of engagement (Shuck et al., 2014; Olden, 2003). Olden (2003) pointed out that three capabilities of stakeholders management consist of (1) identifying the type of stakeholders and their stakes, (2) developing organizational policies and procedure based on stakeholders' interest, (3) establishing workflow and bargaining to achieve missions. Accordingly, in the modern healthcare system, involving the physicians and employees in decision making, appreciating the employees, building trust, communicating effectively, and nurturing innovative thinking are the most policies toward cultivating a healthy organization (Adeyemi, 2017; Scheer, 2016; Herscovitch, 2004).

Thanks to technology, healthcare leaders can collect excellent resources to build strategic initiatives upon the employees' experiences and feedback through electronic survey systems during digital health adoption and implementation. The Electronic Medical Record (EMR) has been one of the achievements of the digital transformation, and an example of significant change initiatives in the hospitals, that has brought up the immense lesson learned for the hospitals (Stensaker & Meyer, 2011). Establishing the EMR has indicated that involving physicians, nurses, and employees in all phases of project (planning, implementation, post-implementation) as well as ensuring their adaption and adoption of digital technology, providing appropriate training, creating transparent processes, and providing financial incentives can improve engagement of employees in the process of change and reduce the resistance to change (Sandhu 2020; Adeyemi, 2017; McCabe & Timmins, 2016; K. H. Cohn, 2015; Aziz, 2007; Rogers 1995).

Notably, physicians and nurses have a critical role in the tangible and intangible outcomes of a hospital. Their engagement in digital transformation can impact on the profitability of the hospital. Cohn (2015) represented a study by Healthcare Consultancy Vital Work-Life and Physician Recruiting Firm

Cejka Search among 1,600 physicians and indicated the top critical factor of physicians' engagement in the process of change is, "feeling that my opinions and ideas are valued and have a voice in how my time is structured and used" (p. 60). Furthermore, technology sustains healthcare outcomes. In this regard, McCabe and Timmins (2016) asserted, "Health care technology has the power to transform the nursing profession through collecting, synthesizing, and analyzing data about nursing care in a way that demonstrates its impact on the provision of quality outcomes in health care" (pp. 105-106).

Individuals Empowerment

According to World Health Organization (WHO), the human resource empowerment is "a process through which people gain greater control over decisions and actions affecting their health, which signifies a process on both the individual and the community level" (Meskó, Radó, & Győrffy, 2019, p. 2). In people leadership, human resources are the most valuable capital in all organizations. Similarly, in healthcare organizations that serve the health and wellbeing of individuals, communities, and society, the employees are the source of productivity and growth of healthcare. Faddis (2018) argued that forecasting the hospitals' future clarifies that " management at the smart hospital can enhance its operational efficiencies by having a clear picture of asset and labor use based on productivity efficiencies, labor accountability, and the ability to identify solutions to low-performance metrics" (p. 37).

The digital era has significantly impacted the patient's-physicians' relationship and tied them to digital technologies. Nowadays, physicians and nurses utilize digital health from simple techniques such as patients' data entry to sophisticated systems such as robotic surgeries. Also, the patients become more informed and involved in their health through internet and m-Health services. They participate in their treatment and decision-making processes toward self-care management. However, leaders and managers have faced the multigenerational workforce that requires special leadership techniques in the fast-paced changing environment. The generational differences with different levels of digital health technology adoption affect the willingness to change; Baby boomers adopted landline phone and TV, Gen X are desktop computers and email friendly, Gen Y keen on social media and tablets, and Gen Z is the geek of smartphones and google glasses (McCabe & Timmins, 2016). Wiedmer (2015) stated, "Identifying and retaining valuable intellectual capital among a multigenerational workforce are critical goals for leaders" (p. 56).

Weiner (2016) and Commins (2014) conducted interviews among healthcare leaders and C-suites and cited physicians' engagement as the most challenging part of hospital management. Moreover, physicians' involvement in transformation requires creating transparency, involving them in decision making, creating work-life balance, improving processes, increasing incentives, and assigning decision-making authority (Ommaya et al., 2018; Weiner, 2016; Commins, 2014). Wade, Taylor, Kidd, and Carati (2016) conducted a qualitative study to identify impact factors of shifting from the small scale of digital health projects (home telehealthcare) to large-scale projects; the results indicated that clinicians' technology acceptance and leadership support are first fundamental factors in starting establishing digital health projects. Moreover, Lloyd & Ferguson (2017) stated that educating nurses in health informatics and nurturing nursing workforce as informatics specialists may make a bridge between nursing and IT concepts toward creating a universal language of digital transformation; it would help to embrace digital health changes smoothly in healthcare organizations and accelerate the adoption and adaption of digital technologies. Hence, ensure the full acceptance of clinicians, nurses, and healthcare professionals in digital transformation before diving to establish massive digital health changes and empowering them in technologies would bring effective outcomes.

Culture Development

The digital technology phenomenon was a fortuitous consequence of academic research that converted to the snowball effect and affected all aspects of our lives, including our culture. The culture encompasses all tangible and intangible behavioral aspects of people, including shared beliefs, values, feelings, thoughts, and behaviors in groups, which shape the organizational behavior (Robbins & Judge, 2017). Buvat et al. (2018) stated that digital culture is the set of seven components: "customer-centricity, innovation, collaboration, digital-first mindset, agility and flexibility, open-culture, and data-driven decision making" (p. 3).

Although organizational culture change is a complicated process and takes a long time, it is possible to happen if the shared values agree with organizational effectiveness, and the organizational climate aligns with the physiological climate. In 2017, Capgemini Digital Transformation Institute conducted a study across France, Italy, Germany, Spine, Netherland, Sweden, UK, and the US in 1700,340 organizations to evaluate top barriers in digital transformation. The results indicated that culture (62%) is the top roadblock in organizations. Kluge (2008) stated that the most challenging task in the digital evolution of health care is not technical or legal issues, it is about how to identify and create a cultural bridge between the present and future that validate the change. Thimbleby (2013) stated that technology is improving things and making the profit for organizations, whereas human culture does not make a tangible return on investment. It might be the reason for more incentive in technology development rather than culture development (Thimbleby, 2013). Likewise, Edgar Schein (2004) in his book, *Organizational Culture, and Leadership* stated that culture is created by shared experience, but it is the leader who initiates this process by imposing his or her beliefs, values, and assumptions at the outset.

Gardenswartz, Rowe, Digh, & Bennett (2003) divided culture into three dimensions. First, the personal culture defined as the integration of personality, skills, interests, and value. Second, organizational culture referred to shared values, knowledge, morals, and believes that shape organizational practice. The third one is the national culture, which is the collection of behaviors, attitudes, norms, and customs of people that derive from their country's history and their origins. Springer, Clark, Faan, Strohfus, and Belcheir (2012) and Smollan and Sayers (2009) argued that the sustainable organizational culture is an intangible factor derived from attitude and reflection of employees, and it can make transformational actions. Bayan (2018) indicated that culture drives continuous knowledge development and talent empowerment to accelerate this process. However, the digital transformation will happen if the employees believe that digital technologies can help them toward their desired, positive future; investment in nurturing digital culture is one of the primary keys to attaining this aim.

The economic indicators of global markets, as well as in the US healthcare market, indicate that not only global healthcare trade is increasing but also other indicators such as multiculturality is elevating due to demographic change, globalization, and digitalization. According to the Pew Research Center report, the US demographic trend is projected to be more diverse (59 million immigrants); 43 percent of the millennial generation is not white, and also, the women's roles in leadership are increasing (D. Cohn & Caumont, 2016). Moreover, digital technologies are crossing the borders and connecting all people around the globe with different cultures and values. Digital technologies such as telehealth have been dramatically expanded, and it is going to create a healthcare system without a wall in the future.

Therefore, healthcare leaders must figure out how to lead multicultural environments and how to meet the stakeholders' expectations beyond geographical borders. It is the managerial innovation to understand the sociocultural background of healthcare stakeholders to encourage complex problem solving and digital technology adoption based on new learnings toward increasing organizational performance.

Exploring the Important Role of Globalization

The digital services have removed barriers such as distance, language, and time zone, and increased the interaction between people, governments, and industries (Hill & Hult, 2018; Friedman, 2007). The technology has opened a door for globalizing individuals rather than globalizing companies (Friedman, 2007). Friedman (2007), in his book, *The World is Flat*, stated the world is getting flatter. In Globalization 1.0, the critical agent of change was human power and his creativity to use wind, steam, and horsepower. The Globalization 1.0 converted the massive world to a common word. The Globalization 2.0 era was the industrial revolution with the manifestation of multinational markets and smaller world. Moreover, now the Globalization 3.0 has flattened the world by developing individualization and empowering people to contribute to global markets individually (pp. 9-10). Jerry Rao stated, "cultures and willing to change have a huge advantage in this world," and "you have to have a strong culture, but also openness to adopt and adapt from others" (as cited in Friedman, 2007, p. 422).

Furthermore, the shift explicitly in the nature of businesses due to digital technologies and global trading has led the organizations to redesign their strategies, leadership, human resource, and performance toward the sustainable effectiveness model. Therefore, managerial innovation leads the organization to embrace sustainable effectiveness organizations (SEM) helps to make a value-creating system. The SEM considers not only economic profit and competitive advantage but also find the stakeholders' value, social values, and environmental values (Lawler, 2014). The value-based organizations align their strategies based on increasing the social responsibility, corporate citizenship, and environmental stewardship in order to minimize harm, maximize benefit, and produce higher standards of living and quality of life for all communities (Eason, 2016; Ross, 2016).

The economic indicators of global markets, as well as in the US healthcare market, indicate that not only global healthcare trade is increasing but also other indicators such as multiculturality is elevating due to demographic change, globalization, and digitalization. According to the Pew Research Center report, the US demographic trend is projected to be more diverse (59 million immigrants); 43 percent of the millennial generation is not white, and also, the women's roles in leadership are increasing (D. Cohn & Caumont, 2016). Moreover, digital technologies are crossing the borders and connecting all people around the globe with different cultures and values. Digital technologies such as telehealth have been dramatically expanded, and it is going to create a healthcare system without a wall in the future. Therefore, healthcare leaders must figure out how to lead multicultural environments and how to meet patients' and stakeholders' expectations beyond geographical borders. Also, physicians must understand the sociocultural background of patients who are affected by processes of care services via technologies such as telehealth and video conferencing examinations

SOLUTIONS AND RECOMMENDATIONS

This chapter indicated that healthcare organizations need to invest in five components to develop hospital operation in the adoption of digital transformation to fortify managerial innovation, including technology, leadership, people, change, and globalization. The digital technologies can mitigate the borders and increase accessibility to healthcare as well as promote the awareness and knowledge of populations. The leadership and their skills to create a shared vision, communicate the vision and goals, and motivate employees to achieve effectiveness are part of managerial innovation. During the digital transformation, the ways to nurturing agents of change, sustain changes, and improve organizational performance are the forte of leaders. To increase the ability of managers and leads in managerial innovation, the author has recommended boosting three skills, including systems thinking, innovative thinking, and global thinking. Moreover, establishing learning systems to empower employees and engaging them in decision making and creating the desired future are important factors that healthcare leaders must develop their strategies based on them, such as triple loop learning in order to learn about learning. Likewise, the author described that healthcare organizations must put globalization in their strategic plans to create sustainable effectiveness organizations and to make a difference in the lives of all people of all ages around the globe. The viewpoint for obtaining global strategy helps organizations think beyond the profit of the organization; it helps them to contribute to the establishment of sustainable effectiveness by linking the individuals(micro-scale) to the global (macro scale).

FUTURE RESEARCH DIRECTIONS

The opportunities for future research can be to explore the organizational factors to engaging individuals in the process of managerial innovation, investigate in leadership models to develop an innovation culture in healthcare organizations, and understand the needs of healthcare organizations to become proactive in innovation rather than only be the users of innovation. Moreover, the qualitative and quantitative research to explore and investigate the stakeholder's perspective and health leaders' strategies can enlighten the road map of the future healthcare landscape and improve innovative approaches.

CONCLUSION

For effective digital transformation, healthcare organizations require to foster their managerial innovation toward creating value for the organization. The value-based organizations align their strategies based on increasing the social responsibility, corporate citizenship, and environmental stewardship to minimize harm, maximize benefit, and produce higher standards of living and quality of life for all communities (Damanpour, 2017). The leaders and managers are responsible for creating a shared vision and communicate the vision of value creation. In managerial innovation, human factors dominate to conquer effective outcomes. While innovation is a multidimensional process, management innovation must alter the changes based on the employee's viewpoints and their incentives. Therefore, the author indicated that engaging the employees, developing the digital culture, and empowering them to facilitate the process of change adoption. Moreover, innovation is a risky action and evaluation of human factors, such as performance, is more sophisticated than technological evaluation due to a variety of behavior, culture, assumption, and response of individuals. Thus, managerial innovation must be multidimensional tactic to address internal change agents such as employes and external factors such as globalization; And, all elements must contribute in maximizing the outcomes focusing on not only the return of investment and competitive advantage, but also value creation for stakeholders, community, and society (Colbert & Kurucz, 2007; Eason, 2016).

REFERENCES

Adeyemi, O. A. (2017). *Organizational change and how it effects healthcare employees: A study on employee resistance to change in electronic medical record implementation (Doctoral dissertation).* (10602795 D.M.), Colorado Technical University, Ann Arbor. ProQuest Dissertations & Theses Global database.

Agarwal, R., DesRoches, C., & Jha, A. K. (2010). The digital transformation of healthcare: Current status and the road ahead. *Information Systems Research*, *21*(4), 796–809. doi:10.1287/isre.1100.0327

Allman, S. (2009). Leadership vs. Management. *Successful Meetings, 58*(11), 12.

Arbab Kash, B., Spaulding, A., Johnson, C. E., & Gamm, L. (2014). Success factors for strategic change initiatives: A qualitative study of healthcare administrators' perspectives. *Journal of Healthcare Management*, *59*(1), 65–81. doi:10.1097/00115514-201401000-00011 PMID:24611428

Argyris, C. (1977). Double loop learning in organizations. *Harvard Business Review, No.77502.*

Atchison, T. A. (1999). Managing change. *Frontiers of Health Services Management*, *16*(1), 3–29. doi:10.1097/01974520-199907000-00002 PMID:10557432

Azemi, A. (2018). Promoting innovation through systems thinking and systems design. In *Proceedings of the 2018 Annual IEEE International Systems Conference (SysCon).* IEEE. 10.1109/SYSCON.2018.8369600

Aziz, L. (2007). *Managing change in healthcare information technology projects: the role of the project manager*. Project Management Institute.

Bauer, J. C. (2002). Rural America and the digital transformation of health care: new perspectives on the future...third annual Southern Illinois Healthcare/Southern Illinois University Health Policy Institute, Rural Health Care: Current Issues and Future Directions, May 31 and June 1, 2001. *Journal of Legal Medicine*, *23*(1), 73–83. doi:10.1080/019476402317276678 PMID:11957332

Bennis, W., Sample, S. B., & Asghar, R. (2015). *The art and adventure of leadership*. San Francisco, CA: John Wiley & Sons.

Billington, C. (2015). A change in mindset - building the digital future. *Public Sector*, *38*(2), 5–9.

Bilodeau, L. (2019). What is whole systems thinking? Retrieved from https://sustain.ok.ubc.ca/whole-systems-plan/what-is-it/

Bleich, M. R. (2014a). Developing leaders as systems thinkers--Part I. *Journal of Continuing Education in Nursing*, *45*(4), 158–159. doi:10.3928/00220124-20140327-13 PMID:24702051

Bleich, M. R. (2014b). Developing leaders as systems thinkers--part II. *Journal of Continuing Education in Nursing*, *45*(5), 201–208. doi:10.3928/00220124-20140424-13 PMID:24801681

Blyth, M. J. (2005). Learning from the future through scenario planning. Retrieved from http://www.fourscenes.com.au/LearningFromScenarios0305.pdf

Boisot, M., & MacMillan, I. (2004). Crossing epistemological boundaries: Managerial and entrepreneurial approaches to knowledge management. *Long Range Planning*, *37*(6), 505–524. doi:10.1016/j.lrp.2004.09.002

Bowersox, D. J., Closs, D. J., & Drayer, R. W. (2005). The digital transformation: Technology and beyond. *Supply Chain Management Review*, *9*(1), 22–29.

Bowman, G. (2016). The practice of scenario planning: An analysis of inter- and intra-organizational strategizing. *British Journal of Management*, *27*(1), 77–96. doi:10.1111/1467-8551.12098

Briscoe, P. (2016). Global systems thinking in education to end poverty: Systems leaders with a concerted push. *International Studies in Educational Administration (Commonwealth Council for Educational Administration & Management (CCEAM)), 43*(3), 5-19.

Buvat, J., Crummener, C., Kar, K., Sengupta, A., Solis, B., Aboud, C., & Aoufi, E. H. (2018). The digital culture challenge: Closing the employee-leadership gap. Retrieved from Cambridge International Examination. (2016). *Global thinking*. Paper presented at the Cambridge Schools Conference, Kuala Lumpur. https://www.cambridgeinternational.org/Images/346617-global-thinking.pdf

Cohn, D., & Caumont, A. (2016). *10 demographic trends that are shaping the U.S. and the world.* Retrieved from https://www.pewresearch.org/fact-tank/2016/03/31/10-demographic-trends-that-are-shaping-the-u-s-and-the-world/

Cohn, K. H. (2015). Training for physician engagement. *Talent Development*, *69*(5), 58–62.

Coile, R. C. Jr. (2000). The digital transformation of health care. *Physician Executive*, *26*(1), 8. PMID:10788126

Colbert, B. A., & Kurucz, E. C. (2007). Three Conceptions of Triple Bottom Line Business Sustainability and the Role for HRM. *Human Resource Planning*, *30*(1), 21–29.

Commins, J. (2014). Addressing physician engagement. *HealthLeaders, 17,* 8-8, 10.

Daily News. (2019). Need for global thinking. *Daily News*.

Damanpour, F. (2017). *Organizational innovation*. Oxford University Press; doi:10.1093/acrefore/9780190224851.013.19

Davis, F. D. (1989). Perceived usefulness, perceived ease of use, and user acceptance of information technology. *Management Information Systems Quarterly*, *13*(3), 319–340. doi:10.2307/249008

Davis, H. P. (2014). Managing black swan uncertainty: A grounded theory examination of portfolio managers' assessment of unknown risk. (3609144 Ph.D.), Capella University, Ann Arbor. ProQuest Central; ProQuest Dissertations & Theses Global database.

Dess, G., McNamara, G., Eisner, A., & Hyun, S. (2019). *Strategic management: Creating competitive advantages* (9th ed.). New York: McGraw-Hill Education.

Domegan, C., McHugh, P., Devaney, M., Duane, S., Hogan, M., Broome, B. J., ... Piwowarczyk, J. (2016). Systems-thinking social marketing: Conceptual extensions and empirical investigations. *Journal of Marketing Management, 32*(11-12), 1123–1144. doi:10.1080/0267257X.2016.1183697

Eason, C. C. (2016). Achieving Sustainable Effectiveness. *The Journal of Applied Management and Entrepreneurship, 21*(2), 119–121. doi:10.9774/GLEAF.3709.2016.ap.00009

Faddis, A. (2018). The digital transformation of healthcare technology management. *Biomedical Instrumentation & Technology, 52*(s2), 34–38. doi:10.2345/0899-8205-52.s2.34 PMID:29775387

Fleming, A., Wise, R. M., Hansen, H., & Sams, L. (2017). The sustainable development goals: A case study. *Marine Policy, 86*, 94–103. doi:10.1016/j.marpol.2017.09.019

Friedman, T. L. (2007). *The world is flat, A brief history of the twenty-first century* (2nd rev ed.). New York: Farrar, Straus, and Giroux Publisher.

Gardenswartz, L., Rowe, A., Digh, P., & Bennett, M. (2003). *The global diversity desk reference: Managing an international workforce*. Pfeiffer.

Gardner, P. J. (2009). Organizational change: All we want is better projects-why so difficult? *AACE International Transactions*, PM.03.01-PM.03.25.

Gill, R. (2003). Change Management or Change Leadership. *Journal of Change Management, 3*(4), 307–318. doi:10.1080/714023845

Hadeel, A., & Sandhu, K. (2021). PLS Model Performance for Factors Influencing Students Acceptance of Elearning Analytics Recommender. *International Journal of Virtual and Personal Learning Environments* (IJVPLE): New Perspectives in Online Education: Environments, Approaches, Authoring Tools. USA

Harris, G. (2014). Key steps of scenario planning and analysis Retrieved from https://www.wecc.org/Administrative/QPG%20Introduction%20to%20Scenario%20Planning.pdf

Henden, A. (2019). Brexit latest: What is Black Swan theory? What does it mean for Brexit? *Express (Online)*.

Herscovitch, L. (2004). *Resistance to organizational change: Toward a multidimensional conceptualization (Doctoral dissertation)*, The University of Western Ontario, Canada. ProQuest Dissertations & Theses Global database.

Hill, C., & Hult, T. (2018). *Global business today* (10th ed.). New York: McGraw-Hill Education.

HIMSS-Elsevier Digital Healthcare Award Asia Pacific. (2017). *HIMSS-Elsevier Digital Healthcare Award Asia Pacific 2017 Winners Include Four Hospitals Recognized for Innovative use of Technology to Improve Patient Care*. PR Newswire.

Kane, G. C. (2015). How digital transformation is making health care safer, faster and cheaper. *MIT Sloan Management Review, 57*(1).

Kelly, S. (2016). Creating a culture of continuous improvement and sustainable management systems at Abbott Diagnostics Longford. *Global Business & Organizational Excellence, 36*(1), 6–24. doi:10.1002/joe.21753

Kluge, E.-H. W. (2008). Ethical aspects of future health care: Globalization of markets and differentiation of societies- ethical challenges. *Studies in Health Technology and Informatics, 134*, 77–87. PMID:18376035

Kotter, J. (1995). Leading change: Why transformation efforts fail. (cover story). *Harvard Business Review, 73*(2), 59–67.

Kotter, J. (2012). *Leading change. Harvard Business Review*. Library of Congress Cataloging in Publication Data.

Kumar, V. (2013). *101 Design Methods* (1st ed.). Hoboken, NJ: John Wiley & Sons.

Kuza, C. C. (2017). Burnout among critical care nurses and the role of ICU telemedicine. (Doctoral thesis), University of Pittsburgh. ProQuest Dissertations & Theses Global database.

Lawler, E. E. (2014). Sustainable Effectiveness and Organization Development. *OD Practitioner, 46*(4), 65–67.

Lin, C.-L., Wang, Y.-N., & Tsai, H.-M. (2013). Innovative Thinking in Nursing Practice. *Hu Li Za Zhi, 60*(2), 97–102. PMID:23588694

Linski, C. M. III. (2013). Process consultation and law enforcement: A practitioner's case study report. *Organization Development Journal, 31*(4), 63–73.

Lloyd, J., & Ferguson, S. L. (2017). Innovative information technology solutions: Addressing current and emerging nurse shortages and staffing challenges worldwide. *Nursing Economics, 35*(4), 211–212.

Manuele, F. A. (2019). Systems/macro thinking a primer. *Professional Safety, 64*(1), 37–42.

Marshak, R. J., & Grant, D. (2008). Organizational discourse and new organization development practices. *British Journal of Management, 19*(s1), S7–S19. doi:10.1111/j.1467-8551.2008.00567.x

Mary Norine, W., & Rumsfeld, J. S. (2017). Leading the digital transformation of healthcare. *Journal of the American College of Cardiology, 70*(21), 2719–2722. doi:10.1016/j.jacc.2017.10.020 PMID:29169479

McCabe, C., & Timmins, F. (2016). Embracing healthcare technology - What is the way forward for nurse education? *Nurse Education in Practice, 21*, 104–106. doi:10.1016/j.nepr.2016.10.007 PMID:27810650

McCullough, J. S., Casey, M., Moscovice, I., & Prasad, S. (2010). The effect of health information technology on quality in U.S. hospitals. *Health Affairs, 29*(4), 647–654. doi:10.1377/hlthaff.2010.0155 PMID:20368594

Mehta, A. (2019, May 16). Innovative thinking needed to defang existing health issues: Ajoy Mehta. *eHealth*.

Meskó, B., Radó, N., & Győrffy, Z. (2019). Opinion leader empowered patients about the era of digital health: A qualitative study. *BMJ Open, 9*(3). doi:10.1136/bmjopen-2018-025267 PMID:30898816

Mintzberg, H. (2009). *Managing*. Oakland, CA: Berrett - Koehler Publication.

Morse, S. (2018). 82% of clinicians, IT leaders blame burnout on interruptions from tech tools. *Healthcare Finance*. Retrieved from https://www.healthcarefinancenews.com/news/82-clinicians-it-leaders-blame-burnout-interruptions-tech-tools

Nahavandi, A. (2015). *The art and science of leadership* (7th ed.). New York: Pearson Education.

Nanaimo News Bulletin. (2017). *Climate change talk encourages global thinking*. Nanaimo News Bulletin.

Odor, H. O. (2019). A literature review on organisational learning and learning organisations. *International Journal of Information Business and Management*, *11*(3), 281–295.

Olden, P. C. (2003). Hospital and community health: Going from stakeholder management to stakeholder collaboration. *Journal of Health and Human Services Administration*, *26*(1/2), 35–57. PMID:15330379

Ommaya, A. K., Cipriano, P. F., Hoyt, D. B., Horvath, K. A., Tang, P., Paz, H. L., . . . Sinsky, C. A. (2018). Care-centered clinical documentation in the digital environment: Solutions to alleviate burnout. *NAM Perspectives. Discussion Paper, National Academy of Medicine, Washington, DC*. doi:10.31478/201801c

Organizational Learning. (2014). Single and double loop learning. *Organizational Learning*, Retrieved from https://organizationallearning9.wordpress.com/single-and-double-loop-learning/

Reifsteck, M. W., & Reese, M. L. (2016). The digital health explosion. A comprehensive strategy is required to remain relevant and responsive. *Healthcare Executive*, *31*(3), 66–67. PMID:27319110

MENA Report. (2017). New Zealand: Techweek17 to encourage global thinking. *MENA Report*.

Robbins, S. P., & Judge, T. A. (2017). *Organizational behavior*. New York: Pearson Education.

Rogers, E. (1995). Diffusion of Innovations, 4th ed. The Free Press New York.

Romme, G. L. A., & Witteloostuijn, A. (1999). Circular organizing and triple loop learning. *Journal of Organizational Change Management*, *12*(5), 439–453. doi:10.1108/09534819910289110

Ross, K. (2016). Corporate Stewardship: Achieving Sustainable Effectiveness. *People & Strategy*, *39*(1), 59–60.

Roth, M., & Keller, B. (2019). Modernization through digital transformation. *TD: Talent Development*, *73*(1), 32.

Sandhu, K. (2020). Digital Systems Innovation for Health Data Analytics. Handbook of Research on Optimizing Healthcare Management Techniques. Editor: Nilmini Wickramasinghe (Swinburne University of Technology, Australia & Epworth HealthCare, Australia). doi:10.4018/978-1-7998-1371-2.ch019

Sargent, K. (2011). Scenario planning. *Contract (New York, NY)*, *52*(5), 60.

Scheer, V. (2016). Process Improvement for Implementing and Sustaining Change in Healthcare. (10125146 M.S.), The College of St. Scholastica, Ann Arbor. ProQuest Dissertations & Theses Global database.

Schein, E. H. (2004). Organizational culture and leadership (3rd ed.). San Francisco, CA: The Jossey-Bass Business & Management Series.

Schwarzmüller, T., Brosi, P., Duman, D., & Welpe, I. M. (2018). How does the digital transformation affect organizations? Key themes of change in work design and leadership. *Management Review*, *29*(2), 114–138. doi:10.5771/0935-9915-2018-2-114

Shafran, R. F. (2017). Leading organizational change: A phenomenological study of the nonlinear strategies used by female leaders in global companies. (Doctoral thesis), Pepperdine University, Ann Arbor. ProQuest Dissertations & Theses Global database.

Shuck, B., Twyford, D., Reio, T. G. Jr, & Shuck, A. (2014). Human resource development practices and employee engagement: Examining the connection with employee turnover intentions. *Human Resource Development Quarterly*, *25*(2), 239–270. doi:10.1002/hrdq.21190

Smollan, R. K., & Sayers, J. G. (2009). Organizational culture, change, and emotions: A qualitative study. *Journal of Change Management*, *9*(4), 435–457. doi:10.1080/14697010903360632

Sobowale, A., & Ortigara, A. R. C. (2017). From MDGs to SDGs: The need for Global Thinking and Local Actions in the Nigerian Water Sector. *Arid Zone Journal of Engineering, Technology, and Environment*, *13*(5), 620–629.

Springer, P. J., Clark, C. M., Faan, A., Strohfus, P., & Belcheir, M. (2012). Using transformational change to improve organizational culture and climate in a school of nursing. *The Journal of Nursing Education*, *51*(2), 81–88. doi:10.3928/01484834-20111230-02 PMID:22201273

Stary, C. (2014). Non-disruptive knowledge & business processing in knowledge life cycles – aligning value network analysis to process management. *Journal of Knowledge Management*, *18*(4), 651–686. doi:10.1108/JKM-10-2013-0377

Stensaker, I., & Meyer, C. (2011). *Change experience and employee reactions: Developing capabilities for change* (Vol. 41).

Tait, I. E. (2013). Comparison of project management software tool use in healthcare and other industries. (Doctoral thesis), Capella University, Ann Arbor. ProQuest Central; ProQuest Dissertations & Theses Global database.

Taleb, N. N. (April 2007). The black swan: The impact of the highly improbable. Retrieved from https://www.nytimes.com/2007/04/22/books/chapters/0422-1st-tale.html?pagewanted=all&_r=0

Thimbleby, H. (2013). Technology and the future of healthcare. *Journal of Public Health Research*, *2*(3), 28. doi:10.4081/jphr.2013.e28 PMID:25170499

Thomas, D. (2014). Leading under pressure, coping practices of executives: A mixed methods study. (Doctoral thesis), Pepperdine University, Ann Arbor. ProQuest Central; ProQuest Dissertations & Theses Global database.

Van Fleet, D. D., & Peterson, T. O. (2016). Improving healthcare practice behaviors. *International Journal of Health Care Quality Assurance*, *29*(2), 141–161. doi:10.1108/IJHCQA-07-2015-0089 PMID:26959895

Venkatesh, V., & Davis, F. D. (2000). A theoretical extension of the technology acceptance model: Four longitudinal field studies. *Management Science*, *46*(2), 186–204. doi:10.1287/mnsc.46.2.186.11926

Wade, V. A., Taylor, A. D., Kidd, M. R., & Carati, C. (2016). Transitioning a home telehealth project into a sustainable, large-scale service: A qualitative study. *BMC Health Services Research*, *16*(1), 1–10. doi:10.118612913-016-1436-0 PMID:27185041

Wade, W. (2012). *Scenario planning- a field guide to the future*. Hoboken, NJ: John Wiley & Sons.

Walsh, M. N., & Rumsfeld, J. S. (2017). Leading the digital transformation of healthcare. *The ACC Innovation Strategy*, *70*(21), 2719–2722. doi:10.1016/j.jacc.2017.10.020 PMID:29169479

Wang, L., & Alexander, C. A. (2013). Automated identification technology and information security in digital hospitals. *International Journal of Information and Network Security*, *2*(5), 348–359. doi:10.11591/ijins.v2i5.3914

Watkins, M. J., Mohr, B. J., & Kelly, R. (2011). Appreciative Inquiry (AI) (2nd ed.). San Francisco, CA: John Wiley & Sons.

Weiner, L. J. (2016). Improving physician engagement. *HealthLeaders, 19,* 8-8, 10.

Weinstein, R. S., Lopez, A. M., Krupinski, E. A., Beinar, S. J., Holcomb, M., McNeely, R. A., ... Barker, G. (2008). Integrating telemedicine and telehealth: Putting it all together. *Studies in Health Technology and Informatics*, *131*, 23–38. PMID:18305320

Weston, S. (2017). The real challenge for digital transformation is not your technology. *CIO (Framingham, MA)*.

Whitehead, N. P., Scherer, W. T., & Smith, M. C. (2015). Systems thinking about systems thinking a proposal for a common language. *IEEE Systems Journal*, *9*(4), 1117–1128. doi:10.1109/JSYST.2014.2332494

Wiedmer, T. (2015). Generations do differ: Best practices in leading Traditionalists, Boomers, and Generations X, Y, and Z. *Delta Kappa Gamma Bulletin*, *82*(1), 51–58.

Zennouche, M., Zhang, J., & Wang, B. W. (2013). Factors influencing innovation at individual, group and organisational levels: A content analysis. *International Journal of Information Systems and Change Management*, *7*(1), 23–42. doi:10.1504/IJISCM.2014.065052

Ziadlou, D. (2019, September). Digital health leadership in healthcare organizations[YouTube]. Switzerland International Society for Telemedicine and e-Health (ISfTeH).

Chapter 9
Digital Entrepreneurship for Immigrants:
Motivations, Strategies, Ecosystems, and Performance

Carson Duan
University of New England, Australia

Bernice Kotey
University of New England, Australia

Kamaljeet Sandhu
https://orcid.org/0000-0003-4624-6834
University of New England, Australia

ABSTRACT

This chapter examines the concept of digital entrepreneurship for immigrants (DEI) through the lens of entrepreneurship research theories and practices. The authors argue that the current focus of digital entrepreneurship (DE) research has centered on successful digital entrepreneurs with proactive innovation strategies and high-performance outcomes, largely being well-known brands. On the other hand, immigrants who are engaging in digital entrepreneurship have been neglected in current related literature. This chapter presents a conceptual framework for combined immigrant and digital entrepreneurship study. The authors propose future research themes for DEI drawing from the results and implications of two Chinese immigrant digital entrepreneurs in NZ and Alibaba e-commerce platforms in China.

INTRODUCTION

In today's digital economy era, there is a general acknowledgement of the importance of digital entrepreneurship (DE) to socioeconomic development among policymakers, researchers and society (Sahut, Luca, & Frédéric, 2019). DE has rapidly become a new interdisciplinary research field (Nambisan,

DOI: 10.4018/978-1-7998-2799-3.ch009

2017); the field has been examined through information systems (Du, Pan, Zhou, & Ouyang, 2018), economics (Richter, Kraus, Brem, Durst, & Giselbrecht, 2017), management, innovation (Nambisan, 2017), policy, education (P. Li, Liu, Zhang, & Li, 2017), strategy and sociology standpoints in conjunction with entrepreneurship disciplines (Zaheer, Breyer, & Dumay, 2019). To date, there are several terms defining DE such as internet entrepreneurship between 2000-2001, cyber entrepreneurship around 2004, information and communication technologies (ICT) entrepreneurship (Bogdanowicz, 2015) and IT associated entrepreneurship (Steininger, 2019). Scholars make use of these terms interchangeably, or with a self-explained definition for the research target. In this research, the concept of DE is "the process of entrepreneurial creation of digital value through the use of various socio-technical digital enablers to support effective acquisition, processing, distribution, and consumption of digital information (Sahut et al., 2019)" or "new venture opportunities presented by new media and internet technologies (Davidson, Vaast, & Ieee, 2010)."

Despite increasing research interest in the digital entrepreneurship sector, only a few studies have devoted attention to immigrants (Melnikova, Grunwald, Ahrens, & Zascerinska, 2019) and small and medium enterprises (SMEs) (Afutu-Kotey, Gough, & Owusu, 2017). Scholars have focused on extensively successful internet platforms, infrastructures and technology companies, whereas routine frontline digital entrepreneurs have not received enough attention from a DE perspective. As a matter of fact, many frontline ICT services and development firms are SMEs. Many entrepreneurs who create economic value through/by the Internet platform are sole traders and start their venture either opportunity-driven or necessity-driven.

While immigrant entrepreneurship (IE) has played a critical role in the socio-economic development of most advanced countries from the 1970s (Zhou, 2004), DE has been in a crucial position since the digital economy era started in the early 1990s (Sussan & Acs, 2017). It is widely regarded that digital entrepreneurship has changed the business landscape and is a major driving force in propelling the economy, however, there is a lack of conceptual framework developed in DE, particularly DE for immigrants. In addition, there is very little research that focuses on DE motivations, or factors influencing a person's capabilities for digital entrepreneurship. This research paper seeks to address this gap; it is a first attempt in discussing digital entrepreneurship for immigrants.

Well-known and often-researched examples with digital service business models include Google, YouTube, Amazon and many others (Sahut et al., 2019). With these newly created models, markets have been witnessing a transition in firms offering strategies from products to services and then servicing as advocated in the service-dominant logic (SDL) (Sussan & Acs, 2017). Drawing on SDL, service exchanges are the only economic activities that exist among economies and societies, Karmarkar (2015) argues that there are two key trends: the move to a service economy and the evolution to an information economy, "Information Services is where these two trends come together… More and more value will continue to accrue to services, to the information sector and to their intersection: information services." Additionally, SDL logic sees service as a process, i.e. doing something for another party, the locus of value creation moves from the "producer" to a collaborative process of co-creation between parties (Vargo, 2008).

This study focuses on China and Chinese immigrants in New Zealand as case studies. First, the Chinese migrants traditionally have an entrepreneurship culture (Zheng, Khavul, & Crockett, 2012; Zhao & Hsu, 2007). Compared to other ethnic groups, Chinese migrants are more likely to set up businesses or undertake self-employment, as being an entrepreneur represents achievement in Chinese culture (Collins, 2002). According to Inland Revenue NZ (2013), 24.1% of Chinese immigrants in New Zealand are entrepreneurs, compared to 18% of entrepreneurs within the total population (NZ, 2013). More importantly, the Chinese economy has developed by leaps and bounds in the last four decades, and research results from traditional Chinese IE studies have not been able to capture the changes in China adequately (Y. Wang & Warn, 2018; Z. Zhang, 2013). For the last twenty years in China ICT infrastructure has been significantly improved, smartphones and PC technology are a staple amongst daily life and internet and e-commerce platforms are widely accepted. By a 2016 global survey, China was the largest digital market including both platforms and customers (Evans, 2016).

Furthermore, there are confused assumptions in literature and government/organization reports. On the one hand, necessity entrepreneurs account for the majority of immigrant entrepreneurs (Zolin, Chang, Yang, & Ho, 2016; Maritz, 2004), engaging in labor intensive, low pay, disserted business activities. On the other hand, scholars and government administrations admit that immigrants make significant contributions to digital and high-tech entrepreneurship, and they suggest that immigrants make up a large portion of digital entrepreneurs (Lan & Zhu, 2014; D. M. Hart & Acs, 2011). In extant literature, scholars classified digital and high-tech entrepreneurs as opportunity entrepreneurs (Chrysostome & Arcand, 2009). This paper attempts to clarify these confused assumptions.

This chapter aims to examine digital entrepreneurs for immigrants through the lens of necessity and opportunity-driven entrepreneurship typologies to analyze their corresponding business strategies, access and use of opportunities and resources from entrepreneurial and digital ecosystems. As a consequence of entrepreneurship, the assessment of firm performance is investigated from different measurement perspectives. This paper will follow entrepreneurship theories with the integration of ICT concepts to examine the process of digital entrepreneurship for immigrants.

The following sections present a literature background on immigrant and DE first, a conceptual framework for DEI is then illustrated, and the key components of the conceptual framework are explained. Following is research methodology and case studies, and later the authors provide research results and implications, solutions and recommendations. The paper finishes with future research directions and the conclusion.

BACKGROUND

IE research is essential not only because of the high percentage of foreign-born populations in advanced countries, but because a significant portion engage in formal and informal entrepreneurial activities (Z. Zhang, 2013; Spoonley & Meares, 2009; Collins, 2002; Le, 2000). Evidence shows some countries have experienced significant socioeconomic gains from international migration and IE (Z. Zhang, 2013; Collins, 2003); for example, immigrants in Australia own 33.6% of all businesses, despite making up only 28.8% of the total population (ABS, 2019). Similarly, in the U.S., although immigrants only make up 13% of the total population, they account for 27.5% of all entrepreneurs (Blanding, 2016; Vandor & Franke, 2016). The Global Entrepreneurship Monitor (GEM) report for the United Kingdom shows that people with migrant backgrounds are twice as likely as their white British counterparts to be early-stage entrepreneurs (M. Hart, Bnner, Levie, & Heery, 2018). Additionally, the number of new immigrant businesses maintain a positive relationship with the total amount of non-immigrant firms in the host country (Li, Isidor, Dau, & Kabst, 2018).

Digital entrepreneurship became an academic topic among scholars and policymakers in the early 2000's – five years after the first wave of digital ventures started (Sahut et al., 2019; Zaheer et al., 2019). Common entrepreneurship theories and frameworks have been applied to explain various aspects of the way DE operates. They include the venture of typology, strategic formulation, categorization of entrepreneurs, entrepreneurial process, opportunity recognition, individual traits, business environment (Zaheer et al., 2019) and entrepreneurship ecosystems (Sussan & Acs, 2017). Sussan and Acs further integrated entrepreneurial and digital ecosystems, proposing the concept of a digital entrepreneurial ecosystem. The most focused theories in digital entrepreneurship literature is the entrepreneurial process, which focuses on sequences of strategic activities and views entrepreneurship as a journey instead of an action (Zaheer et al., 2019). Calculations from Zaheer et al. (2019) find that over 36% of articles in DE literature is based on the "process" theory. It proves that DE research has the same trend as general entrepreneurship and it is considered a subfield of entrepreneurship discipline (He, 2019). Following this trend, some scholars have suggested in-depth studies into the mechanisms of how both digital and entrepreneurial ecosystems influence different entrepreneurial activities (Sussan & Acs, 2017). Scholars recommend studying DE with an EE influence from social, culture and market aspects.

This research will be the first attempt to conceptually investigate the relations between digital entrepreneurship and immigrants' role and DEI impact factors from the perspective of entrepreneurs' internal motivation and external environments. The research will be conducted under a newly created integrated conceptual framework, with constructs of entrepreneurial and digital ecosystem frameworks, necessity and opportunity motivations, strategic activities and firm performances. The findings will contribute to the IE and DE fields with empirical evidence in terms of the relationship among necessity/opportunity motivation, entrepreneurs' strategic activities and ecosystems of the host and home country. Considering that understanding the critical factors for DEI success can favor the development of government policies and programs that affect all immigrants (Sussan & Acs, 2017; Sim, 2015), the results will have significant theoretical and practical implications, helping policymakers improve ecosystems to promote digital entrepreneurship for immigrants (Rosique-Blasco, Madrid-Guijarro, & Garcia-Perez-de-Lema, 2017; Kushnirovich, 2015)

INTEGRATED CONCEPTUAL FRAMEWORK OF DIGITAL ENTREPRENEURSHIP FOR IMMIGRANTS

Drawing on entrepreneurship theories, DEI research can be conceptualized as a study among entrepreneurial motivations, strategies, digital entrepreneurial ecosystem and business performance outcomes in immigrant context as illustrated by Figure 1. This framework is consistent with conventional entrepreneurship theories and the latest arguments of immigration and digital entrepreneurship. For the purpose of causal effect analysis, a few new attempts are added onto the framework including 1) considering DEI in multi-ecosystems, including digital and entrepreneurial ecosystems from both home and host countries, 2) acknowledging DEI effects on entrepreneurial motivations, and 3) measuring growth performances to match the differences of entrepreneurial motivations and strategies. The conceptual framework synthesizes four sets of factors contributing to the outcomes of DEI: entrepreneurs' motivation, business strategies, business performance and digital and entrepreneurial ecosystems. These ecosystems play a significant role for digital innovations because it contains leverageable resources for entrepreneurs' resource assembly, facilitating experimentation, entrepreneurial innovation and value creation (Song, 2019; Du & Li, 2018).

Numerous theories contribute to the framework. First, the theoretical foundation is the well-cited Gartner's (1985) four-dimension framework of entrepreneurship. Gartner describes entrepreneurship as an individual undertaking innovative activities to create a new economic venture relying on the specific business environment, which has been widely referenced from both IE and DE perspectives (Brzozowski et al., 2017; Waldinger, 2013; Drori et al., 2009). Secondly, both intrinsic and extrinsic immigrant digital entrepreneurial motivations positively influence digital entrepreneurial intention. Entrepreneurship orientation such as conscientiousness, openness to experience, and agreeableness positively affect intrinsic digital entrepreneurial motivation; while extraversion, conscientiousness, and neuroticism positively affect extrinsic digital entrepreneurial motivation (Y. S. Wang, Lin, Yeh, Li, & Li, 2016). Thirdly, scholars argue that both IE and DE are influenced by the factors of social, political, economic and technological environments; determinants of netizens, governments, enterprises, and media; and industry associations from digital ecosystems (Xie, Xie, & Martinez-Climent, 2019). Some of them express that one of the crucial strategies for conducting DE research is to consider that digital entrepreneurs are operating in an open innovation ecosystem, and both interaction with consumers and the sharing of information amongst consumers is crucial (Yu, Roy, Quazi, Nguyen, & Han, 2017). Their argument is that e-satisfaction is not only a critical outcome metric, but also a primary predictor of customer e-loyalty, measured in terms of the customer's positive attitude, stickiness and propensity to spread positive word of mouth. Therefore, it is rational to conceptualize that DEI is driven by motivation, operates in entrepreneurial and digital ecosystems, adopts appropriate business strategies and achieves expected performance.

Two recently matured concepts, the entrepreneurship ecosystem (Erik Stam, 2015; Isenberg, 2011) and digital ecosystem (W. J. Li, Du, & Yin, 2017), were merged conceptually by Sussan and Acs (2017) and revisited by Song (2019). This new concept of digital entrepreneurial ecosystems (DEE) provides the living and non-living digital ecosystem components, including "digital infrastructure governance (DIG), digital user citizenship (DUC), digital entrepreneurship and digital marketplace" in the digital economy (Song, 2019; Sussan & Acs, 2017). In the context of DEI, this framework conceptualized a construct of immigrant digital entrepreneurial ecosystem. It contains location-specific immigrants' host and home country entrepreneurial ecosystems and a boundary-less digital ecosystem. The difference between digital entrepreneurial ecosystem and immigrant digital entrepreneurial ecosystem is that the latter covers the immigrants' home country EE.

Lastly, the causal relations among entrepreneurial inputs, outputs and outcomes are significant foundations for the framework. Stam (2015) recognizes that the EE is "a set of interdependent actors and factors coordinated in such a way that they enable productive entrepreneurship." His framework, strongly supported by scholars (Nicotra, Romano, Giudice, & Schillaci, 2018; Freiling & Baron, 2016), illustrates that entrepreneurial activities and strategies are the outputs of an EE and firm performances are outcomes of EE.

IMPORTANT COMPONENTS OF DIGITAL ENTREPRENEURSHIP OF IMMIGRANTS

Motivation

A well-accepted entrepreneurial motivation typology is the notion of necessity and opportunity-driven entrepreneurship (Reynolds, Bygrave, Autio, Cox, & Hay, 2002). Necessity entrepreneurs engage in business ventures out of need, whereas opportunity entrepreneurs start new ventures due to personal ambitions. Another typology classifying entrepreneurs is the argument of routine and high-growth entre-

Figure 1. Integrated concept framework of digital entrepreneurship for immigrants (Own Elaboration)

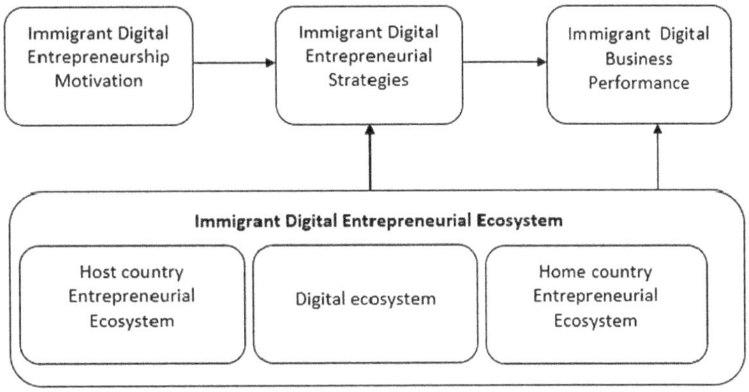

preneurship (Sussan & Acs, 2017). Routine entrepreneurs are business owners involved in executing an established ongoing firm such as restaurants, garages, hairdressers etc. High growth entrepreneurs operate in a lesser established and less clearly defined marketplace in which the products, services, functions, supply chains and users are not completely known (Sussan & Acs, 2017). There is an assumption that startups in digital clusters are opportunity-driven (D. M. Hart & Acs, 2011; Lan & Zhu, 2014) because of the knowledge, investment, market potential and existing risks in digital service innovation (Cavallo, Ghezzi, Dell'Era, & Pellizzoni, 2019; Farani, Karimi, & Motaghed, 2017; Wan, Jiang, & Xie, 2017). In the following sections, necessity and routine, and opportunity and high-growth are used interchangeably.

Although a large portion of entrepreneurship studies do not separate necessity/routine and opportunity/high-growth entry modes, literature does show that the two groups of entrepreneurs are different (van der Zwan et al., 2016; Rubach, Bradley, & Kluck, 2015). In the necessity scenario, entrepreneurs are forced into entrepreneurship due to a variety of reasons resulting in inability to find satisfying employment in the local labor market (Chrysostome & Arcand, 2009). Likewise, it is difficult for new entrepreneurs to receive support from government agents and gain start-up capital through formal financial institutions because their projects may be considered untrustworthy, even though the idea has been proven in other countries or regions. Consequentially, they are forced to enter less economically profitable and deserted business clusters and work long hours with family support to achieve business success (Chrysostome & Arcand, 2009). Research reveals that necessity immigrant entrepreneurs tend to bond with the local community in which people share the same culture in a city or region, either as the market niche or resource provider (Aldrich & Waldinger, 1990). Necessity IE studies dominate traditional entrepreneurship research, following consensus that business ventures were the sole method for economy advancement. However, more recently the DE concept has become heterogeneous, referring to many digital entrepreneurs who start their own business, not because of barriers and drivers in the digital society, but as a result of their eagerness to exploit a business opportunity and pursue a personal dream (Wahee & Bhardwaj, 2014).

An individual's socioeconomic profile, personality and perception of the entrepreneurial environment are the critical determinants of entrepreneurs' engagement in either necessity or opportunity entrepreneurship (van der Zwan, Thurik, Verheul, & Hessels, 2016). Both Serviere's (2010) and Chrysostome's (2010) models identified that low levels of education and a lack of language skills are obstacles to social mobility, which leads to immigrant necessity entrepreneurship. In contrast, local society's socioeconomic status, job opportunities, and institutional and government support are external factors determining individuals' choices. In some situations, entrepreneurship is perceived as the only option for some individuals, particularly those lacking significant opportunity for paid employment (Block, Kohn, Miller, & Ullrich, 2015).

In summary, entrepreneurial motivations have consequences for the way in which a business is managed, such as business aspirations, market entry strategies and business performance (van der Zwan et al., 2016). Current DE motivation studies center on university students and there is a lack of research on digital entrepreneurs (Zhang, 2016; Hu, 2015; Z. W. Li, Millman, Matlay, & Wong, 2008) in the seed stage of the venture creation journey. Notably, there is an inability to locate articles concerning necessity-driven digital entrepreneurial processes. Even though early research has shown that opportunity entrepreneurial intentions are influenced by personal perspectives (major of study, family background, gender), there is a lack of research on the causal effects. Furthermore, the relationship between motivations and the effect of the environment is not yet sufficiently researched in IE and DE literature (Wahee & Bhardwaj, 2014). There is a need to research the relationships mentioned above in separate necessity and opportunity settings from the motivations, strategies, opportunities and resources coming from both ecosystems and performance perspectives.

Business Strategies

The business strategies undertaken by a digital entrepreneur are determined by entrepreneurs' characteristics and the business environment (Bogdanowicz, 2015). Entrepreneurs with more capitals (financial, human, etc.) tend to pursue strategies centered on infrastructure and platform marketplace. In contrast, entrepreneurs with fewer capitals are pursuing digital user focused strategies. While opportunity-driven digital entrepreneurs engage in proactive strategies, necessity entrepreneurs tend to be reactive in business strategies with the ultimate goal of survival. These survival strategies have featured low risk, cash-flow focused and necessary market orientation, and are sometimes supported by government business programs. Literature shows a positive relationship between a firm's proactive strategies and its growth performance (Kotey et al. 2013), whilst reactive firm strategies are practical for the survival of the necessity entrepreneurs. Motivation, the willingness to work long hours with less hourly pay is also treated as one of the strategic resources for business survival and growth (Barney, 1991).

Entrepreneurial strategic activities are recognized as processes in which individuals access resources, exploit opportunities and create businesses for innovation (Erik Stam, 2015; Gartner, 1985). Stam's (2015) EE framework of entrepreneurial activities being the outputs of EE is widely accepted. Some scholars express that opportunity entrepreneurial activities with proactive strategies will benefit more than necessity activities from EE (Erik Stam & Spigel, 2016). Others claim that EE fertilizes both necessity and opportunity entrepreneurs (Sussan & Acs, 2017), and it is the researcher's responsibility to clarify what type of entrepreneurship is the target group (GIZ, 2018).

Overall, the entrepreneurial strategies of digital firms can be presented on a continuum from proactiveness to reactiveness. The strategic activities convert motivation, resources and opportunities to organizational strategies, with economic performance as one of the key outcomes within a given economic environment. Previous studies have focused on individual entrepreneurs, providing content-rich factors related to the strategies they undertake. However, there is limited attention to environmental factors, especially with regards to the relationship between necessity entrepreneurs and ecosystem factors. The sparse research in this area lacks a systematic approach to analyzing how digital and EE variables affect strategic activities and entrepreneurship outcomes. The framework presented above aims to fill these gaps.

Digital Entrepreneurial Ecosystem for Immigrants

Digital Ecosystem

The digital ecosystem is the business environment of DE. The term has a root of "Digital Business Ecosystem" which refers to a socio-economic development environment favoring ICTs (Nachira, 2002). Like a biological ecosystem, a perfect digital ecosystem will solve complex internal problems through automatically evolving the whole system. The digital ecosystem emerged through advancements in ICT, web powered social networks and virtual communities, emergence of business networks through knowledge sharing, and ICT-enabled service.

E-commerce is defined as the economic activities in a digital ecosystem, which is borderless, infinite, multifaceted and timeless. It increases efficiency between producers and buyers, reduces transaction costs (He, 2019) and removes the boundaries of marketspace. Digital artifacts, platforms and infrastructures are three distinct and related factors in the realm of DE (Nambisan, 2017). A platform such as Alibaba, Tencent and Amazon creates a business ecosystem. This platform business model operates differently with a traditional pipeline and hierarchy business structure (He, 2019); its value proposition is created through connectivity and network traffic. In the platform firm, users co-create value for all participants and foster new revenue streams by sustaining the advantage of the platform (He, 2019).

Li et al. (2012) defined DE as "…a self-organizing, scalable and sustainable system composed of heterogeneous digital entities and their international focusing on interactions among entities to increase system utility, gain benefits, and promote information sharing, inner and inter cooperation and system innovation (p119)." Some scholars emphasize the importance of DE research at an extended social network area, drawing from the concept of the loose geographic boundary character of the discipline (Liu, Nandhakumar, & Zachariadis, 2018). As Stone (2017) expressed, digital service companies are likely to upright industries on a global scale (Sussan & Acs, 2017). The European Commission draws a seven-pillar digital ecosystem framework with barriers and drivers to DE, these are the regulatory frameworks, market conditions, access to finance, entrepreneurial capabilities, immigrants and culture (Bogdanowicz, 2015).

Entrepreneurship Ecosystems

The latest improvement in the entrepreneurship discipline is to research the phenomenon within an entrepreneurial ecosystem framework (Erik Stam, 2015; WEF, 2013; Isenberg, 2011). Instead of investigating singular or a few factors, scholars are researching EE as a system of interconnected socioeconomic and political-cultural elements. Scholars believe that the business environment enhances entrepreneurship by providing opportunities and resources, hence creating high growth firms (Nicotra, Romano, Giudice, & Schillaci, 2018; Erik Stam, 2015; Kotey, 2006). EE has become the latest instrument for external environment research in general entrepreneurship studies (Spigel & Harrison, 2018), and has the potential to become an accepted framework to guide IE and DE study after enough examination.

Several definitions of "EE" have been developed for the last two decades (Stam and Spigel, 2016; Spigel, 2015; Mason and Brown, 2014; Vogel, 2013; Isenberg, 2000). Although each definition has considered different entrepreneurial factors that dynamically interact with one another, the universal agreement is that a functional EE provides capitals (financial, knowledge, institutional, social and etc.) to enable entrepreneurs to discover, assess, exploit and take advantage of market opportunities (Nicotra et al., 2018; Erik Stam, 2015; Isenberg, 2011). Recent studies have holistically improved the elements of an ecosystem by extending EE to cover extra dimensions such as cultural, market, human, research institute, government and institutional (Motoyama & Knowlton, 2017; Isenberg, 2011).

Digital entrepreneurs operate in two business environments (Sussan & Acs, 2017) or dual ecosystems. Supported empirical research shows that there is a high percentage of digital entrepreneurs engaging in artifacts, platforms, and infrastructure development activities (Nambisan, 2017). These opportunities not only exist in the marketplace of digital products and services in developed countries, but of developing economies as well (Boojihawon & Ngoasong, 2018). Based on the resource-based view (RBV) and opportunity structure theory, the different capabilities of access resources and opportunities between digital entrepreneurs cause disparities in entrepreneurial activities and business performances.

EE emerged as an accepted framework within policymakers, it is an economic development strategy that creates a supportive environment to foster entrepreneurship. However, existing research on EE has been largely typological and theoretical and it is yet to be explored how they influence entrepreneurial activities (Spigel & Harrison, 2017). Digital entrepreneurs' willingness to participate and be embedded within EE vary drastically (Sussan & Acs, 2017). Although opportunity digital entrepreneurs' willingness and fruitful returns have been sufficiently investigated (Du & Li, 2018; Song, 2019), necessity entrepreneurship research under the framework of digital and EEs have been neglected, regardless of the fact that both entrepreneurship typologies have gained worldwide acceptance among scholars.

To summarize, the interactions between immigrant digital entrepreneurs' motivation, resources and opportunities in digital and entrepreneurial ecosystems determine activities undertaken and ultimate performance outcomes. However, how ecosystem factors, motivations and entrepreneurial activities influence firm outcomes have not been adequately studied in IE (Nicotra, Romano, Giudice, & Schillaci, 2018; E. Stam, 2015) and DE (Song, 2019) sub-disciplines within the necessity and opportunity context. Although IE survival indicators could be mainly influenced by the local EE, growth indicators can be impacted by both host and home ecosystems, while both home and host country EEs affect DEI simultaneously.

Performance Assessment of Digital Entrepreneurship for Immigrants

The performances of immigrant and digital businesses have been a concern because DE has been treated as a key to promoting economic growth (M. T. Chand, Rosalie L., 2011), and IE has been an important pathway to not only enhancing local socioeconomic growth, but integrating immigrants into the host society (Heilbrunn & Kushnirovich, 2008). Although there are numerous conceptual and empirical studies on immigrant business performance, the results are not conclusive. Most of these studies focus on comparing business performance between immigrants and native firms in the low skilled sector (Q. F. Wang & Liu, 2015) with many studies noting that immigrant business income and performance are inferior to their native peers. Nonetheless, these findings vary for different immigrant groups in one country (M. Chand & Ghorbani, 2011; Constant & Shachmurove, 2006; Oliveira, 2010; A. Portes & Yiu, 2013), and the same immigrant group in different countries (Lan & Zhu, 2014; Oliveira, 2010). A framework that effectively integrates all relevant factors to investigate outcomes from immigrant entrepreneurs will be important to promoting and supporting the prosperity of the immigration economy (Kushnirovich & Heilbrunn, 2008; Rosique-Blasco et al., 2017; Yang, Ho, & Chang, 2012).

Acs's (2006) research shows that the economic performance of an immigrant firm depends on the entrepreneur's motivation to start a firm and the institutional conditions in the host country. Necessity-driven firms with survival strategies will have a low economic impact because of lower investments, fewer employees and lack of business skills (Reynolds et al., 2002), which all restrict long-term expansion. On the other hand, opportunity-driven firms have innovative strategies and additional resources to exploit profitable opportunities, which results in a positive impact on economic development (Reynolds et al., 2002).

Some scholars argue that the performance assessment of immigrant businesses needs to be the same as SME (Zhou, 2004; Alejandro Portes & Zhou, 1996). The reasons behind it are 1) most necessity enterprises are SMEs, and 2) although IE focuses on the specific features of ethnic communities, business performance variables of all enterprises look much alike regardless of industry or owner's ethnicity. It is well agreed that necessity and opportunity enterprises are not the same due to different goals and challenges, and they should be assessed separately. If increases in market share and employment are useful measurements of opportunity entrepreneurship, these measurements are not necessarily relevant to necessity entrepreneurship. With survival as the main objective, the indicators for necessity entrepreneurship should focus on sustainable business activities and earnings. Meanwhile, the successes of necessity DE should also include job creation for the owners, reducing competition with cooperate workers, and providing increased earnings.

All in all, it is difficult to assess necessity entrepreneurs in an identical manner to opportunity entrepreneurs. If there is common ground to assess them, business effectiveness approaches would be optimal. The commonly used four effectiveness approaches include 1) systems resources approach, 2) goal approach, 3) internal functioning approach, and 4) strategic constituencies approach. The approach particularly focuses on entrepreneurs' outputs, or on the degree of goal attainment achieved by the entrepreneur. Sarapaivanich (2007) illustrates a suggested typology of entrepreneurship assessment in goal approach (Figure 2), where she split the goals attained by entrepreneurs into objective financial and non-financial goals plus subjective financial and non-financial goals to assess the achievement of a SME owner. Since there are often conflicts between financial and non-financial goals there must be a trade-off among the various goals of the entrepreneurs in order to achieve a balanced performance. This method is applied later in the paper for two DEI cases.

RESEARCH METHODOLOGY

This research adopted a qualitative method due to DE investigation is relatively nascent and under-researched in extant literature (Ngoasong, 2018). Case studies include one case for an e-commerce platform, one case for necessity entrepreneurs and one for opportunity entrepreneurs, each sample being purposely selected. Such an approach has been applied in various studies and approved to be suitable for knowledge development and improvement (Hohenthal, 2006). Ngoasong (2008) cited Yin's (2009) definition for a case study as "an empirical inquiry that investigates a contemporary phenomenon in-depth and within its real-life context, especially when the boundaries between phenomenon and context are not clearly evident (p.18)." Thus, each case representing one of many samples that can be examined as empirical evidence to develop the theoretical framework (Ngoasong, 2018) suits this theoretical research. Two data collection methods were applied: face to face conversion deployed for the necessity entrepreneur and credited secondary data sources applied for the opportunity entrepreneur and digital platform.

China could be a good example for digital platform and DE investigation because 1) China has experienced a rapidly growing digital marketplace (Economist, 2019), 2) China has the largest number of e-commerce platforms with 64 (Alibaba, Tencent, Baidu etc.) (Evans & Gawer, 2016); 3) China has the largest number of e-shop owners whose online sales reached USD36bn (Economist, 2017), and 4) Chinese immigrants living outside of China have a desire to trade with their home country, and are named "born global" immigrant entrepreneurs, (Qiu & Gupta, 2015). The success of China's e-commerce platforms have relied on the advancement of ICT and latest digital service business models.

CASES: ALIBABA E-COMMERCE PLATFORM IN CHINA AND CHINESE IMMIGRANT DIGITAL ENTREPRENEURS IN NEW ZEALAND

The Internet has been widely accepted in China for the last 20 years and has dramatically impacted industry, commercial and even agricultural practices. In recent years China's e-commerce sector has demonstrated explosive growth momentum, becoming a popular trading standard in daily life. New innovations brought about by the application of e-technologies have stimulated entrepreneurship to benefit the micro-entrepreneur and SME owner. The United Nations' report titled "National report on e-commerce development in China" expressed four stages of e-commerce development in China: the initial stage (1996-2000), the accelerated development stage (2000-2008), the standardization stage (2008-2014) and the globalization stage (starting from 2014) (Yue, 2017).

At the same time, China's cross-border e-commerce is opening up with the "Going global strategy and bringing-in", which has become a new engine of China's economic development. Between 2016 and 2017, China's cross-border E-commerce import and export growth was 38.7% with a value reaching 6.6 trillion Chinese yuan, and an average annual growth rate of 33.1% from 2012 to 2017 (Yue, 2017). United Nations' report shows that the cross-border e-commerce transactions have increased more than fivefold (1.2 to 6.6 trillion yuan) from 2010 to 2016 (Yue, 2017).

The selected case of e-commerce platforms is Alibaba in China. Alibaba has designed a business environment with six pillars for micro, small and medium enterprises (MSEMs), and they describe the development as the Alibaba business ecosystem (He, 2019). Alibaba platforms bring multisided marketplaces, including producers, e-shop owners, buyers and logistics partners, with Alibaba e-commerce platforms to co-create value. It acts as a one-stop business center for MSEMs owners by providing operation and management services. First, for e-startups, the platforms provide e-shop operation training through which users are able to design a personalized website. Second, Ali-platforms (Taobao, Tmall, AliExpress etc.) provide effective e-channels between shop owners and consumers. Third, Ali-platforms provide financing services including payments, taxes and refunds. Fourth, business intelligent services offer customer and product analysis. Fifth, Alibaba cooperates with logistics partners to provide delivery, warehouse, product return and claims services through Ali-platforms. Last but not least, the Ali-platform provides scaling up capabilities to enable e-shops to grow by providing API, cloud-computing and online-2-offline services. Through this service, e-shop owners can create their own platform but still get required services from Ali-platform. The concept of the Alibaba business ecosystem is illustrated by Figure 3. This virtual cluster, or self-contained entrepreneurial ecosystem, facilities the participants to share data and improve economic performance for each party by one-stop services. A combination of available technology such as PC, laptop, smartphone, the Internet and high-speed connection encourages millions of MSMEs to open e-shops on Alibaba e-commerce platforms. In 2014, the Taobao and Tmall platforms have over nine million e-shops and one-third of them are actively trading daily (ChinaNews, 2014).

The first digital entrepreneur case selected for this study is an e-shop owner, James. He is a Chinese immigrant who moved to Auckland four years ago. Currently he is running his digital trading business on Taobao and Tmall selling NZ goods such as wine and dairy products to Chinese customers. By all definitions he is a digital entrepreneur, since the economic value he creates mainly comes from e-commerce activities. He started his e-shops on his home country e-commerce platforms because he could not find a satisfying job in the NZ labor market within his major, but he has ultimately achieved his goal of supporting family. He can be classified as a necessity-driven immigrant digital entrepreneur (Sahut, Luca, & Frédéric, 2019; Reynolds, Bygrave, Autio, Cox, & Hay, 2002). There is a lack of statistics on how many Chinese immigrants are running e-shops on Alibaba e-commerce technologies to support their lifestyle. We can only assume there are many of them in Australia and New Zealand based on the news of "Never been so disgusted: Coles customer's viral photo of baby formula 'express lane' causes outrage" on News.com.au (Chung, 2019), and one of the largest NZ supermarkets (Countdown) setting up limitations for the purchase of baby formula products due to private export to China (Williams, 2019).

The second digital entrepreneur case is an opportunity entrepreneur, Sean, who is the founder and CEO of PayPlus NZ which is an Alibaba technology enabled enterprise. The case details were published on Stuff.co.nz by Anuja Nadkarni in Oct 2018 based on ANZ bank small business research. The PayPlus case showed that Alibaba is not only providing e-shop operation platforms for MSMEs, but offering finances, payment services and technologies to opportunity entrepreneurs by providing an integrated payment service that offers Chinese tourists and students the ability to pay through Alipay and WeChat Pay. Sean came to NZ as a student in early 2000, and fifteen years later he became a business owner of six 2degree stores in NZ. In 2016, he sold his stores and embraced Alipay, a technology from Alibaba allowing smartphone payments. He recognized the demand for easy payment with increasing Chinese migrants, and local businesses and shops needed to accept and adapt to the popular payment methods in China. He connected New Zealand businesses with Chinese customers through Alipay and the whole business has been supported by Alibaba since then. He is a typical example of an opportunity-driven immigrant as a digital entrepreneur with home country technology and support. As he indicated, home country network capital enabled him to build up a trusted relationship with suppliers. He also admitted that the Chinese state policy of "one road one belt", encouraging local enterprises to expand overseas, has positively impacted his business partnership with Alibaba.

RESULTS AND AND IMPLICATIONS

Chrysostom and Arcand (2009) argue necessity immigrant businesses cannot be assessed in the same way as opportunity entrepreneurs by their economic performance, market share and employment expansion. In order to assess these two cases of DEI, a goal approach was applied. The perspective of the goal approach purports that all businesses seek to achieve some ultimate and recognizable goal, defining business performance as being centered on goal achievement (Etzioni, 1960). As new immigrant digital entrepreneurs, in these two cases James and Sean are both achieving their financial and non-financial goals, even though the context of the goals might be hugely different. Both immigrant digital entrepreneurs are satisfied with their achievements so far based on secondary data and conversation.

The cases presented are successful stories of DE by Chinese immigrants living in New Zealand. The research reveals the following theoretical and practical implications in the context of DEI.

Digital technologies enable DE with "less bounded entrepreneurial initiative (Nambisan, 2017)" and immigrants are able to take these advantages, particularly home country digital infrastructure and platform facilities. Although several home country IE impact factors have been investigated, such as finance (Kushnirovich & Heilbrunn, 2008), human capital (Van Tubergen, 2005), market (Gould, 1994), culture (Pruthi, Basu, & Wright, 2018) and government (You & Zhou, 2019), home country technology and infrastructure impacts on DE studies are scarce. This case studies indicates that home country entrepreneurial factors such as technology development, infrastructure advancement and business support capability do impact immigrants' entrepreneurship propensity and their business performance. Immigrant digital entrepreneurs are simultaneously influenced by both home and host country ecosystems, as a digital ecosystem bridges two country-specific business environments.

With DE, individual entrepreneurs have become less predefined and more distributed. A collection of entrepreneurs with diverse objectives and motivations engage in the entrepreneurial initiative (Nambisan, 2017) to co-operate a business cluster and co-create value for each party and aggregated values. In our two cases, necessity- and opportunity-driven immigrant entrepreneurs are working on different aspects of one platform to achieve self-defined goals. For the necessity case, Alibaba platforms opened up his home country market for him by allowing him to be an interconnected node of the home country entrepreneurial ecosystem. The traditional view of necessity immigrant entrepreneurs can only engage in the co-ethnic oriented or local neighbourhood focused business, to an extent, has become a false statement in a digital platform enabling business environment. For the opportunity case, Alibaba not only provides opportunities and resources to him to pursue his personal ambitions but also commercialise Alibaba's technologies in immigrants' host country. While the e-commerce platforms are being the entrepreneurial enablers of these two immigrants, they also create value for the platforms.

Immigrants are keen to undertake cross-border economic activities between their host and home countries (Drori et al., 2009; X. Lin & Tao, 2012) when they appear to have different resource advantages, and immigrants are able to take these benefits from both their host and home countries. Immigrants have cultural and social advantages when they trade with their home country based on theories of "bounded solidarity and enforceable trust" (Alejandro Portes & Sensenbrenner, 1993). As a Chinese immigrant trading on a China-based e-shop platform in which the majority of customers are Chinese, immigrant digital entrepreneurs have transnational competitive advantages of market knowledge, products differentiation and social-cultural proximity advantages.

China's state policy of "mass entrepreneurship and innovation" which intend to encourage individuals living in China to pursue micro-business startups have indirectly impacted Chinese immigrants living overseas. This state policy has forced the government agencies to loosen microfinance rules and allowed digital platforms to act as financial service providers with the purpose of promoting entrepreneurship. The policy also helped the improvement of national and international logistics which indirectly encourages micro-international-business startups for domestic individuals and people living overseas. These rule changes and service enhancements also served immigrant entrepreneurs well.

DE is an entrepreneurial pathway for both opportunity and necessity immigrant entrepreneurs. One of the advantages for sole traders sharing one large e-commerce platform (e.g. Alibaba) with all others is that they share identical services from customer services, accounting to business intelligence. The platform takes enormous administration, marketing, accounting, inventory management burdens off entrepreneurs' duties. Therefore, micro-business owners only need to focus on critical activities such as strategic activities and resource management; otherwise, these advanced business services would be scarcely available to MSMEs.

SOLUTIONS AND RECOMMENDATIONS

Currently, entrepreneurship research has been shifting to systematic evaluations of entrepreneurship practices and engage in research on multi-factors; scholars believe that systematically exploring the discipline in a multivariate approach can produce valuable results (D. M. Lin, Zheng, Lu, Liu, & Wright, 2019; Reis Catarina Oliveira, 2010; Baycan-Levent & Nijkamp, 2009). It has been expressed that the regional and cultural settings need to be addressed (Baycan, Sahin, & Nijkamp, 2012). Lin et al. (2019) emphasized that home country background settings such as political, socioeconomic and cultural factors impact IE research, and should not be ignored (D. M. Lin et al., 2019) for IE study. In fact, the socioeconomic, cultural and political factors used in existing IE research are primarily based on direct experiences of researchers and lack theoretical and empirical evidence of cause-effects (Nicotra et al., 2018). Scholars note that the EE framework provides a comprehensive framework to holistically examine entrepreneurship sub-disciplines regardless of geographic region, which will result in comparable research findings (GIZ, 2018). The knowledge created through EE framework will be more generalizable

DEI studies need to consider three business environments: digital ecosystem, host country and home country entrepreneurial ecosystems. The current framework of the digital entrepreneurial ecosystem less emphasises the importance of immigrants' home country. The same issues occur in IE studies, this research suggests engaging IE research within the concept of dual EEs which cover immigrants' host and home country ecosystems. With this approach, we are able to understand IE influential factors from a holistic view. Finally, this paper recommends the concept of the immigrant digital entrepreneurial ecosystem, as Figure 1 shows, for DEI investigation.

FUTURE RESEARCH DIRECTIONS

This article proposes that DEI research can follow the route of necessity (push) and opportunity (pull) motivation dichotomy. These two types of entrepreneurs should be studied separately. Although opportunity or high-growth digital entrepreneurship has been studied to a certain extent, necessity or routine digital entrepreneurship has been neglected. Digital entrepreneurship for immigrants has not been attracting much scholars' interests, even though DEI can be a pathway to improve immigrants' socioeconomic condition and community cohesion. In filling these gaps, the authors propose an entrepreneurship process-orientated conceptual framework by integrating entrepreneurial motivations, digital business development strategies, digital and entrepreneurial ecosystems and digital firm outcomes. Based on the proposed framework, we derived our directions for future DEI research.

Although empirical results and theoretical research so far demonstrates that there are expected differences between entrepreneurship in the digital world and in general (Nambisan, 2017), the expected performance and impacts are still hypotheses that require serious examination (Bogdanowicz, 2015). This study conceptualizes the role of DEI in the rapidly growing global field of digital entrepreneurship. Although both IE and DE have differences from conventional entrepreneurship aspects (Bogdanowicz, 2015; Brzozowski, Cucculelli, & Surdej, 2014, 2017), the main research frameworks are still identical.

Entrepreneurs' personal motivations and intentions, business strategies, entrepreneur-friendly ecosystems and goal-achieving performances are still essential constructs for DEI research from the entrepreneurial process perspective.

This investigation and related work on digital entrepreneurship for immigrants open a few research directions, with four areas of focus to pursue. The first area is about DEI opportunity recognition. Do immigrants exploit opportunities that are available only to them? What are the reasons that natives could not recognize these opportunities? Are these opportunities from the immigrants' host country or home country? The second area of interest is related to immigrant community development: how is DEI helping immigrants' socioeconomic development and community integration? The third important area is the impact DEI has on regional and national socioeconomic change. The last research direction highlighted centers on how immigrants can take the opportunities provided by this wave of DE: What type of opportunities can be taken by resource-scarce immigrants? For capital-rich immigrants, what are their opportunities?

CONCLUSION

Immigrant entrepreneurs are culturally and socially close to their home country entrepreneurial ecosystem while being geographically located in a host country ecosystem. Research shows that they are able to take business opportunities and resources from both ecosystems. Furthermore, the boundary-less digital ecosystem bridges home and host country ecosystems to facilitate DE by providing digital infrastructure, digital user citizenship and digital marketplace (Sussan & Acs, 2017). The result is consistent with Steininger's (2018) proposal that four major roles of ICT play in digital entrepreneurial operations: a facilitator, a mediator, an outcome and a ubiquity (Steininger, 2019). It is also coherent with classifications by Sahut et al. (2019) of digital technology's functionalities: 1) digital technologies as enablers, through which to research on the process of new venture creation or transformation through ICT; and 2) digital technologies as enablers and outputs, through which to research entrepreneurial opportunities generated by digital innovation and new venture creation.

The study thus adds value to DE literature in 1) elevating necessity DE study as a complementary subject of current literature, 2) emphasizing the differences between necessity and opportunity DE, 3) analyzing the demographic and psychological factors affecting two entrepreneurial motivations, 4) introducing the capabilities of opportunities and resource access between two types of entrepreneurship, 5) presenting the differences between necessity and opportunity motivated entrepreneurs through innovation and survival strategies, and 6) arguing the heterogeneity of performance assessment. The study adds value to IE literature in 1) ICT technology bring immigrants closer to their home country, by which immigrant entrepreneurs are able to organize entrepreneurial resources and exploit opportunities from both host and home countries, 2) Immigrants have transnational competitive advantages when applying home country infrastructure and technology in their host country or vice versa, and 3) goal approach can be deployed to assess necessity and opportunity entrepreneurial performance, in the case the economic performance approach is hard to achieve.

The results are, to a certain extent, inconsistent with the traditional view of IE regarding necessity entrepreneurs, in which necessity immigrant entrepreneurs are described without capitals (finance, human, social, culture, etc.) to engage in transnational or international entrepreneurial activities. The case studies show that these capital shortage assumptions may not be able to explain the DEI phenomenon in a boundary-less digital EE. Digital e-commerce platforms bring immigrants closer to their home country markets, finance, human, government, business support, social and cultural environments. These platforms enable necessity entrepreneurs to enact their business activities identical to other types of entrepreneurs. The e-commerce platform turned the immigrants' disadvantages in their host country into advantages in the transnational economy, particularly to Chinese immigrants since China's e-commerce platform and business models are leading the trend. The traditional assumption of only opportunity entrepreneurs engaging in the digital business may no longer be true in platform service business model.

ACKNOWLEDGMENT

This research was funded by the Australian Commonwealth Government.

REFERENCES

Afutu-Kotey, R. L., Gough, K. V., & Owusu, G. (2017). Young Entrepreneurs in the Mobile Telephony Sector in Ghana: From Necessities to Aspirations. *Journal of African Business*, *18*(4), 476–491. doi:10.1080/15228916.2017.1339252

Barney, J. B. (1991). Firm resources and sustained competetitive advantage. *Journal of Management*, *17*(1), 99–120. doi:10.1177/014920639101700108

Block, J. H., Kohn, K., Miller, D., & Ullrich, K. (2015). Necessity Entrepreneurship and Competitive Strategy, IZA Discussion Papers, No. 8219. *Small Business Economics*, *44*(1).

Bogdanowicz, M. (2015). *Digital Entrepreneurship Barriers and Drivers*. In D. K. Boojihawon, & Z. M. Ngoasong (Eds.), Emerging digital business models in developing economies: The case of Cameroon. *Strategic Change-Briefings in Entrepreneurial Finance*, *27*(2), 129–137. doi:10.1002/jsc.2188

Brzozowski, J., Cucculelli, M., & Surdej, A. (2014). Transnational ties and performance of immigrant entrepreneurs: The role of home-country conditions. *Entrepreneurship and Regional Development*, *26*(7-8), 546–573. doi:10.1080/08985626.2014.959068

Brzozowski, J., Cucculelli, M., & Surdej, A. (2017). The determinants of transnational entrepreneurship and transnational ties' dynamics among immigrant entrepreneurs in ICT sector in Italy. *International Migration (Geneva, Switzerland)*, *55*(3), 105–125. doi:10.1111/imig.12335

Cavallo, A., Ghezzi, A., Dell'Era, C., & Pellizzoni, E. (2019). Fostering digital entrepreneurship from startup to scaleup: The role of venture capital funds and angel groups. *Technological Forecasting and Social Change*, *145*, 24–35. doi:10.1016/j.techfore.2019.04.022

Chand, M., & Ghorbani, M. (2011). National culture, networks and ethnic entrepreneurship: A comparison of the Indian and Chinese immigrants in the US. *International Business Review*, *20*(6), 13. doi:10.1016/j.ibusrev.2011.02.009

Chand, M. T., & Rosalie, L. (2011). Global competitiveness, consumer choice and 'country of origin' effect: An exploratory East-West study. *Asia Pacific Journal of Management*, *17*(3), 15.

Chrysostome, E., & Arcand, S. (2009). Survival of Necessity Immigrant Entrepreneurs: An Exploratory Study. *Journal of Comparative International Management, 12*(2).

Chung, F. (2019). 'Never been so disgusted': Coles customer's viral photo of baby formula 'express lane' causes outrage. *News.com.au*. Retrieved from https://www.news.com.au/finance/business/retail/never-been-so-disgusted-coles-customers-viral-photo-of-baby-formula-express-lane-causes-outrage/news-story/11f9a42ce383329864cd09365d48e1ce

Constant, A., & Shachmurove, Y. (2006). Entrepreneurial ventures and wage differentials between Germans and immigrants. *International Journal of Manpower, 27*(3), 208–229. doi:10.1108/01437720610672149

Davidson, E., & Vaast, E. (2010). Digital Entrepreneurship and its Sociomaterial Enactment. In *43rd Hawaii International Conference on Systems Sciences Vols 1-5* (pp. 2978-+). Los Alamitos, CA: IEEE Computer Soc.

Du, W. Y., & Li, W. J. (2018). *Digital Entrepreneurship Ecosystem as a New Form of Organizing: A Case of Zhongguancun*. Singapore: World Scientific Publ Co Pte Ltd.

Du, W. Y., Pan, S. L., Zhou, N., & Ouyang, T. H. (2018). From a marketplace of electronics to a digital entrepreneurial ecosystem (DEE): The emergence of a meta-organization in Zhongguancun, China. *Information Systems Journal, 28*(6), 1158–1175. doi:10.1111/isj.12176

Economist, T. (2017). The everywhere stores: Online retail is booming in China. *The Economist, Oct. 26*.

Economist, T. (2019). Red Moon Rising: How China could dominate science. Should the world worry? *The Economist, Jan. 12*.

Evans, C. P., & Gawer, A. (2016). *The Rise of the Platform Enterprise: A Global Survey*. Retrieved from https://www.thecge.net/app/uploads/2016/01/PDF-WEB-Platform-Survey_01_12.pdf

Farani, A. Y., Karimi, S., & Motaghed, M. (2017). The role of entrepreneurial knowledge as a competence in shaping Iranian students' career intentions to start a new digital business. *European Journal of Training and Development, 41*(1), 83–100. doi:10.1108/EJTD-07-2016-0054

GIZ. (2018). *Guide for Mapping the Entrepreneurial Ecosystem: Observe – Analyse – Visualise*. Retrieved from https://www.coursehero.com/file/41633904/GuideforMappingtheEntrepreneurialEcosystempdf/

Gould, D. (1994). Immigration Links to the Home Country: Empirical Implications for US Bilateral Trade Flow. *The Review of Economics and Statistics, 76*(2), 302. doi:10.2307/2109884

Hart, D. M., & Acs, Z. J. (2011). High-Tech Immigrant Entrepreneurship in the United States. *Economic Development Quarterly, 25*(2), 116–129. doi:10.1177/0891242410394336

Hart, M., Bnner, K., Levie, J., & Heery, L. (2018). *GEM UK 2017 report: enterprising immigrants boosting prosperity in the UK*. Retrieved from https://www.gemconsortium.org/about/news/66

He, X. H. (2019). Digital Entrepreneurship Solution to Rural Poverty: Theory, Practice, and Policy Implications. *Journal of Developmental Entrepreneurship, 24*(1), 32. doi:10.1142/S1084946719500043

Heilbrunn, S., & Kushnirovich, N. (2008). The impact of policy on immigrant entrepreneurship and businesses practice in Israel. *International Journal of Public Sector Management, 21*(7), 693–703. doi:10.1108/09513550810904514

Hohenthal, J. (2006). Integrating qualitative and quantitative methods in research on international entrepreneurship. *Journal of International Entrepreneurship, 4*(4), 175–190. doi:10.100710843-007-0010-6

Kushnirovich, N., & Heilbrunn, S. (2008). Financial Funding of Immigrant Businesses. *Journal of Developmental Entrepreneurship, 13*(2), 167–184. doi:10.1142/S1084946708000910

Lan, T., & Zhu, S. J. (2014). Chinese apparel value chains in Europe: Low-end fast fashion, regionalization, and transnational entrepreneurship in Prato, Italy. *Eurasian Geography and Economics, 55*(2), 156–174. doi:10.1080/15387216.2014.948471

Li, P., Liu, S. W., Zhang, S. B., & Li, C. H. (2017). A Study on Factors Affecting Students' Entrepreneurial Intentions in HIT. In M. A. Tao (Ed.), *2017 24th Annual International Conference on Management Science & Engineering* (pp. 358-362). New York: IEEE.

Li, W. J., Du, W. Y., & Yin, J. M. (2017). Digital entrepreneurship ecosystem as a new form of organizing: The case of Zhongguancun. *Frontiers of Business Research in China, 11*(1), 21. doi:10.118611782-017-0004-8

Liu, J. Y., Nandhakumar, J., & Zachariadis, M. (2018). When guanxi meets structural holes: Exploring the guanxi networks of Chinese entrepreneurs on digital platforms. *The Journal of Strategic Information Systems, 27*(4), 311–334. doi:10.1016/j.jsis.2018.10.003

Melnikova, J., Grunwald, N., Ahrens, A., & Zascerinska, J. (2019). *Immigrants and Asylum Seekers' Digital Entrepreneurship Competence: Evaluation of the Theoretical Framework.* Paper presented at the 9th Balkan Region Conference on Engineering and Business Education and 12th International Conference on Engineering and Business Education Sibiu, Romania.

Nachira, F. (2002). *Towards a Network of Digital Business Ecosystems Fostering The Local Development European Commission.*

Nambisan, S. (2017). Digital Entrepreneurship: Toward a Digital Technology Perspective of Entrepreneurship. *Entrepreneurship Theory and Practice, 41*(6), 1029–1055. doi:10.1111/etap.12254

Ngoasong, M. Z. (2018). Digital entrepreneurship in a resource-scarce context: A focus on entrepreneurial digital competencies. *Journal of Small Business and Enterprise Development, 25*(3), 483–500. doi:10.1108/JSBED-01-2017-0014

Nicotra, M., Romano, M., Giudice, D. M., & Schillaci, E. C. (2018). The causal ralation between entrepreneurial ecosystem and productive entrepreneurship: A measurement framework. *The Journal of Technology Transfer, 43*(3), 33. doi:10.100710961-017-9628-2

Oliveira, R. C. (2010). *The determinants of immigrant entrepreneurship and employment creation in Portugal: Chapter 4 of Migrant Entrepreneurship in OECD Countries.*

Portes, A., & Sensenbrenner, J. (1993). Embeddedness and Immigration: Notes on the Social Determinants of Economic Action. *American Journal of Sociology, 98*(6), 1320.

Portes, A., & Yiu, J. (2013). Entrepreneurship, transnationalism, and development. *Migration Studies, 1*(1), 75–95. doi:10.1093/migration/mns036

Pruthi, S., Basu, A., & Wright, M. (2018). Ethnic ties, motivations, and home country entry strategy of transnational entrepreneurs. *Journal of International Entrepreneurship, 16*(2), 210–243. doi:10.100710843-017-0223-2

Qiu, R., & Gupta, V. (2015). A conceptual study of immigrant born globals and their multinationalization process. *Journal of Global Entrepreneurship Research, 5*(1), 1–18. doi:10.118640497-015-0018-7

Reynolds, D. P., Bygrave, D. W., Autio, E., Cox, W. L., & Hay, M. (2002). *Global Entrepreneurship Monitor: 2002 Executive Report.*

Richter, C., Kraus, S., Brem, A., Durst, S., & Giselbrecht, C. (2017). Digital entrepreneurship: Innovative business models for the sharing economy. *Creativity and Innovation Management, 26*(3), 300–310. doi:10.1111/caim.12227

Sahut, J.-M., Luca, I., & Frédéric, T. (2019). The age of digital entrepreneurship. *Small Business Economics*, 1–11. doi:10.100711187-019-00260-8

Song, A. K. (2019). The Digital Entrepreneurial Ecosystem-a critique and reconfiguration. *Small Business Economics, 53*(3), 569–590. doi:10.100711187-019-00232-y

Spigel, B., & Harrison, R. (2017). Toward a process theory of entrepreneurial ecosystem. *Strategic Entrepreneurship Journal, 12*, 17.

Spigel, B., & Harrison, R. (2018). Toward a process theory of entrepreneurial ecosystems. *Strategic Entrepreneurship Journal, 12*(1), 151–168. doi:10.1002ej.1268

Stam, E. (2015). Entrepreneurial Ecosystems and RegionalPolicy: A Sympathetic Critique. *European Planning Studies, 23*(9), 10. doi:10.1080/09654313.2015.1061484

Stam, E., & Spigel, B. (2016). Entrepreneurial Ecosystems.

Steininger, D. M. (2019). Linking information systems and entrepreneurship: A review and agenda for IT-associated and digital entrepreneurship research. *Information Systems Journal, 29*(2), 363–407. doi:10.1111/isj.12206

Sussan, F., & Acs, Z. J. (2017). The digital entrepreneurial ecosystem. *Small Business Economics, 49*(1), 55–73. doi:10.100711187-017-9867-5

van der Zwan, P., Thurik, R., Verheul, I., & Hessels, J. (2016). Factors influencing the entrepreneurial engagement of opportunity and necessity entrepreneurs. *Eurasian Bus Rev, 6.*

Van Tubergen, F. (2005). Self-Employment of Immigrants: A Cross-National Study Study of 17 Western Societies. *Social Forces, 84*(2), 23. doi:10.1353of.2006.0039

Wahee, S. J., & Bhardwaj, B. (2014). Entrepreneurial Motivation and Intentions: The Antecedent of Cyber Entrepreneurship. In V. Ribiere, & L. Worasinchai (Eds.), *Proceedings of the 2nd International Conference on Innovation and Entrepreneurship* (pp. 289-296). Nr Reading: Acad Conferences Ltd.

Wan, J. P., Jiang, Q. W., & Xie, L. Q. (2017). *Research on Risk Factors of Entrepreneurship in Internet Industry*. Calgary, Canada: University Calgary Press.

Wang, Q. F., & Liu, C. Y. (2015). Transnational activities of immigrant-owned firms and their performances in the USA. *Small Business Economics*, *44*(2), 345–359. doi:10.100711187-014-9595-z

Wang, Y. S., Lin, S. J., Yeh, C. H., Li, C. R., & Li, H. T. (2016). What drives students' cyber entrepreneurial intention: The moderating role of disciplinary difference. *Thinking Skills and Creativity*, *22*, 22–35. doi:10.1016/j.tsc.2016.08.003

Williams, K. (2019). Chinese formula demand won't see Kiwi babies go without. *Stuff.co.nz*. Retrieved from https://www.stuff.co.nz/business/101470144/chinese-formula-demand-wont-see-kiwi-babies-go-without

Xie, X. F., Xie, X. M., & Martinez-Climent, C. (2019). Identifying the factors determining the entrepreneurial ecosystem of internet cultural industries in emerging economies. *The International Entrepreneurship and Management Journal*, *15*(2), 503–522. doi:10.100711365-019-00562-z

You, T. L., & Zhou, M. (2019). Simultaneous Embeddedness in Immigrant Entrepreneurship: Global Forces Behind Chinese-Owned Nail Salons in New York City. *The American Behavioral Scientist*, *63*(2), 166–185. doi:10.1177/0002764218793684

Yu, X. Y., Roy, S. K., Quazi, A., Nguyen, B., & Han, Y. Q. (2017). Internet entrepreneurship and "the sharing of information" in an Internet-of-Things context The role of interactivity, stickiness, e-satisfaction and word-of-mouth in online SMEs' websites. *Internet Research*, *27*(1), 74–96. doi:10.1108/IntR-02-2015-0060

Yue, H. (2017). *National Report on E-Commerce Development in China*.

Zaheer, H., Breyer, Y., & Dumay, J. (2019). Digital entrepreneurship: An interdisciplinary structured literature review and research agenda. *Technological Forecasting and Social Change*, 148.

Chapter 10
Understanding Digital Transformation of Procurement Through E-Procurement Systems Implementation:
Business Partner Relationship Perspective

Nergiz Ilhan
Monash University, Australia

Md. Mahbubur Rahim
Monash University, Australia

ABSTRACT

E-procurement systems require a rigorous implementation process for attracting their acceptance within buying organisations and generating benefits. However, despite a rich body of literature on e-procurement systems, little is known on how buying organisations generally implement their e-procurement system. Even though an important role is played by the procurement function, limited studies have been reported to explain the details of the activities undertaken by buying organisations for implementing an e-procurement system, and in what ways those activities are influenced. To address these gaps, this chapter proposes an association between the type of business relationship a buying organisation intends to maintain with its suppliers when deciding to implement an e-procurement system, and the e-procurement implementation process it follows. The chapter further presents empirical evidence in support of one particular type of business relationship by analysing the e-procurement system implementation experience of an Australian organisation.

INTRODUCTION

In recent years, the notion of digital procurement is emerging. Digital procurement represents a concept

DOI: 10.4018/978-1-7998-2799-3.ch010

that involves automation of procurement process using e-procurement systems that incorporate smart analytical tools like big data and data mining solutions(Srai & Lorentz, 2019). Digital procurement supported by electronic procurement (E-procurement) further represents a key element of Business-to-Business (B2B) e-commerce (Gunasekaran et al., 2009). In particular, e-procurement systems reflects an application of e-commerce technologies to support a buying organisation's purchasing activities (Garrido-Samaniego et al., 2010). In recent years, however the notion of e-procurement systems is undergoing an evolution as it is emphasising on the incorporation of "smart" IT systems (Schmidt et al., 2015). According to Glas & Kleemann (2016), smart systems automatically recognise demand for a certain material and from that, independently generate a purchase order that is transmitted to the respective supplier without any necessary human interference. Hence, digital e-procurement extends the traditional e-procurement systems. These systems can provide operational as well as strategic benefits to buying organisations and their suppliers (Rahim & Kurnia, 2014; Tai et al., 2010). However, mere implementation of e-procurement systems does not guarantee those benefits (Pearcy & Giunipero, 2008). This is because realisation of benefits is largely influenced by the way e-procurement systems implementation process is undertaken by buying organisations (Ilhan & Rahim, 2017). Hence, the importance of e-procurement implementation process cannot be underestimated. Although, there currently exists a rich body of literature on e-procurements systems, a large portion of the literature is concerned with several issues like identifying those factors that affect a buying organisation's decision to implement these systems (Hardy & Williams, 2008), e-procurement implementation challenges (Hardy & Williams, 2008), e-procurement use (Muriuki et al., 2019; Soliman et al., 2005), and the benefits arising from e-procurement systems implementation (Oliech & Mwangangi, 2019; Rahim & Bantwal, 2012). In contrast, understanding the activities involved in implementing an e-procurement system in organisations has received inadequate attention from the eBusiness scholars. A comprehensive literature review on e-procurement systems conducted by Schoenherr & Tummala (2007) further supports this view. It is thus not clearly known: a) what activities buying organisations generally undertake when implementing an e-procurement system to link with suppliers, and b) how those activities are influenced (Rahim & Bantwal, 2012).

Against this backdrop, this book chapter reports a research project that seeks to address the following research question:

RQ: What influences an organisation to choose the key activities involved in digitally enabled e-procurement systems implementation process?

This research question is answered by examining the e-procurement systems implementation experience of an Australian organisation, analyse its experience in terms of seven activities involved in e-procurement systems implementation process undertaken by that organisation, and explain selection of these activities using business relationship as a theoretical lens.

BACKGROUND LITERATURE

- Digital procurement: Digital procurement represents a concept that involves e-procurement system incorporating smart analytical tools like big data analytics, the Internet of Things, cloud computing, data mining tools. According to Srai & Lorentz (2019), digital procurement practices involve "the use of advanced 'digital technologies' for procurement purposes". Digital procurement can quicken the pace of procurement transformation and essentially decrease the grinding of progress management (Platform, 2019).

- E-procurement systems: Multiple definitions of e-procurement exist in the e-commerce literature (Gunasekaran et al., 2009). However, in its simplest terms, e-procurement systems refer to the web-enabled solutions designed at automating and streamlining the key activities (e.g. ordering, sourcing, supplier evaluation, receiving) involved in an organisation's procurement process (Rahim et al., 2011). A range of technologies qualify as e-procurement systems (Tatsis et al., 2006). Examples include: Electronic Data Interchange (EDI), online auctions, procurement modules of Enterprise Resource Planning (ERP) systems, packaged e-procurement solutions, e-catalogues, and e-marketplaces. However, web-enabled e-procurement solutions have largely replaced traditional EDI systems (Smart, 2010). Some systems (e.g. package solutions) provide greater support to the procurement ordering process than others (e.g. EDI and online auctions). As the scope of procurement function is too broad encompassing such activities as negotiation, sourcing, ordering, supplier performance evaluation, and receipt among others (Croom & Brandon-Jones, 2007), a wide range of IT-enabled e-procurement systems are now available in support of those activities. The research reported in this book chapter however focuses on those e-procurement solutions that: a) automate ordering process, b) are initiated by the buying organisations, c) operate based on the IT infrastructure of buying organisations, and d) send procurement related documents to suppliers over the Internet. According to this definition, EDI, online auctions and third-party marketplaces are outside the scope of e-procurement definition (adopted for this research), because they do not generally support the order requisition and ordering activities associated with purchasing. However, packaged e-procurement solutions (e.g. SAP EBP, Starightbuy, and Oracle iProcurement) and procurement modules of Enterprise Resource Planning (ERP) systems are included within the scope of e-procurement considered relevant for the research reported in this book chapter. A high-level conceptual diagram showing the components of an e-procurement system and its interactions with suppliers' back-end IT systems is shown in Figure 1.

Figure 1. An e-procurement system: A conceptual diagram

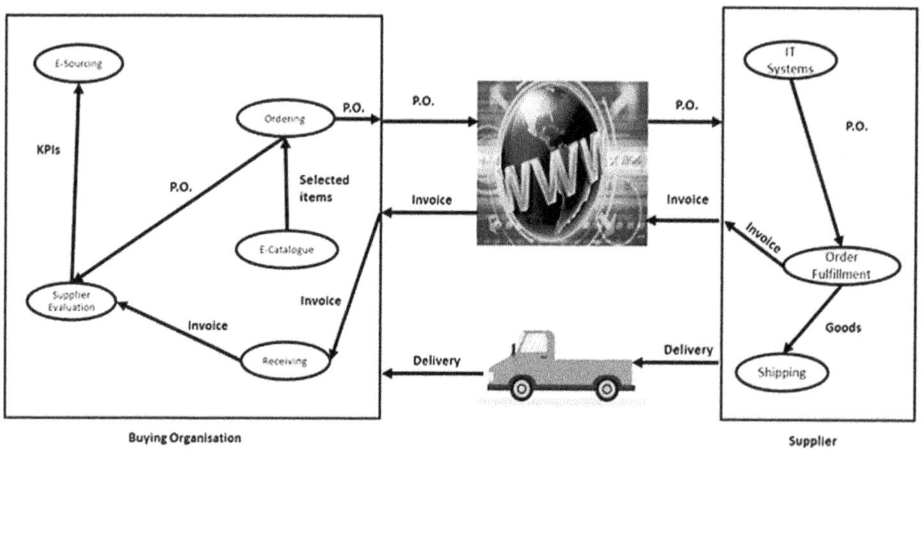

- E-procurement systems literature: There exists a rich body of literature on e-procurement. In this research, an attempt has been made to divide the existing e-procurement literature into three broad topical areas drawing on the stage model concept (Rogers, 1995): e-procurement adoption, e-procurement implementation, and e-procurement post-implementation. Drawing on literature review, a set of eight key themes encompassing these areas have been identified (Figure 2) and are briefly discussed below.

- Theme 1 (Factors affecting e-procurement adoption): Considerable literature exists on e-procurement adoption in many countries (Gunasekaran et al., 2009). Adoption studies primarily focus on identifying various factors influencing the decision of buying organisations to adopt an e-procurement system. This has been acknowledged by Schoenherr & Tummala (2007) and Pani et al. (2011) in their structured literature review on e-procurement. Examples of studies focusing on adoption factors include the works of Hardy & Williams (2008), Gunasekaran & Ngai (2008), Altayyar & Beaumont-Kerridge (2016), and Afolabi et al. (2019), among others. The adoption factors are generally classified into three categories: technological, organisational, and external. One particular issue associated with the factor-based studies is that they have not systematically examined how the factors differ based on such organisational characteristics as size and type (public, private).

In recent years, some scholars have identified the level of maturity of the buying organisation's purchasing department as a requirement (i.e. organisational contextual factor) necessary for the adoption of digitally enabled e-procurement systems (Fröhlich et al., 2018). According to them five levels of maturity can be observed: ad-hoc stage, independent stage, supporting stage, integrative stage, and extended or digital stage. Most digitally enabled e-procurement solutions connect buying organisations with ICT systems of their suppliers. As a result, digital readiness is required on both sides of the buyer–supplier dyad (Kosmol et al., 2019). However, very limited research has been reported on how buyer-supplier readiness (as a factor) can affect the decision of buying organisations to adopt such systems.

Figure 2. Key themes within three broad topical areas of e-procurement literature

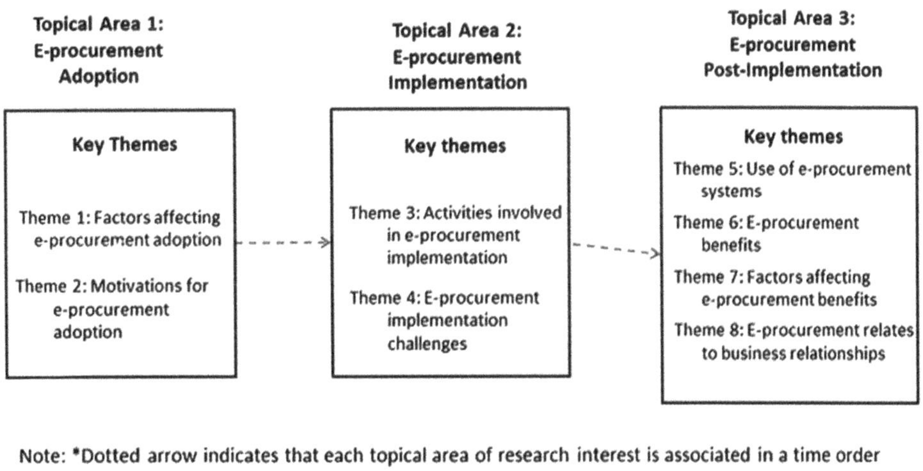

- Theme 2 (Motivations for e-procurement adoption): Adoption motivation too has received attention from several scholars who identified economic motive to be the primary driver for oganisations to adopt e-procurements systems. For example, Macmanus (2002) examined the rate of e-procurement implementation in the US public sector, remarked that motivation for implementation was based on expectations of lower purchase prices, reduced transaction costs, and increased speed. Likewise, Croom & Brandon-Jones (2007) reported for the UK public sector the significant motivation for e-procurement adoption was considered to be the economic benefits. Similar observations are also noted by Chang and Wong (2010), who in a survey noted two major types of economic motivations of Chinese companies to adopt an e-procurement system: lowering cost and conforming to a standard procurement process. However, the existence of non-economic motivation has also been reported by other scholars. For instance, Rahim & As-Saber (2011) analysed the experiences of an Australian city municipality and found that economic motives were not the only driving forces influencing e-procurement adoption in that municipality. The municipality's desire to meet governmental service targets through improved relationships with suppliers was also found to have influenced the municipality's adoption of e-procurement. The notion of motivations for e-procurement context has also been investigated in several other countries like Sri Lanka and UK (Eadie et al., 2010, 2011; Eadie et al., 2007; Wimalasena & Gunatilake, 2018). Thus, in summary, the presence of economic and socio-political (e.g. government compliance) has been identified in the e-commerce literature for e-procurement adoption context. Interestingly, no studies have reported on how e-procurement adoption factors could differ based on type of adoption motivations.
- Theme 3: (Activities involved in e-procurement implementation process): No studies exist that report a detailed discussion of various activities involved in e-procurement systems implementation process. Specific activities involved in e-procurement systems implementation are generally mentioned in a fragmented manner across several studies. However, in recent years several authors have begun describing e-procurement implementation at a higher level for the public sector. For example, Ahmed et al. (2019) explained the role of bureaucracy for e-procurement implementation in Bangladesh. In particular, they emphasised on the institutional capacity building (technical, legal) supported by strong bureaucracy for facilitating e-procurement implementation. In another recent study, Ahmad et al. (2019) reported the implementation experience of public e-procurement for Jamaica context. They have identified how such activities like training and certification of procurements staff, training of suppliers, communication and awareness of new procurement law, enactment of new procurement law, and building e-catalogues contributed to the successful e-procurement implementation initiative.

Thus, drawing on a review, a set of seven implementation activities are now identified: streamlining procurement process of buying organisations (Puschmann & Alt, 2005), internal integration of e-procurement systems with back-end IT systems (Davila et al., 2003; Neef, 2001), external integration of e-procurement system with suppliers' IT systems (Ahmed et al., 2019; Min & Galle, 2003; Rahim & As-Saber, 2011), joint planning with suppliers (Ageshin, 2001), building e-catalogs for suppliers (Ahmed et al., 2019; Ronchi et al., 2010; Trkman & Mccormack, 2009), and even integrating multiple supplier e-catalogs into a single buyer-managed view of e-catalogue (Gunasekaran et al., 2009), organising training for suppliers (Ahmed et al., 2019; Neef, 2001), and arranging training for employees of buying organisations (Ahmed et al., 2019; Naseebullah et al., 2010; Panayiotou et al., 2004).

- Theme 4 (E-procurement implementation challenges): A number of barriers have been discussed in the literature that may impede e-procurement implementation initiatives. For example, Christian Tanner (2008) report that Swiss organisations regularly experience barriers in grasping e-procurement implementation process. In other studies like Rahim et al. (2008) and Dhillon Dhillon (2008) found lack of organisational competency affect implementation initiatives for e-procurement context. E-procurement systems are also known to engender IT-enabled organisational change which could generate certain level of resistance from staff. This in turn can exert negative influence on how implementation of these systems is negotiated within organisational contexts (Hardy & Williams, 2008). Further, low satisfaction among internal customers may also significantly reduce the benefits of digitalised procurement (Croom & Johnston, 2003). In a recent study, Chang (2019) discussed implementation failure of an e-procurement system among a Taiwanese manufacturing company, its buying center in China, and four suppliers located in China. The implementation and usage of the e-procurement system required mutual investment in the system shaped by relationships. However, relationship was not strongly supported owing to absence of knowledge about the system, sharing of activities, and governance mechanism. In another recent study, Ahmed et al. (2019) identified inadequate capacity building as a major barrier for implementing public e-procurement systems among four ministries for Bangladesh context.
- Theme 5 (Use of e-procurement systems): A number of studies discussed use of e-procurement systems in organisations. For example, Rahim (2008) identified two sets of factors affecting employees' use of e-procurement systems in an Australian organisation. These include comprising perceived usefulness, perceived ease of use, employee involvement, reliability, customised training, vendor support and management support are identified. In their study, Davila et al. (2003) further discussed use of e-procurement systems in organisations and differentiate between conservative and aggressive users. In another study, Soliman et al. (2005) discuss various aspects involved in e-procurement systems use by buyer organisations for a supply chain context. The use of e-procurement systems by employees within organisations has begun to receive attention again in recent years. For example, Brandon-Jones & Kauppi (2018) examined the acceptance of e-procurement systems by individual employees at a university in the Netherlands using Technology Adoption Model (TAM) lens. In another study, Muriuki et al. (2019) conducted a descriptive survey among 211 procurement staff among energy sector state corporations in Kenya, and highlighted the lack of technical knowledge possessed by them as a factor for their e-procurement use.
- Theme 6 (E-procurement benefits): Many benefits from successful implementation of e-procurement systems have been reported for both commercial and public sectors (Bartezzaghi & Ronchi, 2003). For example, Moon (2005) discussed such benefits as reduced transaction costs, faster ordering, wider range of vendor choices, greater control on procurement spending, and less paperwork, among others. Recent empirical studies of Pule & Bakkabulindi (2017) have illustrated significant increase in benefits for organisations using e-procurement. Drawing on Granger causality econometrics method, Shkolnyk Shkolnyk et al. (2019) confirmed that use of e-procurement systems in Ukraine public sector has created savings in public procurement. One important strategic benefit that results from successful implementation of e-procurement systems is buyer satisfaction as reported by Gupta & Tripathi (2018). In another recent study, Ahmad et al. (2019) identified benefits for both government organisations and suppliers for the public e-procurement in Jamaica context. Government organisations reported at least a 2% reduction in transaction costs. On the other hand, suppliers reported greater convenience and time savings as a result of

using e-procurement systems. In their systematic literature review on sustainable public procurement, Adjei-Bamfo et al. (2019) reported e-procurement benefits for three layers of public sector (e.g. national, central government, and local government), and reported government's use of e-procurement creates such benefits as reduced corruption, reduced cost of operation, increased efficiency and enhanced accountability. In their study, Oliech & Mwangangi (2019) discussed e-procurement under the umbrella of strategic procurement initiative for the context of five large hospitals in Kenya. Drawing on a survey among procurement users, they reported that use of an e-procurement contributed in improving in financial performance of hospitals.

Scholars have classified these benefits in two categories: operational and strategic. Operational benefits are internal focused and can help organisations to gain immediate short-term benefits. Typical examples include reduced transaction costs, shorter order cycle, reduced inventory, and less maverick buying practices. Reduction in transaction costs occur due to avoidance of paper work (Gunasekaran et al., 2009). In contrast, strategic benefits are often intangible in nature and take considerable time to achieve. They include such benefits as greater choice of suppliers and increased understanding of procurement needs.

- Theme 7 (Factors affecting e-procurement benefits): In addition to describing benefits, several attempts have been made to determine the factors that influence realisation of benefits from e-procurement systems implementation. Typical studies focusing on this theme include those of Rahim & As-Saber (2011) and Makosa & Mwangangi (2018). In their study, Rahim et al. (2011) identified 4 factors (e.g. senior management support, training), which managers should consider for wanting to experience benefits from the implementation of e-procurement systems. In another study, As-Saber et al. (2014) reported integration of social media with e-procurement could help organisations to broadcast about tenders more efficiently and quickly reaching a wider audience, facilitate stakeholder feedback on procurement related issues, and, at the same time, contribute to establishing transparency in the process. This in turn could improve efficiency and effectiveness of the e-procurement process. In a more recent study, Makosa & Mwangangi (2018) reported training, capacity building, and adequacy of ICT framework were found to influence benefits of e-procurement system.
- Theme 8 (E-procurement related business partner relationships): Several studies report how the implementation of these systems may influence buyer-supplier relationships. So far mixed findings are reported on how implementation of e-procurement systems can influence buyer-supplier relationships. For example, scholars like Carr & Smeltzer (2002) report an improvement in business relationship between buying organisations and their suppliers as a result of e-procurement systems implementation. In contrast, Leek et al. (2003) raise reservations about improving business relationships with suppliers through e-procurement systems. However, most studies generally view buyer-supplier relationship as a dependent variable and discusses how the introduction of e-procurement systems can have an effect on changing the nature of such relationships. The association between e-procurement systems and business partner relationship has also been reflected in several recent studies. For example, Makosa & Mwangangi (2018) reported an improvement in supplier relationship for the use of a public e-procurement system for the context of South Sudan. In another study, Oliech & Mwangangi (2019) reported an improvement in relationship between five hospitals in Kenya with its suppliers due to the use of an e-procurement system. For the context of Jamaica's public e-procurement system, Ahmad et al. (2019) confirmed an efficient

relationship with suppliers as a result of introducing e-procurements system because the system allowed suppliers to submit their proposals for evaluations at their convenience.

- Gaps in the background literature: In summary, a thematic analysis of three topical areas encompassing the existing e-procurement literature indicates that despite receiving considerable attention from e-commerce and supply chain scholars, there still exists several issues (Figure 3) that deserve further research attention. However, as this research is concerned with e-procurement implementation, it is noted that inadequate research has so far been reported in the existing literature on two specific issues associated with e-procurement implementation. First, in-depth case studies detailing how organisations undertake e-procurement implementation process involving multiple activities are not regularly reported. Hence, e-procurement literature focusing on implementation process is scarce and requires more attention. Second, it is unclear in what ways factors would influence how buying organisations undertake e-procurement implementation process. The research reported in this book chapter thus seeks to address these concerns using partner relationship as a suitable theoretical lens.

Figure 3. Research gaps in e-procurement literature: A summary

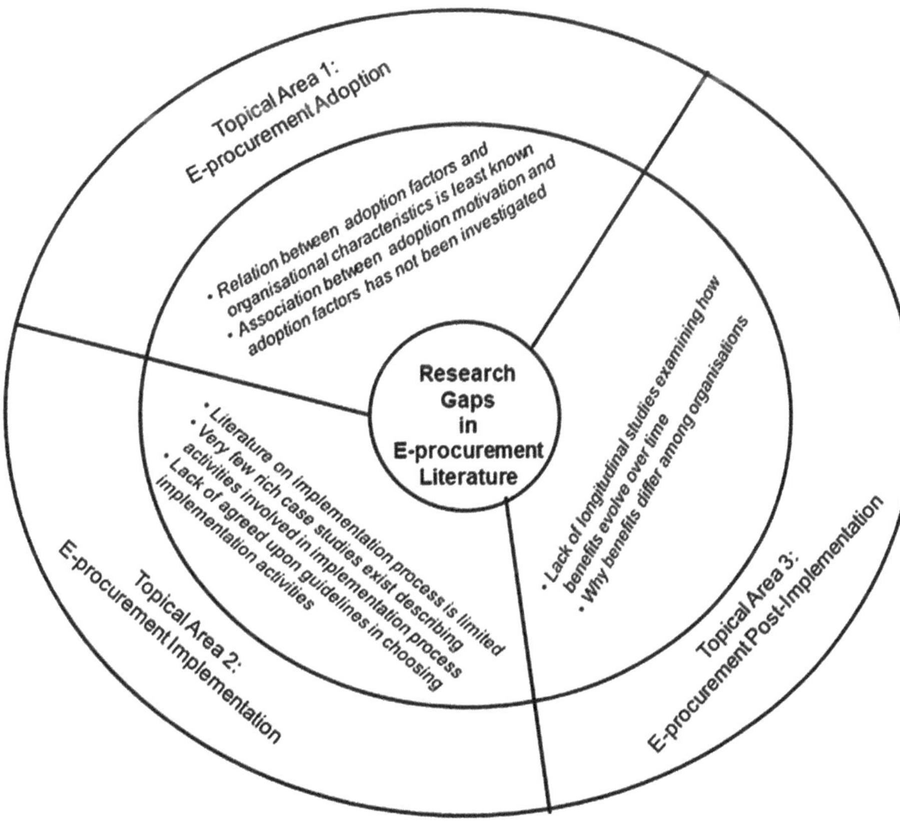

RESEARCH PROPOSITIONS

In this book chapter, it is proposed that the type of business relationship a buying organisation intends to maintain with its suppliers bears a potential to serve as a useful theoretical lens to determine how a digitally enabled e-procurement system would be implemented by that organisation. This is because e-procurement is an inter-organisational in nature. Hence, implementation of an e-procurement system cannot be done without the participation of suppliers. Hence, business relationship involving two-way interactions is a necessity for successful e-procurement implementation. Drawing on e-commerce literature, two types of business relationships are identified that buying organisations may seek to establish and maintain when dealing with e-procurement systems: arms-length relationship and collaborative partnership. It is thus proposed that, when the influence of contextual factors is controlled, the activities involved in a given e-procurement systems implementation process are likely to differ among organisations depending on the type of business relationships (i.e. arms-length, and collaborative partnership) that those organisations intend to establish with their suppliers.

It is argued that business relationships can be described in terms of two dimensions: 'Type of resource sharing' and 'Type of e-procurement investment'. Type of resource sharing is characterised by sharing of information, knowledge and tangible resources (e.g. employee, capital, hardware, software) that can possibly occur in a relationship between an organisation and its suppliers when implementing an e-procurement system. Sharing of resources has received considerable discussion within the supply chain management literature. For example, exchange of relevant information between buyers and suppliers is emphasised by such scholars as Morgan & Hunt (1994)) and Sin et al. (2005). Likewise, exchange of knowledge and tangible resources among supply chain partners in support of implementing B2B e-business systems has been highlighted by scholars like Dale (2003), Johnson & Whang (2002). On the other hand, the term 'Type of e-procurement investment' represents an organisation's financial commitment or willingness for capital injection in support of implementing an e-procurement initiative for sustaining its intended type of business relationship. Investments can be made in the e-procurement software itself, development of internal and external integration, sponsoring design of internal and external training programs, creation of e-catalogs, and supplier education. In principle, the notion of 'Type of e-procurement investment' is quite consistent with 'investments made by buying companies' that according to Bensaou (1999) is considered to be a dimension describing buyer-supplier relationship. 'Type of e-procurement investment' can be described in terms of four increasing categories: basic, moderate, considerable, and intensive. Basic, moderate and considerable categories are all concerned with e-procurement investments made within the buying organisations (hence, they have internal investment focus), however they differ in terms of breadth of items covered for capital injection. Conversely, intensive category represents e-procurement investment made in both internally and externally focused items (e.g. supplier education, external integration). An investment is considered: a) Basic - when an organisation invests in e-procurement software systems and training needed to implement and use that system, b) Moderate - when an organisation invests in e-procurement software systems, internal integration, and training needed to implement and use that system, c) Considerable: when an organisation invests in e-procurement software systems, internal integration, e-catalogs, and training needed to implement and use that system, d) Intensive: when an organisation invests in e-procurement software systems, internal integration, e-catalogs, and training needed to implement and use that system, supplier education, and external integration with suppliers.

Table 1. A set of research propositions

Types of Business Relationships	Propositions Linking E-Procurement Enabled Business Relationship Types With Implementation Activities
Type A Arms-length Relationship	P1: An organisation representing 'efficiency seeking relationship' would streamline procurement process P2: An organisation representing 'efficiency seeking relationship' would perform internal e-procurement integration P3: An organisation representing 'efficiency seeking relationship' would not perform external e-procurement integration P4: An organisation representing 'efficiency seeking relationship' would not engage in joint planning activities P5: An organisation representing 'efficiency seeking relationship' would not build e-catalogs P6: An organisation representing 'efficiency seeking relationship' would not organise supplier training P7: An organisation representing 'efficiency seeking relationship' would arrange training for its own employees
Type B Collaborative Partnership Relationship	P1: An organisation representing 'partnership relationship' would streamline procurement process P2: An organisation representing 'partnership relationship' would perform internal e-procurement integration P3: An organisation representing 'partnership relationship' would perform external e-procurement integration P4: An organisation representing 'partnership relationship' would engage in joint planning activities P5: An organisation representing 'partnership relationship' would build e-catalogs P6: An organisation representing 'partnership relationship' would organise supplier training P7: An organisation representing 'partnership relationship' would arrange training for its own employees

An effort is now made to derive a set of seven research propositions (Table 1) that predict those e-procurement activities that are likely to be performed by organisations representing those two relationship types.

- Arms-length Relationship (Type A): It is characterised by 'information only' sharing between a buying organisation and its suppliers and by 'moderate' category of e-procurement investment made by the organisation to support its e-procurement implementation initiative. This scenario may occur when an organisation intends to use an e-procurement system as a tool to enhance its internal procurement related efficiency. E-procurement implementation under this scenario is not directed at improving bonding with suppliers but is aimed at enhancing the effectiveness of the procurement function of an organisation. One way to address this is to reduce transaction costs by removing fragmented and duplicate non-value adding activities. As such, implementation efforts tend to be directed towards simplifying procurement process (P1). However, such improvement generally involves modest level of changes in procurement related activities and does not trigger a radical transformation of all key activities involved in procurement process. Hence, procurement process does not undergo a reengineering effort. Efficiency can also be increased by improving data quality by avoiding multiple data entry. In particular supplier and invoice related data are often stored in backend financial application system. Without such information, the users of e-procurement systems would need to re-enter those data which would result in increased processing time and introduction of potential purchasing errors. Hence, establishing an internal integration with e-procurement systems would be performed (P2). The pure focus on internal efficiency means that the e-procurement system is unlikely to represent a strategic investment, and any changes to procurement processes are directed towards automated transaction processing rather than business process transformation. As such, these organisations would not share procurement related knowledge and other tangible resources with their suppliers, and are unlikely to engage in joint planning activities (P4). As enhancing bonding with suppliers is not the primary reason to invest in e-procurement systems, these organisations would not develop e-catalogues in consultation with their suppliers (P5). They care less for organising e-procurement training for suppliers (P6) and would not seek changes in suppliers' fulfillment systems; thus no external integration is required (P3). However, these organisations recognise the importance of arranging training for their own employees to facilitate e-procurement integration, implementation and acceptance within their organisations (P7).

- Collaborative Partnership Relationship (Type B): It is characterised by all three types of resources (i.e. information, knowledge, and tangible resources) sharing between a buying organisation and its suppliers and by 'intensive' category of e-procurement investment made by the organisation to support its e-procurement implementation initiative. This scenario can be observed when an organisation contributes a great deal of assets into implementing an e-procurement system and believes in long term strategic partnership development with its suppliers by agreeing to share relevant procurement related information (e.g. budget, scheduling and forecasting), knowledge (e.g. procurement, metrics, benchmarks), and even tangible resources (e.g. interface source code) with them. The buying organisation thus undergoes a change from the traditional operating procurement function to strategic one. This line of thinking is supported by Gupta & Narain (2012). Thus, it expects to attain sustained operational and strategic benefits as a result of maintaining the long-term partnership relationship which it believes could be facilitated through successful e-procurement systems implementation. As a result, these organisations are more likely to reengineer their internal procurement processes (P1) and introduce radical changes. This line of argument is largely consistent with the views of Pillutla et al. (2014) who advocate for strong support for reengineering procurement process to expect strategic benefits. Such a relationship encourages buyer organisations to develop an integrated e-procurement system by linking them with backend applications (P2).

As e-procurement systems tend to be strategic in nature, greater focus is likely to be made on improving the depth and quality of information available to buying organisations' managers by ensuring a long-term bonding with suppliers. This means that implementation approach selected by these organisations would involve sharing appropriate information, knowledge and even resources with suppliers through procurement related joint planning (P4) to ensure smooth purchasing and reduce uncertainty involved in purchasing items from suppliers. This is in line with the views of Cai et al. (2006) who argue for the need to establish a strong business partnership relationship between buyers and suppliers for sharing accurate and customised information to meet strategic and operational needs of buyers with suppliers. In order to gain trust from the suppliers, and convince them of the merits of joining electronic relationships with the buying organisations, these organisations would cooperate with suppliers for building e-catalogues (P5) and will establish external interface by integrating the e-propcurement systems with suppliers' back-end IT systems (P3). As the attainment of the buying organisations' strategic objectives is influenced by suppliers' proactive participation in the electronic relationship, these organisations are expected to aggressively promote the notion of e-procurement systems to their suppliers, and hence would not hesitate to offer training opportunities to their suppliers on how to join the electronic relationships with organisations via the e-procurement systems (P6). These organisations also recognise training to be an important facilitator to implement and promote the acceptance of e-procurement systems by their own employees (P7)

RESEARCH DESIGN

- Research approach: An exploratory case study research approach was chosen because of the intentions to discover the existence of a possible relationship between two sets of theoretical constructs (i.e. e-procurement enabled business relationship types and e-procurement systems implementation process). This choice is in line with the recommendations provided by Yin (2003).
- Selection of case organisation: The experience of a single case organisation from the Australian municipal sector is reported to explore the applicability of one particular type of business relationship. The case organisation, which is a large city council, was chosen because of its rich experience (over 5 years) in using an e-procurement system. In addition, being a large government organisation with considerable purchasing capability, it is wanted to know how the council views its e-procurement system in connection to developing relationship with its suppliers. It is acknowledged that use of a single case organistaion restricts generalisability of findings. Despite this, as this research is still in the formative stage, a deliberate choice has been made to restrict research attention to a single case organisation to explore the existence of core constructs for the context of one particular type of business relationship and its associated the research propositions.
- Unit of analysis: The unit of analysis is an individual organisation because each organisation is likely to have one single e-procurement system implemented in support of its procurement process.
- Data collection and analysis: Data were gathered from multiple sources, including semi-structured interviews, e-mail communications, e-procurement related documents maintained by the council and information posted on the council website. Four senior managers who had an involvement with the e-procurement systems implementation initiative at the council were interviewed: Purchasing manager, IT manager, Senior business manager, and Account manager. An interview protocol was developed that included questions on the background of interviewees, business relationships, information, knowledge and resource sharing between council and its suppliers, e-procurement investment, and e-procurement systems implementation activities. A set of principles and indicators were then devised to operationalise each attribute of 'Type of resource sharing' and 'Type of e-procurement investment'. For example, four principles were used to identify 'Type of resource sharing': diversity of information is shared, direction of information exchange, what knowledge is shared?, and what resources are shared?. Likewise, 'Type of e-procurement investment' was evaluated using such principles as: investment made in e-procurement software, investment made in e-procurement hardware, investment made in e-procurement training, investment made in internal integration, investment made in e-catalogs, investment made in external integration, and investment made in supplier education. A Drawing on suggestion of Neuman (2003), a coding scheme was prepared to extract themes from the interview transcripts and data sources (e.g. emails, interview transcripts), to match those indicators. In order to evaluate the research propositions (Table

1), data collected from the council were analysed using the pattern matching logic described by Yin (2003).
- Methodological rigour: Established methodological guidelines provided by Yin (2003) and Sarker & Lee (2000) were applied to generate reliable and valid findings. Details could not be provided due to page limitations.

CASE ORGANISATION

A large city council, with over 500 staff and A$270 million budget in 2011 from Australia participated in this research project. The council currently uses Oracle ERP system that operates within Unix environment. For the past five years, the council has been using Oracle Internet Procurement (IP5) solution that is internally integrated with other modules of the Oracle ERP system. Currently, about 300 users have access to IP5, they can use it for sending purchase orders to suppliers in either e-mail or internet fax format generated by the IP5 solution. A total of 5500 suppliers are registered in the system. However, these suppliers are not directly connected with the IP5 solution. In other words, the IP5 solution facilitates one-way communication between the council and the suppliers. The council generates 17,000 purchase orders in electronic format annually.

CASE STUDY FINDINGS

Classifying business relationship: Guided by a set of principles and indicators, and using pattern-matching technique (Yin, 2003), empirical evidence has been identified in support of establishing 'Type of resource sharing' and 'Type of e-procurement investment' relating to the IP5 e-procurement system implementation initiative at the participating council. Drawing on the evidence collected from the council, it is found that 'Type of resource sharing' to be 'information only', and 'Type of e-procurement investment' to be 'Moderate' in nature. Therefore, the e-procurement implementation initiative undertaken by the council reflects 'Arms-length (Type A)' business relationship. The council fundamentally wanted to maintain an arms-length relationship with its suppliers despite introducing an e-procurement system. Investment in e-procurement systems was internally focused. Another observation is that the council did not recognise innovative web enabled e-procurement systems as a mechanism to revolutionise its existing relationships with suppliers. Hence, the sheer absence of interest of the council management to establish external integration is observed. This sentiment is echoed by the IT manager as follows:

"I don't really think e-procurement system has any role in facilitating good long-term relationships with our suppliers. There are no dialogs happening in between our Oracle system and suppliers' IT systems. Oracle e-procurement system is just a mechanism to send information from us; it is a one-way traffic."

The council is found to share various types of procurement related information with its suppliers through the Oracle IP5 e-procurement system. No tacit or explicit knowledge is shared between the council and its suppliers with regard to procurement policy. For example, council's purchasing performance metrics and forecasts are neither openly discussed with the suppliers nor are transmitted electronically to the suppliers. Likewise, the council has no intention to share any tangible resources with its suppliers.

Therefore, 'Type of resource sharing' is restricted to information only as the suppliers were considered outsiders. On the matter of 'Type of e-procurement investment' the council acknowledged that considerable amount of funds (about A$2 million) it had poured into implementing an Oracle based e-procurement system. The deal the council made with Oracle includes costs of e-procurement software module, training of key council staff (we call them power users) by Oracle consultants, and middleware linking the e-procurement module with back-end Oracle ERP system to ensure seamless data transfer. In addition, the council had spent considerable staff time for evaluating the accuracy of internal back-end integration and in performance acceptance testing. The council further invested in staff training as its internal staff members were trained in the use of the e-procurement system. A customised training program was designed by key council staff (power users) with considerable inputs from Oracle consultants and these key staff were in turn responsible for training the remaining council staff. Given this evidence, Type of e-procurement investment' made by the council could be considered as 'Moderate' in nature. Classifying the e-procurement initiative at the council as 'Arms-length (Type A)' relationship is in agreement with the observations of Croom & Brandon-Jones (2007). In a case study of e-procurement systems adoption in a UK based public sector organisation, they found that although e-procurement solution was recognised to have a potential for enhancing business relationship with suppliers but the solution was not considered to be a pre-requisite for building close relationships with suppliers and was intended to improve internal efficiency and productivity only. Without further empirical investigation, it is however difficult to ascertain whether this observation applies to all councils in Australia.

- Identifying e-procurement systems implementation activities: Drawing on case study data, indicators were identified for each of the seven implementation activities undertaken by the council to implement Oracle IP5 e-procurement system. An activity was considered to have been performed when evidence for at least one of its indicators was found in the case study data.
- Evaluating research propositions: The research propositions for the 'Arms-length (Type A)' relationship were evaluated by comparing the activities predicted by the research propositions proposed in Table 1 with the activities identified from the case study. This is highlighted in Table 2. While making comparisons, empirical evidence from the council was reviewed to better understand the underlying reasons expressed by the council for performing or not performing each of the seven e-procurement systems implementation activities. These were then compared with the logical arguments used in deriving the research propositions for the 'Arms-length Efficiency (Type A)' relationship. This in turn helped in deciding the outcomes of research propositions (i.e. supported, unsupported) for that type of business relationship. A strong agreement is found between the predicted activities and the observed activities undertaken by the council for reasons that are largely in line with the arguments used in deriving the propositions for the Type A relationship. The implementation activities undertaken by the council are now discussed with some details.

On the matter of streamlining procurement process (P1), the council was found to have decentralised its procurement process, redesigned parts of its procurement procedures, and allocated responsibilities to its employees (rather than to their supervisors) who placed purchase orders through the IP5 system. Three reasons were cited by the council for its decision to streamline procurement process: a) to minimise inefficiency inherent in the existing procurement system, b) to improve data quality, and c) to protect procurement system from abuses by employees. This finding is thus quite consistent with the underlying

Table 2. A comparison of predicted and observed e-procurement implementation process

Type of Business Relationship and Case Organisation	Activities Involved in E-Procurement Systems Implementation Process							
	Predicted/ Observed	**A1**	**A2**	**A3**	**A4**	**A5**	**A6**	**A7**
Type B: Efficiency Seeking Relationship	Predicted	Y	Y	N	N	N	N	Y
Case Council	Observed	√Y	√(Y)	√(N)	√(N)	√(N)	√(N)	√(Y)

Note: tick (√) means agreement with the predicted activity, x means disagreement with the predicted activity, Y indicates an activity was performed by the council, N indicates an activity was not performed by the council

arguments used in deriving proposition P1 for the 'Arms-length (Type A)' business relationship. It is therefore argued that proposition P1 is supported for the context of arms-length business relationship.

With regard to integrating e-procurement systems with internal backend applications (P2), it is observed that integration was conceived during the e-procurement adoption decision making stage when the council planned to have seamless data transfer between the Oracle based ERP system and the IP5 e-procurement system. Actual integration was performed by the consultants of the Oracle company but the council actively took part in the evaluation of the integration process. Integration was carefully planned to improve procurement efficiency by saving time spent on data re-entry and to enhance data quality. These reasons are thus in line with the arguments used in deriving proposition P2 for the Type A business relationship. Hence, proposition P2 is supported for the 'Arms-length (Type A)' business relationship.

Concerning external integration (P3), it is noted that the decision of the council for not integrating the IP5 system with the suppliers' IT systems is in agreement with the prediction for 'Arms-length (Type A)' business relationship. Managers at the council acknowledged that without senior management commitment and collaboration with suppliers, external integration cannot be pursued. This in turn requires the council and the suppliers to view e-procurement with strategic significance. However, the council had deliberately chosen a system that required little involvement on the part of the suppliers as they were considered outsiders. Hence, proposition P3 is supported for the 'Arms-Length (Type A)' business relationship.

Relating to joint planning activities (P4), little enthusiasm has been demonstrated by the council. It does not seek interaction with the suppliers by inviting them to participate in joint planning program. This is because: a) of the lack of strategic importance given to the e-procurement initiative, b) the council likes to maintain arms-length relationship and wants to share limited information with its suppliers, and c) the council also takes responsibility of performing planning activities by itself and is quite reluctant to plan its procurement activities closely with suppliers due to its concerns for privacy and trust issues. Even though the council collaborates with other government agencies for preparing its budget and plan for the delivery of its key services within a financial year, it does not consider jointly working with suppliers for better planning its procurement forecasts and identifying a means to satisfy procurement needs. In summary, the experience of the council for not performing joint planning with suppliers is found to be consistent with the arguments used in deriving proposition P4 for the Type A relationship. Hence, proposition P4 is supported for Type A relationship.

The council did not invest resources in developing e-catalogues (P5) in consultation with its suppliers. Council management does not consider suppliers to be strategic partners for their procurement function and is not willing to make any investment in building e-catalogues. Management rather expects the suppliers to demonstrate their reciprocal commitment to electronic relationships by developing their own e-catalogues. This sentiment is echoed by the Account Manager as follows:

"Currently we are just using manual catalogs. No e-catalogs are used at our side. I think – it is the suppliers who should develop their own e-catalogs and place our negotiated price in them for our access. That will reduce invoice mismatch problems, improve data quality, and reduce my work." [Account Manager]

These sorts of reasons are broadly in line with the arguments used in developing proposition P5 for the 'Arms-length (Type A)' business relationship. Hence, proposition P5 is supported for Type A business relationship.

The council was quite unwilling to organise necessary training on e-procurement systems for its suppliers (P6). Supplier training was not considered to be a component of the council's overall e-procurement implementation strategy due to two reasons: a) the e-procurement system was chosen in such a way that it would not invite major changes in the order fulfillment practices of the suppliers, and b) the council management likes to see pro-activeness of its suppliers to trade electronically with the council. These observations too are quite consistent with the arguments used in developing proposition P6 for the 'Arms-length (Type A)' business relationship. Hence, proposition P6 is supported for Arms-length business relationship.

Finally, the council took considerable efforts for organising council staff training in using the IP5 system (P7). The council made investment in four ways: a) the overall deal with Oracle covered expenses for the Oracle consultants to provide training to the power users of the council, b) the power users then, in consultation with the Oracle consultants, designed a training program that is specifically customised for the council context, c) power users were supported by the council to train those council staff who would be interacting with the e-procurement system, and d) council staff were given work off from their daily routine activities to attend mandatory training provided by the power users. This finding is aligned with the arguments used in the development of proposition P7 for the Type A relationship. Hence, proposition P7 is supported for Type A business relationship. This finding is consistent with the findings reported in the existing e-procurement literature.

In summary, the activities undertaken by the council to implement the IP5 e-procurement system are not only in agreement with the predictions expressed in the research propositions (Table 1) for the 'Arms-length (Type A)' business relationship, but the reasons expressed by the council for undertaking each of those activities too are found to be largely consistent with the arguments used in deriving those research propositions. As a result, it is concluded that that the evidence provides good support for the propositions (P1 to P7) with regard to the 'Arms-length (Type A)' business relationship.

DISCUSSION

A reflection is now made on the findings reported above by discussing them in terms of what is reported in the literature.

The notion of streamlining procurement process (P1) undertaken by the participating council is consistent with what has also been reported by several authors for e-procurement context. For example, authors like Sambasivan et al. (2010) and Ahmad et al. (2019) reported an introduction of changes made in the procurement process when digitising procurement function using e-procurement systems.

The notion of internal integration (P2) has also been recognised by other e-procurement scholars. For example, for the context of public e-procurement system implementation in Jamaica Ahmad et al. (2019) reported that internal ICT plans were developed and communicated among government agencies in order to facilitate an electronic interface between the e-procurement system and their in-house IT systems. The idea of establishing internal integration is also frequently mentioned in the IS literature when organisations seek to improve efficiency by reducing double data entry. For example, in their study Wang & Zhang (2005) mentioned about establishing integration between EDI and eSCM systems when organisations prefer to avoid data entry errors by linking their supply chain systems with supplier facing EDI systems. In another study, Themistocleous et al. (2001) reported that business organisations did establish an electronic integration between their EDI systems with other applications supporting several business critical business functions in order to minimise data entry errors.

The notion of not spending efforts for establishing an external integration (P3) is found to be in line with Corini (2000) and Min & Galle (2003), as they argued that without cooperation between buyers and sellers an external integration is unlikely to take place for e-procurement context.

Likewise, on the matter of council's unwillingness to engage in joint planning activities with suppliers (P4) has also been recognised in the literature. In particular, the concerns expressed by the council about trust and privacy of suppliers are also recognised in the broader e-commerce literature. For example, Neef (2001) acknowledges that implementation of e-procurement systems requires a new level of trust due to increased information exchange between buying organisations and their suppliers. According to Day & Hubbard (2003), the impact of internet technologies on supplier relationships raises concerns associated with issues of trust and reliability in an online setting.

The participating city council did not organise training (P6) on e-procurement system for the suppliers. The reasons cited by the council make sense for the type of relationship they like to maintain with suppliers. However, this finding is the opposite of what Ahmad et al. (2019) has reported for the implementation of e-procurement system for the Jamaican public sector context. This is because the Jamaican government emphasised on building an efficient relationship with the suppliers so that suppliers can perceive fairness in public procurement process. Hence, suppliers' training was organised to place greater emphasis on promoting ICT-enabled usage and interactions between government to business (G2B) context. Likewise, according to Panda et al. (2010), e-government initiatives in India also involved providing training to multiple stakeholders including suppliers in the use of public e-procurement. This is because Indian government agencies sought to establish their caring image to external stakeholders.

The importance of training e-procurement users within the participating city council (P7) is fully consistent with the widely held view reported in the existing literature. For example, in their work on e-procurement systems implementation in the Jamaica's public sector, Ahmad et al. (2019) reported on organising considerable training of government employees on how to use the system as well as the changes in procurement process, procurement guidelines, and procurement legal obligations triggered by the system. The importance of staff members training on e-procurement systems has also been found to be important for the capacity development of properly using within the legal frameworks for the context of South Sudan (Makosa & Mwangangi, 2018).

SOLUTIONS AND RECOMMENDATIONS

Drawing on the findings and a reflection on them (as presented in earlier sections), it can be argued that business partner relationship does bear a potential to serve as a suitable theoretical lens to help explain the selection of undertaking certain activities to facilitate digitally enabled e-procurement systems implementation in organisations. However, to help improve this claim a strong recommendation is provided to replicate this study across other sectors (e.g. manufacturing, tourism, service).

FUTURE RESEARCH DIRECTIONS

Drawing on the limitations of the research work implicitly mentioned in this book chapter (e.g. limited scope in terms of choosing a single type of business partner relationship), the following three directions for further research are recommended.

First, the research finding cannot be generalised because the research propositions associated with Type B (Collaborative Partnership relationship) have not yet been empirically evaluated. Therefore, additional case studies are to be undertaken to improve the validity of research finding.

Second, the moderating influence of contextual factors (e.g. management norms, organisation culture) on digitally enabled e-procurement systems implementation activities has not been examined. This particular aspect needs further research attention.

Third, it is possible to identify other type of business partner relationships that may involve additional activities in the implementation process of digitally enabled e-procurement systems. Future research is needed to further explore this aspect.

CONCLUSION

Even though e-procurement systems have been widely discussed in multiple literature streams (e.g. IS/IT, supply chain, e-commerce), relatively less is known on the involvement of key activities in e-procurement systems implementation process initiated by buying organisations. In this book chapter, this particular concern has been addressed by arguing the existence of a possible association (subject to controlling the influence of other contextual factors) between the type of business relationship between a buying organisation and its suppliers, and the e-procurement implementation process undertaken by that buying organisation. This association is evaluated by deriving a set of seven propositions that relate 'Type of business relationship' with seven activities involved in e-procurement systems implementation process. The research assertion on a possible association between 'Type of business relationship' and 'Activities involved in e-procurement implementation process' is then empirically evaluated by analysing empirical evidence collected from a large Australian organisation's experience with implementing an e-procurement system. The empirical findings of this case study are indicative to support the research assertion. This supportive empirical evidence is however limited to the 'Arms-length (Type A)' business relationship only.

A modest contribution of this research is the emergence of an empirical support (although partial) for the proposed research assertion that type of e-procurement enabled business relationship would play an important role for explaining e-procurement systems implementation activities. This finding helps alert senior management to focus their attention to those implementation activities, which they would need to initiate when they consciously decide to establish a particular type of business relationship with their suppliers.

REFERENCES

Adjei-Bamfo, P., Maloreh-Nyamekye, T., & Ahenkan, A. (2019). The role of e-government in sustainable public procurement in developing countries: A systematic literature review. *Resources, Conservation and Recycling*, *142*, 189–203. doi:10.1016/j.resconrec.2018.12.001

Afolabi, A., Ibem, E., Aduwo, E., Tunji-Olayeni, P., Oluwunmi, O., & Ayo-Vaughan, E. (2019). Gauging parameters for e-procurement acquisition in construction businesses in Nigeria. International Journal of Construction Management, 1-10.

Ageshin, E. A. (2001). E-procurement at work: A case study. *Production and Inventory Management Journal*, *42*(1), 48–53.

Ahmad, T., Aljafari, R., & Venkatesh, V. (2019). The Government of Jamaica's electronic procurement system: Experiences and lessons learned. *Internet Research*, *29*(6), 1571–1588. doi:10.1108/INTR-02-2019-0044

Ahmed, F., As-Saber, S., Fry, S., & Smith, R. (2019). Bureaucracy and E-government: A study of e-procurement initiatives in Bangladesh.

Altayyar, A., & Beaumont-Kerridge, J. (2016). External Factors Affecting the Adoption of E-procurement in Saudi Arabian's SMEs. *Procedia: Social and Behavioral Sciences*, *229*, 363–375. doi:10.1016/j.sbspro.2016.07.147

As-Saber, S., Rahman, S., Rahim, M., Cairns, G., Härtel, C., & Waheduzzaman, W. (2014). Promises and Pitfalls of Using Social Media in Public E-procurement: An Appraisal. *AJIS. Australasian Journal of Information Systems*, *18*(3). doi:10.3127/ajis.v18i3.1097

Bartezzaghi, E., & Ronchi, S. (2003). Internet supporting the procurement process: Lessons from four case studies. *Integrated Manufacturing Systems*, *14*(8), 632–641. doi:10.1108/09576060310503447

Bensaou, M. (1999). Portfolios of buyer-supplier relationships. *MIT Sloan Management Review*, *40*(4), 35.

Brandon-Jones, A., & Kauppi, K. (2018). Examining the antecedents of the technology acceptance model within e-procurement. *International Journal of Operations & Production Management*, *38*(1), 22–42. doi:10.1108/IJOPM-06-2015-0346

Cai, S., Jun, M., & Yang, Z. (2006). The impact of interorganizational internet communication on purchasing performance: A study of Chinese manufacturing firms. *The Journal of Supply Chain Management*, *42*(3), 16–29. doi:10.1111/j.1745-493X.2006.00014.x

Carr, A. S., & Smeltzer, L. R. (2002). The relationship between information technology use and buyer-supplier relationships: An exploratory analysis of the buying firm's perspective. *IEEE Transactions on Engineering Management*, *49*(3), 293–304. doi:10.1109/TEM.2002.803389

Chang, H.-L. (2019). Factors Affecting the Offshore Deployment of Interorganizational Systems in China: A Case Study Analysis. *Journal of Organizational Computing and Electronic Commerce*, *29*(3), 163–189. doi:10.1080/10919392.2019.1583811

Christian Tanner, R. W. (2008). Current Trends and Challenges in Electronic Procurement: An Empirical Study. *Electronic Markets*, *18*(6), 6–18. doi:10.1080/10196780701797599

Corini, J. (2000). Integrating e-procurement and strategic sourcing. *Supply Chain Management Review*, *2*(3), 70–75.

Croom, S., & Brandon-Jones, A. (2007). Impact of e-procurement: Experiences from implementation in the UK public sector. *Journal of Purchasing and Supply Management*, *13*(4), 294–303. doi:10.1016/j.pursup.2007.09.015

Croom, S., & Johnston, R. (2003). E-service: Enhancing internal customer service through e-procurement. *International Journal of Service Industry Management*, *14*(5), 539–555. doi:10.1108/09564230310500219

Dale, C. (2003). The competitive networks of tourism e-mediaries: New strategies, new advantages. *Journal of Vacation Marketing*, *9*(2), 109–118. doi:10.1177/135676670300900201

Davila, A., Gupta, M., & Palmer, R. (2003). Moving procurement systems to the internet: The adoption and use of e-procurement technology models. *European Management Journal*, *21*(1), 11–23. doi:10.1016/S0263-2373(02)00155-X

Day, G. S., & Hubbard, K. J. (2003). Customer relationships go digital. *Business Strategy Review*, *14*(1), 17–26. doi:10.1111/1467-8616.00240

Dhillon, G. (2008). Organizational competence for harnessing IT: A case study. *Information & Management*, *45*(5), 297–303. doi:10.1016/j.im.2008.01.008

Eadie, R., Perera, S., & Heaney, G. (2010). A cross-discipline comparison of rankings for e-procurement drivers and barriers within UK construction organisations. *Journal of Information Technology in Construction,* 217-233.

Eadie, R., Perera, S., & Heaney, G. (2011). Analysis of the use of E-Procurement in the Public and Private Sectors of the UK construction Industry. *Journal of Information Technology in Construction*, *16*, 669–686.

Eadie, R., Perera, S., Heaney, G., & Carlisle, J. (2007). Drivers and barriers to public sector e-procurement within Northern Ireland's construction industry. *Journal of Information Technology in Construction*, *12*, 103–120.

Gunasekaran, A., McGaughey, R. E., Ngai, E. W., & Rai, B. K. (2009). E-Procurement adoption in the Southcoast SMEs. *International Journal of Production Economics*, *122*(1), 161–175. doi:10.1016/j.ijpe.2009.05.013

Gunasekaran, A., & Ngai, E. W. (2008). Adoption of e-procurement in Hong Kong: An empirical research. *International Journal of Production Economics*, *113*(1), 159–175. doi:10.1016/j.ijpe.2007.04.012

Gupta, M., & Narain, R. (2012). A survey on supplier relationship in e-procurement in Indian organisations. *International Journal of Logistics Systems and Management*, *12*(1), 89–121. doi:10.1504/IJLSM.2012.047060

Gupta, M., & Tripathi, S. (2018). A framework for buyer satisfaction in e-procurement in Indian scenario: An integrated ISM and SEM approach. *International Journal of Procurement Management*, *11*(6), 748–776. doi:10.1504/IJPM.2018.095657

Hardy, C. A., & Williams, S. P. (2008). E-government policy and practice: A theoretical and empirical exploration of public e-procurement. *Government Information Quarterly*, *25*(2), 155–180. doi:10.1016/j.giq.2007.02.003

Ilhan, N., & Rahim, M. (2017). Benefits of e-procurement systems implementation: Experience of an Australian municipal council. In Digital Governance and E-Government Principles Applied to Public Procurement (pp. 249–266). Hershey, PA: IGI Global. doi:10.4018/978-1-5225-2203-4.ch012

Johnson, M. E., & Whang, S. (2002). E-business and supply chain management: An overview and framework. *Production and Operations Management*, *11*(4), 413–423. doi:10.1111/j.1937-5956.2002.tb00469.x

Kosmol, T., Reimann, F., & Kaufmann, L. (2019). You'll never walk alone: Why we need a supply chain practice view on digital procurement. *Journal of Purchasing and Supply Management*, *25*(4). doi:10.1016/j.pursup.2019.100553

Leek, S., Turnbull, P. W., & Naude, P. (2003). How is information technology affecting business relationships? Results from a UK survey. *Industrial Marketing Management*, *32*(2), 119–126. doi:10.1016/S0019-8501(02)00226-2

MacManus, S. A. (2002). Understanding the incremental nature of e-procurement implementation at the state and local levels. *Journal of Public Procurement*, *2*(1), 5–28. doi:10.1108/JOPP-02-01-2002-B001

Makosa, P., & Mwangangi, P. (2018). Factors influencing e-procurement implementation public sector in South Sudan. *International Journal of Supply Chain and Logistics*, *2*(2), 1–23.

Min, H., & Galle, W. P. (2003). E-purchasing: Profiles of adopters and nonadopters. *Industrial Marketing Management*, *32*(3), 227–233. doi:10.1016/S0019-8501(02)00266-3

Moon, M. J. (2005). E-procurement management in state governments: Diffusion of e-procurement practices and its determinants. *Journal of Public Procurement*, *5*(1), 54–72. doi:10.1108/JOPP-05-01-2005-B003

Morgan, R. M., & Hunt, S. D. (1994). The commitment-trust theory of relationship marketing. *Journal of Marketing*, *58*(3), 20–38. doi:10.1177/002224299405800302

Muriuki, J. I., Guyo, W., Odhiambo, R., & Kinoti, J. (2019). Effect of electronic procurement technical support staff on procurement performance in energy sector state corporations in Kenya. *International Journal of Supply Chain Management, 4*(1), 20-38.

Naseebullah, M., PDD, D., & Hassan, M. (2010). The role of factors in business-to-business e-procurement implementation.

Neef, D. (2001). E-Procurement: From strategy to implementation: FT press.

Neuman, W. L. (2003). *Social Research Methods: Qualitative and Quantitative Approaches*. Massachusetts: Allyn & Bacon.

Oliech, C. O., & Mwangangi, P. (2019). Effect of strategic procurement management on performance of level five hospitals in Kenya. *International Journal of Supply Chain Management*, *4*(1), 39–62.

Panayiotou, N. A., Gayialis, S. P., & Tatsiopoulos, I. P. (2004). An e-procurement system for governmental purchasing. *International Journal of Production Economics*, *90*(1), 79–102. doi:10.1016/S0925-5273(03)00103-8

Panda, D., Sahu, G., & Gupta, P. (2010). *Promoting transparency and efficiency in public procurement: E-procurement initiatives by government of India.* Paper presented at the 7th International Conference on E-government (ICEG).

Pani, M. R., Agrahari, A., De, S., & Sahoo, G. (2011). Literature review and research issues in e-procurement. *Management and Labour Studies*, *36*(3), 225–246. doi:10.1177/0258042X1103600302

Pearcy, D. H., & Giunipero, L. C. (2008). Using e-procurement applications to achieve integration: What role does firm size play? *Supply Chain Management*, *13*(1), 26–34. doi:10.1108/13598540810850292

Pillutla, S., Yao, D.-Q., & Li, X. (2014). Data mining the adoption intention of e-procurement system for Chinese companies. *International Journal of Management and Decision Making*, *13*(2), 207–219. doi:10.1504/IJMDM.2014.060872

Platform, P. (2019). Digital procurement and procurement transformation. Retrieved from https://www.procuredesk.com/digital-procurement/

Pule, S., & Bakkabulindi, A. (2017). E-procurement implementation: Analysing the applicability of the EPOS method on the efficiency of public sector organisations-a comparative study. *International Journal of Procurement Management*, *10*(2), 135–149. doi:10.1504/IJPM.2017.082783

Puschmann, T., & Alt, R. (2005). Successful use of e-procurement in supply chains. *Supply Chain Management*, *10*(2), 122–133. doi:10.1108/13598540510589197

Rahim, M. (2008). Identifying factors affecting acceptance of e-procurement systems: An initial qualitative study at an Australian City Council. *Communications of the IBIMA*, *3*(1), 7–17.

Rahim, M., & As-Saber, S. N. (2011). E-procurement adoption process: An arduous journey for an Australian city council. *International Journal of Electronic Finance*, *5*(2), 150–168. doi:10.1504/IJEF.2011.040458

Rahim, M., & Bantwal, A. (2012). Perceived benefits from a local government public procurement initiative: A diffusion of innovation perspective. In Inter-organizational information systems and business management: Theories for researchers (pp. 251–271). Hershey, PA: IGI Global. doi:10.4018/978-1-60960-768-5.ch016

Rahim, M., Chen, X., & Kurnia, S. (2008). Understanding benefits of electronic procurement systems adoption: experience of an Australian City Council.

Rahim, M., Ilhan, N., & Chen, X. (2011). *Factors affecting benefits of e-procurement systems: Experience of an Australian City Council.* Paper presented at the 11th International Conference on Electronic Business (ICEB 2011).

Rahim, M., & Kurnia, S. (2014). *Understanding E-Procurement System Benefits using organisational Adoption Motivation Lens: a Case Study.* Paper presented at the PACIS.

Rogers, E. M. (1995). Diffusion of Innovations (4th ed.). New York: The Free Press.

Ronchi, S., Brun, A., Golini, R., & Fan, X. (2010). What is the value of an IT e-procurement system? *Journal of Purchasing and Supply Management, 16*(2), 131–140. doi:10.1016/j.pursup.2010.03.013

Sambasivan, M., Wemyss, G. P., & Che Rose, R. (2010). User acceptance of a G2B system: A case of electronic procurement system in Malaysia. *Internet Research, 20*(2), 169–187. doi:10.1108/10662241011032236

Sarker, S., & Lee, A. S. (2000, January). *Using a positivist case research methodology to test three competing theories-in-use of business process redesign.* Paper presented at the International Conference on Information Systems (ICIS), Brisbane, Australia.

Schmidt, R., Möhring, M., Härting, R.-C., Reichstein, C., Neumaier, P., & Jozinović, P. (2015). Industry 4.0-potentials for creating smart products: empirical research results. In *Proceedings of the International Conference on Business Information Systems.* 10.1007/978-3-319-19027-3_2

Schoenherr, T., & Tummala, V. R. (2007). Electronic procurement: A structured literature review and directions for future research. *International Journal of Procurement Management, 1*(1-2), 8–37. doi:10.1504/IJPM.2007.015353

Shkolnyk, I., Bukhtiarova, A., & Horobets, L. (2019). *The efficiency of electronic public procurement system in Ukraine.*

Sin, L. Y., Alan, C., Yau, O. H., Chow, R. P., Lee, J. S., & Lau, L. B. (2005). Relationship marketing orientation: Scale development and cross-cultural validation. *Journal of Business Research, 58*(2), 185–194. doi:10.1016/S0148-2963(02)00493-9

Smart, A. (2010). E-procurement and its impact on supply management–evidence from industrial case studies. *International Journal of Logistics: Research and Applications, 13*(6), 423–440. doi:10.1080/13675567.2010.486760

Soliman, K. S., Janz, B. D., Puschmann, T., & Alt, R. (2005). Successful use of e-procurement in supply chains. *Supply Chain Management.*

Srai, J. S., & Lorentz, H. (2019). Developing design principles for the digitalisation of purchasing and supply management. *Journal of Purchasing and Supply Management, 25*(1), 78–98. doi:10.1016/j.pursup.2018.07.001

Tai, Y.-M., Ho, C.-F., & Wu, W.-H. (2010). The performance impact of implementing web-based e-procurement systems. *International Journal of Production Research, 48*(18), 5397–5414. doi:10.1080/00207540903117915

Tatsis, V., Mena, C., Van Wassenhove, L. N., & Whicker, L. (2006). E-procurement in the Greek food and drink industry: Drivers and impediments. *Journal of Purchasing and Supply Management*, *12*(2), 63–74. doi:10.1016/j.pursup.2006.04.003

Themistocleous, M., Irani, Z., & O'Keefe, R. M. (2001). ERP and application integration: Exploratory survey. *Business Process Management Journal*, *7*(3), 195–204. doi:10.1108/14637150110392656

Trkman, P., & McCormack, K. (2009). Estimating the benefits and risks of implementing e-procurement. *IEEE Transactions on Engineering Management*, *57*(2), 338–349. doi:10.1109/TEM.2009.2033046

Wang, M., & Zhang, S. (2005). Integrating EDI with an E-SCM system using EAI technology. *Information Systems Management*, *22*(3), 31–36. doi:10.1201/1078/45317.22.3.20050601/88743.5

Wimalasena, N. N., & Gunatilake, S. (2018). The readiness of construction contractors and consultants to adopt e-tendering: The case of Sri Lanka. *Construction Innovation*, *18*(3), 350–370. doi:10.1108/CI-03-2017-0025

Yin, R. K. (2003). *Case Study Research: Design and Methods*. Thousand Oaks, CA: Sage.

ADDITIONAL READING

Garrido-Samaniego, M. J., Gutiérrez-Arranz, A. M., & San José-Cabezudo, R. (2010). Assessing the impact of e-procurement on the structure of the buying centre. *International Journal of Information Management*, *30*(2), 135–143. doi:10.1016/j.ijinfomgt.2009.08.004

Glas, A. H., & Kleemann, F. C. (2016). The impact of industry 4.0 on procurement and supply management: A conceptual and qualitative analysis. *International Journal of Business and Management Invention*, *5*(6), 55–66.

Gunasekaran, A., McGaughey, R. E., Ngai, E. W., & Rai, B. K. (2009). E-Procurement adoption in the Southcoast SMEs. *International Journal of Production Economics*, *122*(1), 161–175. doi:10.1016/j.ijpe.2009.05.013

Rahim, M., & Kurnia, S. (2014). Understanding E-Procurement System Benefits using organisational Adoption Motivation Lens: a Case Study. Paper presented at the PACIS.

KEY TERMS AND DEFINITIONS

Business: An organisation that seeks to meet its established vision using resources and dealing with various stakeholders.

Business Partner Relationship: The nature of association between business entities.

City council: A public sector organisation that operates at the local government level.

E-Procurement: Represents a class of software systems that can automate the key tasks involved in procurement process.

Implementation: Represents a stage of an IT system in which several activities are carried out for its introduction for use within an organisational context.

Methodology: A research approach (qualitative, quantitative, hybrid) that is adopted to empirically evaluate a research model to find answers.

System: A collection of various parts or components (including ICTs) to help a business to carry out certain activities by automating a business process.

Chapter 11
Education and Training for Digital Health:
Establishing Evidence Relating to Digital Health Education and Training (E&T)

Sisira Edirippulige
University of Queensland, Australia

Buddhika Senanayake
https://orcid.org/0000-0002-2278-2889
University of Queensland, Australia

ABSTRACT

Digital health is fast becoming an integral part of healthcare services. Research evidence suggests that digital health can benefit stakeholders involved in healthcare, including patients and care providers. As digital health continues to integrate into routine healthcare, practitioners may require new knowledge, skills, and competencies to make the best use of it and to be able to communicate with an increasingly digitally-enabled consumer. Much effort has been made to systematic education and training relating to digital health, an important aspect in developing the digital health workforce.

INTRODUCTION

Digital health is growing. The intentions of digital health include providing *administrative advantage* (e.g. to improve efficiency, and reduce costs through digitisation); *clinical advantage* (e.g. electronic prescribing to improve patient safety, electronic health records (EHR) to facilitate sharing amongst practitioners); to *overcome impediments* to service delivery (e.g. remote consultation, or remote-access to medical imaging); and to *help patients to self-manage* their health (e.g. apps for monitoring chronic conditions, or sharing information amongst providers using a personally-controlled EHR) (Sandhu, 2020; Greenes & Shortliffe, 1990).

DOI: 10.4018/978-1-7998-2799-3.ch011

There is a strong rationale for seeking alternative methods for delivering healthcare using innovative technologies (Hadeel & Sandhu, 2021; Sandhu, 2020; Venkatesh & Davis, 2000; & Davis, 1989). Traditional business models and infrastructure of healthcare organisations are ill-equipped to sufficiently adapt to, or finance, the global health trends of aging populations, declining birth rates, increasing non-communicable chronic disease, poverty, migration and increasing rates of infectious disease. These factors all impose significant logistical and financial hardships for those striving to provide ubiquitous effective and ethical healthcare. The digital transformation of healthcare with information communication technology provides one avenue to better enable patient centred care. Although digital health is growing rapidly in the healthcare industry the education and training component plays a critical role to uplift the future uptake of digital innovation. This chapter explores the current evidence regarding education and training for digital health and discovers the potentials to strengthen digital health in the field of learning and training.

While digital health is not new, advances in technology, in particular, the availability of mobile devices and ubiquitous connectivity, means that services are no longer physically limited to a health facility, or limited only to clinicians (McConalogue, Davis, & Connolly, 2019). Modern healthcare systems increasingly comprise digital hospitals providing services to digital health-consumers. Many governments and health systems have made attempts to introduce telehealth services, a few of which have become large and well-integrated, such as the Ontario Telemedicine Network in Canada (Edirippulige & Armfield, 2017) and Veteran's Affairs Telehealth in the USA (Kampmeijer, Pavlova, Tambor, Golinowska, & Groot, 2016). The interest in telehealth to improve health care services is global. A number of European countries have implemented various telehealth projects; and some countries have conducted large research studies to establish an evidence base (Dattakumar, Gray, Henderson, Maeder, & Chenery, 2012; Marschang, 2014). Global organisations such as the World Health Organisation (WHO) have been promoting telehealth, particularly in low and middle income countries in Asia, Africa, and Latin as a new way to address critical health challenges (Gray, Dattakumar, Maeder, Butler-Henderson, & Chenery, 2014; Smith, 2011).

Some governments have made important policy changes to encourage and incentivise the use of digital health. For example, the Australian federal government introduced reimbursement for certain video-based consultations involving general practitioners, specialists, Aboriginal health workers, nurse practitioners and midwives (Bryant, 2014). In the USA, forty-six states, and Washington DC provide reimbursement for some video-based consultations under their Medicaid programs which supports the healthcare costs of those on low incomes and those living with certain disabilities (Lannan, 2015).

Alongside the increase in the use of telehealth, a flourishing industry of medical, technology and clinical service businesses has grown. The size of the market is difficult to reliably estimate. One group of analysts have suggested that it could be as large as USD $43.4 billion by 2019 (Cope, 2011) while another group were more conservative in their estimate of USD $36.3 billion by 2020 (Gill, 2014).

As digital health continues to integrate into routine healthcare, practitioners may require new knowledge, skills, and competencies to make the best use of it and to be able to communicate with an increasingly digitally-enabled consumer. Almost thirty years ago in 1990, Greenes *et al.* foresaw this need: '*horizontal integration of eHealth into healthcare practices requires the integration of eHealth into medical education*' (Greenes & Shortliffe, 1990). Meanwhile, the lack of education and training (E&T) has been highlighted as an impediment for health professionals to use digital health (Sandhu 2020; Edirippulige & Armfield, 2017; Kampmeijer et al., 2016). While universities may be a logical place to provide digital health E&T, a study involving thirty Australian universities in 2012 (largely by survey, response rate ~25%; supplemented by interviews) concluded that the future health-professional workforce was not being adequately educated to achieve competence to work with digital health (Dattakumar et al., 2012; Gray et al., 2014).

While much effort has been made to establish the evidence base relating to the clinical use of digital health, little has been done to understand the E&T needs for practicing digital health within clinical services. Lack of proper understanding of E&T needs may be a reason for the lack of systematic E&T in this field. The recent national study highlighted that no Australian medical curriculum provides formal E&T to future medical practitioners about digital health (Edirippulige, Brooks, et al., 2018).

However, in recent years, there has been an increased interest in digital health E&T in various sectors. For example, the Australian Digital Health Agency has highlighted the importance of workforce development and E&T has been identified as one of the strategic priorities.

LITERATURE REVIEW

The concept of 'digital health' is broad and complex. There are many different definitions associated with digital health and they may vary depending on the context. However, the common aspects of all these definitions are the delivery of health care services and information and communication technologies (ICT). As a result, there are a number of different terms associated with this disciple. Some commonly used terms include telemedicine, mHealth, telehealth, health informatics, medical informatics. (Iyawa, Herselman, & Botha, 2016).

There are many definitions of digital health and e-health. The World Health Organization defines digital health as 'the use of information and communication technologies (ICT) for health including treating patients, conducting research, educating the health workforce, tracking diseases and monitoring public health' (World Health Organisation, 2016). This has evolved from the original World Health Organization (WHO) definition of e-health which is as 'the transfer of health resources and health care by electronic means'.

Human resources for health is an essential component of an efficient health system. Policymakers should rethink their health workforce strategies to build up competencies and skill mix to meet changing health needs in the future. Countries are taking many actions in health workforce education, retention and reorientation of work processes and of cadres. Many national stakeholders and international development agencies are involved to incorporate digital health into health workforce enhancement (Nancarrow, 2015).

The development of Internet-based technologies, which have significantly changed the way people engage with the data that surrounds their lives. The Internet was considered to be a place that had the potential to change society in fundamental ways. In the context of health care, these expectations were reinforced by views that foresaw the capacity of digital technology to create a more cost-effective way of delivering health care. Moreover, the proliferation of digital technologies across all sectors of society would lead to increased utilization of and reliance upon digital solutions within the healthcare sector (Rich & Miah, 2014).

Technology is the main factor in the emergence and the development of e-Health. As we have learned, the following are the major technical aspects that have made an impact:

- developments in computing, the Internet and information technology
- development and expansion of networks and telecommunication infrastructure

- the rapid adoption of technology by society (Rogers, 1995)
- the technological innovation of medical devices, digital cameras, and physiological sensors

There are plenty of health benefits of Information, Communication, and Technology (ICT). It improves health service delivery, increasing efficiency, reduces operating costs, reduces administrative costs, and enabled entirely novel ways of going about health care management. This definition includes clinical, educational, administrative and research health related activities that are carried out at a distance (Bashshur, Reardon, & Shannon, 2000). Therefore, when we place the word 'clinical' in front of 'telehealth', we are talking about the application of ICTs for medical consultations, diagnosis, treatment, monitoring, patient follow up, case discussion, etc. Digital health consisted of several components to provide health services: i.e. telemedicine/ telehealth, health informatics, mobile health, etc.

Telemedicine

Telehealth is defined as 'the use of information and communication technology (ICT) to deliver health services, expertise and information over distance.' (5) Many, including the World Health Organisation (WHO), use the terms 'telehealth' and 'telemedicine' interchangeably, while others draw distinctions between the two. A Cochrane review concluded that the terms overlap and that 'telemedicine is considered to be the use of communication and information technologies to deliver clinical care where the individuals involved are not at the same location. They can either be two healthcare professionals or a healthcare professional and a patient. Telehealth includes this definition, and also covers telecommunication to deliver non-clinical services such as research and health education promotion' (Currell, Urquhart, Wainwright, & Lewis, 2000). Telemedicine includes telephone and email contact between patients and health professionals, videoconferencing between patients and health professionals and remote patient monitoring, where devices are used in the patient's home to send information to health professionals. There are two main approaches used in telehealth – the real-time approach and the store-and-forward approach. The real-time ('synchronous') approach means that the parties involved interact at the same time using information and communication technologies (ICTs).

Health Informatics

The history of electronic health information exchange (health informatics) is in part matched by the evolution of telemedicine. Clinicians have sought to exploit technological development and progress to overcome everyday logistical issues they face with delivering their services. The ability to electronically transmit text and image-based anatomical pathology, clinical pathology and radiology diagnostic data greatly enhanced the evolution of telemedicine.

Mobile Health

From the world-wide infiltration of mobile communication devices and innovation offerings by advanced networking technologies, a specific field within e-Health has emerged, i.e. mobile health or m-Health. m-health is the fastest growing segment of ICT-based healthcare; devices include smartphones, tablets, global positioning system (GPS) devices, mobile telehealth, and physiological monitoring devices. The latest statistics from April 2019 reveal that mobile devices and transmitters are now in the hands of more people and communities than any other type of communication device. The number of mobile connections exceeds the total population in the world (7.7 billion) and the average number of mobile connections per unique mobile user is 1.52 (We are social, 2019).

World Health Organization (WHO) defines mHealth as medical and public health practice supported by mobile devices, such as mobile phones, patient monitoring devices, personal digital assistants (PDAs), and other wireless devices. mHealth involves the use of voice, video, images and, short messaging services (SMS), general packet radio service (GPRS), third/ fourth and fifth-generation mobile telecommunications (3G/4G and 5G systems), global positioning system (GPS), and Bluetooth technology (World Health Organization, 2011). Digital technologies are becoming an important resource for health services delivery and public health. Mobile wireless technologies are particularly relevant, due to their ease of use, broad reach, and wide acceptance (Hadeel & Sandhu, 2021; Venkatesh & Davis, 2000; & Davis 1989).

SMS

This is the most accessible technology as all mobile phones are SMS capable. SMS is often used to send reminders and motivational prompts (keeping health related appointments, reminders about set goals); the recipient can then use SMS to send back data relative to the intervention.

Cameras

Photos for dermatology review (lesions, etc), photographic evidence of progress.

Native Applications

Via programming interfaces, developers can build special purpose applications – users can log journals relating to their health and health-related goals and participate in online communities for progress monitoring, peer support and motivation. Chronic disease sufferers have a variety of mobile-application programs available to assist them in managing their diseases, i.e. asthma, diabetes, hypertension, heart failure.

Automated Sensing

Mobile phones act as receivers, storage devices and transmitters of data from medical and wellness devices like measuring scales, blood pressure monitors, heart rate monitors, glucometers, portable ECGs, pedometers, ultrasounds.

Interfacing With the Internet

The capacity of mobile devices to be continuously connected to the internet means that the transfer of data to care providers can be automatic and immediate. This allows for the timely review of data by the remote care provider enabling earlier intervention if warranted (World Health Organization, 2011).

Digital health comprise providing administrative advantage (e.g. to improve efficiency, and reduce costs through digitisation); clinical advantage (e.g. electronic prescribing to improve patient safety, electronic health records (EHRs) to facilitate sharing amongst practitioners); to overcome impediments to service delivery (e.g. remote consultation, or remote access to medical imaging); and to help patients to self-manage their health (e.g. apps for monitoring chronic conditions, or sharing information amongst providers using a personally-controlled EHR).

METHODOLOGY

A thorough literature review was performed to identify facts for the research topic. The electronic databases such as "PubMed", "Google Scholar", Medline, Scopus, Emtree and "HINARI" were used to gather data. Keywords such as education and training, digital health, mHealth, health workforce were entered in order to get comprehensive coverage of relevant literature. Citation information along with notes and abstracts were downloaded to Endnote X6 software. Citation and reference lists were made using Endnote X6 software in Harvard APA style –sixth edition.

The exploratory review was conducted on peer-reviewed publications published between January 2000 and April 2019. The reason for limiting searches to this narrow period was due to change in technology and teaching and learning styles rendering findings of earlier studies unsuitable for current contexts. As the study aimed to collate all evidence regarding education and training for digital health, all the potential studies that strengthen digital health in the field of learning and training were included. The review also considered any type of study (quantitative, or qualitative) if adequate information (to assess successful implementation or utilization) on the service delivered was described. Articles related to healthcare models, technical feasibility studies, case studies, letters, and published comments were excluded (inadequacy of information to reliably assess usefulness). Publications with no abstracts and non-English articles were also excluded. Grey literature was not searched.

SOLUTIONS AND RECOMMENDATIONS

Digital Health and Patient-Centred Healthcare

Patient-centred care is a healthcare model that places the patient at the centre of the care instead of the disease or illness. It considers a patient's cultural tradition, personal preferences, and values, family situations, social circumstances and lifestyles as well as the clinical aspects of diagnosis, potential treatment, rehabilitation or advice on healthy behaviours. 'Patient-centred care is an approach to the planning, delivery, and evaluation of health care that is grounded in mutually beneficial partnerships among healthcare providers, patients, and families' (Australian Commission on Safety and Quality in Health Care, 2011).

In the traditional, disease-centered model, physicians make almost all treatment decisions based largely on clinical experience and data from various medical tests. In a patient-centered model, patients become active participants in their own care and receive services designed to focus on their individual needs and preferences, in addition to advice and counsel from health professionals. This model was first described by the Institute of Medicine (US) in its landmark publication 'Crossing the quality chasm: A new health system for the 21st century' (Institute of Medicine, 2001).

Patient-centred care directly improves quality and safety and empower patients. Empowered healthcare consumers require access to high quality medical information and an up to date personal electronic health record. Having a personal electronic health record is vital for the future vision of timely, effective and up-to-date personalised care.

Just like patients, healthcare teams need access to the best possible and most recent information about their patients' health status (European Commission, 2004). A single electronic repository for such information, that all participants in a person's healthcare can contribute to and access, is viewed as critical for the success of many e-health innovations. Consider the efficiencies that can be gained if all relevant information from ongoing disease monitoring (including patient well- being and quality of life indicators), diagnostic and treatment data can be made available to all participants in the multi-disciplinary care team including the patient. A vision for the future e-Health patient is to provide enhanced self-management interventions and individually tailored case management; problems can be quickly identified and clinical information and treatment goals shared between the healthcare team and the patient (Australian Health Minister's Conference, 2008).

Digital health provides equal access to health irrespective of location: Digital health is viewed as an eloquent resolution to the inequity of healthcare access for those that dwell in underserved or geographically disadvantaged communities. This also applies to metropolitan dwellers disadvantaged by age, illness or disability. In particular electronic health care records and telemedicine applications are viewed as a tool that will assist in 'closing the gap' of Indigenous Australian life expectancy, mental health, maternity services, paediatric services of ear, nose, and throat and aged care healthcare inequities (Dods, Alem, & Nepal, 2012).

In summary, digital health provides greater patient engagement through health related websites, apps, tools and wearable technology with gamification and incentives. It promotes healthier and safer communities: the Internet of things, smarter policy, regulation, and response. Digital health technology is productive and provides targeted care: Genomic and precision medicine, digital therapeutics. Finally, it secures the systematic high quality and safer care with electronic medical records and digital hospitals, improved decision support systems, information sharing, and interoperability smart systems and integration and automation (Queensland Health, 2017).

The Changing Role of Health Workers in the Digital Era

Nurses play a critical role in almost every sphere of healthcare such as primary, community and tertiary care. Nurses, regardless of specialty or work setting, treat patients, educate patients and the public about various medical conditions, and provide advice and emotional support to patients' family members. Nurses are responsible for recording patients' medical histories and symptoms, help perform diagnostic tests and analyse results, operate medical machinery, administer treatment and medications and help with patient follow-up and rehabilitation. Research has shown that there is a clear relationship between the number of nurses and the quality of care. Shortage and mal-distribution of nurses is a serious problem (Yogesan, Bos, Brett, & Gibbons, 2009). There is a significant disproportional distribution of nurses within and among countries. For example, while the United States have 773 nurses per 100,000 population, the number of nurses in Uganda for the same size population is 6 (International Council of Nurses, 2005). The disparity of nurses in urban and rural communities is a significant problem in both developing and developed countries.

These shortcomings can be addressed through nursing informatics. According to the official definition of the American Nursing Association (ANA), nursing informatics is the 'specialty identifying, collecting, processing, analysing and managing data'. Although this definition solely concentrates on the information, it is believed that nursing informatics should also incorporate the aspects of the delivery of care. It is apparent that the definitions are continuing to evolve. In broader terms, telehealth is a tool for healthcare delivery, diagnosis, consultation, treatment, and transfer of medical data/images and education by using ICT (i.e. interactive audio, visual, and data communications) at a distance. Multi-Sensory Stimulation Environment (MSSE) was designed with the assistance of robotic technology and mobile applications for children with Autism Spectrum Disorder identifying their need for rehabilitation and special educational purposes (Vladimir et al., 2018).

In general terms, telenursing is the use of ICT for delivering health care services at distance by nurses. Telenursing can be used in both healthcare setting or a non-institutional setting – that is at home or in an assisted-living facility (Wootton, Kvedar, & Dimmick, 2006). Telenursing applications may include real time techniques, for example videoconferencing or telephony and store and forward techniques, for example, email or web-based applications. Through interactive video systems, patients can contact on-call nurses and arrange for a video consultation. This kind of application is particularly helpful for children and adults with chronic conditions and long-term illnesses. Telenursing can also help patients and families to be active in care, such as self-management of chronic conditions. It is useful to support patients in the home, to support remote nurses, to promote and support self-care, to improve medication management and compliance, and cost-saving method.

Health care professionals (doctors, nurses, pharmacists, and other allied healthcare professionals) can utilize digital health for their clinical advantage (e.g. electronic prescribing to improve patient safety, electronic health records (EHRs) to facilitate sharing amongst practitioners and to *overcome impediments* to service delivery (e.g. remote consultation, or remote-access to medical imaging); and to *help patients to self-manage* their health (e.g. apps for monitoring chronic conditions, or sharing information amongst providers using a personally-controlled EHR).

Vision and Strategic Goals for Digital Health

The WHO has prepared a five-year global strategy for digital health for from 2020-2024. There are four main goals. First, it is essential to promote global collaboration and advance the transfer of knowledge on digital health. It provides the platform to advance the implementation of national digital health strategies and strengthen governance for digital health at the global and national levels. The final goal is to advocate for people centred health systems that are enabled by digital health (World Health Organisation, 2019).

A Ten-year vision is developed for digital health in Queensland, Australia. The first priority is to build a consistent and sustainable capability. It strengthens the systems, infrastructure, and enabling tools as well as establishing the frameworks, standards, and governance required to support greater sharing and linkage of trusted information *(to be achieved within three years) (Queensland Health, 2017).*

The second priority is to integrate, grow and expand digital health and digital workforce capabilities. Apply the technology required to integrate, optimise, grow and expand digital health and digital workforce capabilities. This focus will shift capabilities to the creation of the next generation of digital health solutions including integrated technologies, devices and workflows and the integration of big data to enhance decision support, health system planning and performance reporting *(to be achieved within 5 years) (Queensland Health, 2017).*

The third priority is to scale digital health that can enable new and alternative models of healthcare, delivering treatment outcomes that are beyond the current realms of possibility. It promotes taking full advantage of technology and building capability to change the way that patients access care, where that care is provided, and the manner in which care is centred on patients.

Evidence for Digital Health on Workforce

Understanding the E&T Needs for Different Clinical Populations/Stakeholders

A study conducted to explore the progress of eHealth in the Australian medical curriculum by the University of Queensland showed that there is consensus on the importance of eHealth to current and future clinical practice, but there are other priorities and no strong drivers for change. Systemic problems inhibit the inclusion of eHealth in medical education, in which the curriculum is described as 'crowded' and with competing demands, and because accrediting bodies (such as Australian Medical Council) do not expect eHealth competence in medical graduates. Therefore, there is no external pressure for its inclusion. Unless and until accrediting bodies recognise and expect competence in eHealth, it is unlikely that it will enter the curriculum; consequently, the future workforce will remain unprepared (Edirippulige, Brooks, et al., 2018).

A qualitative study of the careers and professional practices of graduates from an e-health postgraduate programme (e-health study at UQ) declared that the digital health programme helped in their career development and more importantly, the current occupation of the majority of respondents was related to health or medicine, and most had worked in an e-health related job since graduating. The overall, level of satisfaction regarding the learning experience doing e-health study at UQ was also very high (Edirippulige, Armfield, & Smith, 2012).

A literature review was conducted to describe the delivery of education and training in telehealth/telemedicine in 2016, which identified nine studies (out of 388 reviewed studies) for final review. Programs from five countries were represented and articles were spread across telemedicine and clinically oriented journals. Education and training in telehealth has been provided as both university level and vocational courses using conventional classroom based delivery methods and e-learning. Reported curriculum items included terminology, clinical applications, the evidence-base, and technological aspects. Currently most recognized universities, public and private organizations have offered telehealth courses within their formal degree program. In addition, Continuous Professional Development (CPD) courses have incorporated telehealth and digital health as essential topics during their curriculums. These CPD courses targeted a range of professionals including health policy makers; telehealth service coordinators; medical doctors; engineers; clinical psychology trainees; nurses; public health professionals; private sector informatics professionals; and information technology professionals (Edirippulige & Armfield, 2017).

Key Topic Areas Relevant to Professional Practices

A study conducted in the UK developed and define domains of competence and associated learning outcomes needed by healthcare graduates to commence working in a digital healthcare environment. It generally applies to all healthcare professionals including doctors, nurses, and allied health professionals. Six key domains of competence were identified.

The agreed domains were: 1) Digital Health: work as a practitioner in the digital healthcare environment; 2) Accessing Data: access and interpret patient data to inform clinical decision-making; 3) Communication: communicate effectively with healthcare professionals and patients in the digital environment; 4)Generating data: generate data for and about patients within the EPR; 5) Multidisciplinary working: work with healthcare professionals with and alongside EPRs; and 6) Monitoring and audit: monitor and improve the quality and safety of healthcare. These domains can be used by academics to guide the integration of EPRs into undergraduate healthcare programmes (Pontefract & Wilson, 2019).

The Association of Faculties of Medicine of Canada and Infoway (2015) launched physician in–training project with the objective of improving clinical practice and patient care by supporting medical students and doctors in their adoption (Rogers, 1995) and use of digital health technologies. The next generation competency based medical education model identified digital health as an essential component for better patient outcomes. They developed 33 virtual patient cases by incorporating the use of digital health to improve patient safety (Baker, Charlebois, Lopatka, Moineau, & Zelmer, 2016).

Suitable Delivery Modes to Provide Relevant Knowledge, Skills, and Competencies

The clinical informatics curriculum was designed for psychiatric residents to improve their skills and competencies for EMR and tele-psychiatry. The e-curriculum in clinical informatics included electronic diagnostics and treatments, mobile apps and devices, EMR, big data, and data science, etc. The exposure therapy through virtual and augmented reality devices and optimizing technology based learning and self care are identified as key competencies of education and training (Torous, Chan, Luo, Boland, & Hilty, 2018).

Another study was conducted in Australia to assess students' perceptions of the value of an eHealth practicum. The majority of students agreed that the practicum allowed them to gain necessary skills in eHealth applications and provided students with an opportunity to explore ways of using different eHealth tools for the delivery of health care at a distance. The study shows that a practical component in eHealth teaching was well received by students. While online teaching is appropriate for providing knowledge, the opportunity to develop practical skills may encourage students to use eHealth techniques in their future practices (Edirippulige, Smith, Armfield, Bensink, & Wootton, 2012).

Another study evaluating the Education ''Clinical Telemedicine/e-Health'' in the Curriculum of Medical Students at the University of Zurich showed that the overall satisfaction and understanding of telemedicine as a supplement in traditional medical consultations increased among students. Most of the students indicated that they would use telemedicine for chronically ill and elderly patients. The positive assessment of the module indicates the successful introduction of the topics of telemedicine and e-health to medical students at the University of Zurich. It was well evident that the students gained confidence and basic skills throughout the course (Brockes, Grischott, Dutkiewicz, & Schmidt-Weitmann, 2017).

Another study examined self-reported knowledge and perception changes associated with an e-health course offered to university undergraduate students in Australia. Students believed the course helped them to improve their knowledge regarding key aspects of e-health. It is important to design an e-health curriculum targeting competencies to provide the necessary knowledge and skills to help students practice e-health in their professional careers (Edirippulige, Samanta, & Armfield, 2018).

Levels of Qualification and Competencies

e-Health is steadily integrating into modern health services, making significant changes in the way health services are traditionally delivered. To work in this new environment, healthcare workers are required to have new knowledge, skills, and competencies specific to e-Health. New research has found that the main motivating factors for studying e-Health were 'relevance to current practice' and 'e-Health is the future'. The participants highlighted the opportunity to 'acquiring relevant knowledge' and 'having exposure' were key benefits offered by the e-Health education program. Participants believed that the postgraduate e-Health program had an influence on their professional practice and the majority of them had worked in the field of e-Health since their graduation. Their professional roles mainly involved the 'implementation of e-Health in health service settings' and the 'use of e-Health in clinical practice'. While e-Health may be steadily integrating into modern health services, e-Health specific job opportunities are still relatively limited. e-Health workforce development must be given priority (Edirippulige, Smith, Wickramasinghe, & Armfield, 2018).

FUTURE RESEARCH DIRECTIONS

It is quite obvious that Information and communication technologies have not been harnessed systematically to improve the health of populations in developing countries. Although developed countries are using modern technology to empower those who use information by providing them with a choice of information to be accessed in their own time and by allowing them to put their own information on the web. Developing countries are par below the standard level compared to developed countries. The current digital divide is more dramatic than any other inequity in health or income. Therefore, it is essential to encourage policymakers in developing countries to embrace digital health by investing in research related to health informatics and e-health. The way forward for these developing countries is to exploit the full interactivity of the internet, which allows rapid feedback and change to continuously mold information into useful knowledge (Edejer, 2000). Although knowledge-based economy with digital health is a novel concept, it tends to neglect human values, the impact of the natural environment, and holism which are essential to nurturing a humanitarian health system (Ženko, Mulej, & Potocan, 2017).

The rapidly rising aging population is a challenge for most of the developed countries as they consume more and more healthcare resources for long periods. The cost of healthcare spending is increasing rapidly and the governments are searching for cost-effective healthcare models to maintain their quality of care. Mental health, substance abuse, and domestic violence are emerging challenges for health systems. More importantly, digital health is a vital tool that can be utilized for most of these new health challenges. Poor communication between health authorities and medical curriculum experts have created a barrier to incorporate digital health in a timely manner, as it is essentially required at health service delivery. Some health managers are not geared enough to rethink and redesign healthcare models and resist innovative approaches to new challenges. Less sustainable funding for digital health research to improve systems and patient outcomes is another challenge (Queensland Health, 2017). Personal data privacy and security are another challenge for digital health limiting its implementation. This should be addressed sensitively to uplift the growth of digital health (Filkins et al., 2016). The future direction of the healthcare industry is determined by the quality of education and training provided for healthcare professionals and to general public on digital health and its practical applicability in day-to-day practice. There should be a uniform agreement between all stakeholders regarding the core education components of digital health and its practical implications similar to fundamental disciplines in medical, para-medical and health science streams.

CONCLUSION

Digital health is experiencing steady growth. Innovators and investors alike see the potential in technologies that can transform the face of healthcare by leveraging data to improve diagnosis, treatment, outcomes and the overall patient experience. The new opportunities offered by digital health are generous, with a variety of uses ranging from disease detection and diagnosis, self-management of chronic conditions, psychological support and clinical decision-making. This novel, patient-centered and technology-driven approach has also created openings for new players to break into the market.

While evidence relating to the benefits and effectiveness of digital health is growing, the workforce development relating to digital health has been paid little or no attention. The utility of digital health in clinical practice requires significant changes in the traditional way of practice. Systematic education and training is needed for delivering essential knowledge and skills for healthcare providers to use digital health in their routine practice. Education research has a major role to play in establishing evidence relating to the needs, content, and methods of delivery of digital health education and training for different groups. The development of digital health curriculum must be evidence-based.

REFERENCES

Australian Commission on Safety and Quality in Health Care. (2011). *Patient-centred care: Improving quality and safety through partnerships with patients and consumers*. Retrieved from https://www.safetyandquality.gov.au/wpcontent/uploads/2012/03/PCC_Paper_August.pdf

Australian Health Minister's Conference. (2008). *National e-Health Strategy Summary*. Melbourne, Australia: Australian Government.

Baker, C., Charlebois, M., Lopatka, H., Moineau, G., & Zelmer, J. (2016). Influencing Change: Preparing the Next Generation of Clinicians to Practice in the Digital Age. *Healthcare Quarterly*, *18*(4), 5–7. doi:10.12927/hcq.2016.24569 PMID:27009699

Bashshur, R. L., Reardon, T. G., & Shannon, G. W. (2000). Telemedicine: A new health care delivery system. *Annual Review of Public Health*, *21*(1), 613–637. doi:10.1146/annurev.publhealth.21.1.613 PMID:10884967

Brockes, C., Grischott, T., Dutkiewicz, M., & Schmidt-Weitmann, S. (2017). Evaluation of the Education "Clinical Telemedicine/e-Health" in the Curriculum of Medical Students at the University of Zurich. *Telemedicine Journal and e-Health*, *23*(11), 899–904. doi:10.1089/tmj.2017.0011 PMID:28498779

Bryant, A. (2014). *An interpretative phenomenological analysis of student engagement in learning at the middle level.* (Education Doctoral Theses Paper), Northeastern University, USA. Retrieved from https://repository.library.northeastern. edu/files/neu:336553 (208)

Cope, J. (2011). Entrepreneurial learning from failure: An interpretative phenomenological analysis. *Journal of Business Venturing*, *26*(6), 604–623. doi:10.1016/j.jbusvent.2010.06.002

Currell, R., Urquhart, C., Wainwright, P., & Lewis, R. (2000). Telemedicine versus face to face patient care: Effects on professional practice and health care outcomes. *Cochrane Database of Systematic Reviews*, (2). doi:10.1002/14651858.Cd002098 PMID:10796678

Dattakumar, A., Gray, K., Henderson, K. B., Maeder, A., & Chenery, H. (2012). We are not educating the future clinical health professional workforce adequately for e-health.

Davis, F. D. (1989). Perceived usefulness, perceived ease of use, and user acceptance of information technology. *MIS quarterly*, 319-340. Competence: Findings of an Australian study. *Studies in Health Technology and Informatics*, *178*, 33–38.

Dods, J., Alem, L., & Nepal, S. (2012). *Caring for the last 3%: Telehealth potential and Broadband implications for remote Australia. CSIRO ICT Centre; 2012*. Retrieved from Canberra.

Edejer, T. T.-T. (2000). Disseminating health information in developing countries: The role of the internet. *BMJ (Clinical Research Ed.)*, *321*(7264), 797–800. doi:10.1136/bmj.321.7264.797 PMID:11009519

Edirippulige, S., & Armfield, N. R. (2017). Education and training to support the use of clinical telehealth: A review of the literature. *Journal of Telemedicine and Telecare*, *23*(2), 273–282. doi:10.1177/1357633X16632968 PMID:26892005

Edirippulige, S., Armfield, N. R., & Smith, A. C. (2012). A qualitative study of the careers and professional practices of graduates from an e-health postgraduate programme. *Journal of Telemedicine and Telecare*, *18*(8), 455–459. doi:10.1258/jtt.2012.gth107 PMID:23209263

Edirippulige, S., Brooks, P., Carati, C., Wade, V. A., Smith, A. C., Wickramasinghe, S., & Armfield, N. R. (2018). It's important, but not important enough: eHealth as a curriculum priority in medical education in Australia. *Journal of Telemedicine and Telecare*, *24*(10), 697–702. doi:10.1177/1357633X18793282 PMID:30343657

Edirippulige, S., Samanta, M., & Armfield, N. R. (2018). Assessment of Self-Perceived Knowledge in e-Health Among Undergraduate Students. *Telemedicine Journal and e-Health*, *24*(2), 139–144. doi:10.1089/tmj.2017.0056 PMID:28708457

Edirippulige, S., Smith, A. C., Armfield, N. R., Bensink, M., & Wootton, R. (2012). Student perceptions of a hands-on practicum to supplement an online eHealth course. *Journal of Medical Internet Research*, *14*(6). doi:10.2196/jmir.2029 PMID:23246840

Edirippulige, S., Smith, A. C., Wickramasinghe, S., & Armfield, N. R. (2018). Examining the Influence of E-Health Education on Professional Practice. *Journal of Medical Systems*, *42*(11), 215. doi:10.100710916-018-1084-5 PMID:30276486

European Commission. (2004). E-Health, Making Healthcare Better for European Citizens: An Action Plan for a European E-health Area. Retrieved from https://ec.europa.eu/digital-single-market/en/news/e-health-making-healthcare-better-european-citizens-action-plan-european-e-health-area

Filkins, B. L., Kim, J. Y., Roberts, B., Armstrong, W., Miller, M. A., Hultner, M. L., ... Steinhubl, S. R. (2016). Privacy and security in the era of digital health: What should translational researchers know and do about it? *American Journal of Translational Research*, *8*(3), 1560–1580. PMID:27186282

Gill, M. J. (2014). The Possibilities of Phenomenology for Organizational Research. *Organizational Research Methods*, *17*(2), 118–137. doi:10.1177/1094428113518348

Gray, K., Dattakumar, A., Maeder, A., Butler-Henderson, K., & Chenery, H. (2014). *Advancing ehealth education for the clinical health professions: Final report 2014.*

Greenes, R. A., & Shortliffe, E. H. (1990). Medical informatics. An emerging academic discipline and institutional priority. *Journal of the American Medical Association*, *263*(8), 1114–1120. doi:10.1001/jama.263.8.1114

Hadeel, A., & Sandhu, K. (2021). PLS Model Performance for Factors Influencing Students Acceptance of Elearning Analytics Recommender. *International Journal of Virtual and Personal Learning Environments* (IJVPLE): New Perspectives in Online Education: Environments, Approaches, Authoring Tools.

Health, Q. (2017). *Digital Health Strategic Vision for Queensland 2026.*

International Council of Nurses. (2005). *Summary- Global shortage of registered nurses.*

Iyawa, G. E., Herselman, M., & Botha, A. (2016). Digital Health Innovation Ecosystems: From Systematic Literature Review to Conceptual Framework. *Procedia Computer Science*, *100*, 244–252. doi:10.1016/j.procs.2016.09.149

Kampmeijer, R., Pavlova, M., Tambor, M., Golinowska, S., & Groot, W. (2016). The use of e-health and m-health tools in health promotion and primary prevention among older adults: A systematic literature review. *BMC Health Services Research*, *16*(S5Suppl 5), 290. doi:10.118612913-016-1522-3 PMID:27608677

Lannan, K. (2015). *Interpretative phenomenological analysis: Exploring the formative experiences of community college students enrolled in remedial courses. (Doctoral dissertation)*, Colorado State University. Retrieved from https://dspace.library.colostate.edu/handle/10217/170325

Marschang, S. (2014). *Health inequalities and eHealth: 2014 eHealth Forum, Report of the e-Health stakeholder Group*. Retrieved from Athens: http://www.ehealth2014.org/wpcontent/

McConalogue, E., Davis, P., & Connolly, R. (2019). Health Technology Assessment: The Role of Total Cost of Ownership. *Business Systems Research*, *10*(1), 180–187. doi:10.2478/bsrj-2019-0013

Nancarrow, S. A. (2015). Six principles to enhance health workforce flexibility. *Human Resources for Health*, *13*(1), 9. doi:10.1186/1478-4491-13-9 PMID:26264184

Pontefract, S. K., & Wilson, K. (2019). Using electronic patient records: Defining learning outcomes for undergraduate education. *BMC Medical Education*, *19*(1), 30. doi:10.118612909-019-1466-5 PMID:30670000

Rich, E., & Miah, A. (2014). Understanding digital health as public pedagogy: A critical framework. *Societies (Basel, Switzerland)*, *4*(2), 296–315. doi:10.3390oc4020296

Sandhu. (2020). Digital Systems Innovation for Health Data Analytics. In N. Wickramasinghe (Ed.), *Handbook of Research on Optimizing Healthcare Management Techniques.* (Swinburne University of Technology, Australia & Epworth HealthCare, Australia). IGI Global.

Smith, J. A. (2011). Evaluating the contribution of interpretative phenomenological analysis. *Health Psychology Review*, *5*(1), 9–27. doi:10.1080/17437199.2010.510659

Torous, J., Chan, S., Luo, J., Boland, R., & Hilty, D. (2018). Clinical Informatics in Psychiatric Training: Preparing Today's Trainees for the Already Present Future. *Academic Psychiatry*, *42*(5), 694–697. doi:10.100740596-017-0811-4 PMID:29047074

USA Institute of Medicine. (2001). *Crossing the Quality Chasm: A New Health System for the 21st Century*. Washington, DC: The National Academies Press.

Venkatesh, V., & Davis, F. D. (2000). A theoretical extension of the technology acceptance model: Four longitudinal field studies. *Management Science*, *46*(2), 186–204. doi:10.1287/mnsc.46.2.186.11926

Vladimir, R.-B., Jorge, G.-M., Martín, L.-N., Yaroslava, R.-B., Javier, C.-R., Diego, Q.-P., ... Verónica Cevallos, L.-W. (2018). A Hybrid System to Support the Psychological Intervention for Children With Autism: A Proposal Based on Knowledge Modeling and Intelligent Technologies. [IJESMA]. *International Journal of E-Services and Mobile Applications*, *10*(4), 74–100. doi:10.4018/IJESMA.2018100105

We are social. (2019). Digital 2019, Q2 Global Digital Statshot. Retrieved from https://wearesocial.com/blog/2019/04/the-state-of-digital-in-april-2019-all-the-numbers-you-need-to-know

Wootton, R., Kvedar, J., & Dimmick, S. L. (2006). *Home Telehealth: Connecting care within community*. London, UK: Royal Society of Medicine Press.

World Health Organisation. (2016). *Global diffusion of eHealth: Making universal health coverage achievable. Report of the third global survey on eHealth*. Retrieved from Geneva, Switzerland: http://apps.who.int/iris/bitstream/10665/252529/1/9789241511780- eng.pdf?ua=1

World Health Organisation. (2019). *Global Strategy on Digital Health 2020-2024*.

World Health Organization. (2011). *mHealth: New horizons for health through mobile technologies; Global Observatory for eHealth series*.

Yogesan, K., Bos, L., Brett, P., & Gibbons, M. C. (2009). Handbook of Digital Homecare. Berlin, Gernmany: Springer-Verlag.

Ženko, Z., Mulej, M., & Potocan, V. (2017). Knowledge-cum-values Management belongs to the Way out from Global Crisis. *Business Systems Research*, *8*(1), 113–123. doi:10.1515/bsrj-2017-0009

ADDITIONAL READING

Brockes C, Grischott T, Dutkiewicz M, et al. Evaluation of the Education "Clinical Telemedicine/e-Health" in the Curriculum of Medical Students at the University of Zurich. *Telemedicine journal and e-health: the official journal of the American Telemedicine Association* 2017; 23: 899-904.

Dattakumar, A., Gray, K., Henderson, K. B., & … . (2012). We are not educating the future clinical health professional workforce adequately for e-health competence: Findings of an Australian study. *Studies in Health Technology and Informatics*, *178*, 33–38. PMID:22797016

Edirippulige, S., Brooks, P., Carati, C., Wade, V. A., Smith, A. C., Wickramasinghe, S., & Armfield, N. R. (2018). It's important, but not important enough: eHealth as a curriculum priority in medical education in Australia. *Journal of Telemedicine and Telecare*, *24*(10), 697–702. doi:10.1177/1357633X18793282 PMID:30343657

Gray, K., Dattakumar, A., Maeder, A., & .... (2014). *Advancing ehealth education for the clinical health professions: Final report 2014*. Sydney: Australian Government Office for Learning and Teaching.

Nancarrow, S. A. (2015). Six principles to enhance health workforce flexibility. *Human Resources for Health*, *13*(1), 9. doi:10.1186/1478-4491-13-9 PMID:26264184

Pontefract, S. K., & Wilson, K. (2019). Using electronic patient records: Defining learning outcomes for undergraduate education. *BMC Medical Education*, *19*(1), 30. doi:10.118612909-019-1466-5 PMID:30670000

Wootton, R., Kvedar, J., & Dimmick, S. L. (2006). *Home Telehealth: Connecting care within community*. London: Royal Society of Medicine Press.

World Health Organisation. Global diffusion of eHealth: Making universal health coverage achievable. Report of the third global survey on eHealth. 2016. Geneva, Switzerland

KEY TERMS AND DEFINITIONS

eHealth: The use of information and communications technology in support of health and health-related fields.

Electronic Health Record: A digital version of a patient's paper chart. It is real-time, patient-centered records that make information available instantly and securely to authorized users.

Health Informatics: The interdisciplinary study of the design, development, adoption, and application of IT-based innovations in healthcare services delivery, management, and planning.

Health System: All activities whose primary purpose is to promote, restore, and maintain health.

mHealth: Medical and public health practice supported by mobile devices, such as mobile phones, patient monitoring devices, personal digital assistants (PDAs), and other wireless devices.

Telehealth: Use of telecommunication techniques for the purpose of providing telemedicine, medical education, and health education over a distance.

Telemedicine: The delivery of health care services, where distance is a critical factor, by all health care professionals using information and communication technologies for the exchange of valid information for diagnosis, treatment and prevention of disease and injuries, research and evaluation, and for the continuing education of health care providers, all in the interests of advancing the health of individuals and their communities.

Chapter 12
Digital Transformation in the Hospitality Industry in an Emerging Country

Kanitsorn Terdpaopong
https://orcid.org/0000-0002-7978-3030
Rangsit University, Thailand

ABSTRACT

Emerging technologies have an undeniably significant impact on the hospitality industry. Hotel owners and general managers face many challenges to compete in an ever-increasingly competitive market. Many hotels are quick to adopt some of the emerging technologies to maximize brand visibility, gain operational efficiencies, and enhance customer experiences. The digital transformation is well underway. Hotels need to keep up with technological changes; otherwise, they are taking risks being left behind. This chapter encompasses the digital transformation of the hospitality industry to examine the impact of it on operational competence and organizational performance in an emerging country like Taiwan. Five case studies are used in this chapter. In-depth interviews with the executives of the leading hotels in Taiwan are a source of rich information and reflect the gradual transformation of technology in the hotel industry of Taiwan.

INTRODUCTION

Hospitality is a general term that refers to a segment of the service industry that includes hotels, restaurants, entertainment, sporting events, cruises, and other tourism-related services. The hospitality industry is essential not only to societies as a whole but also to a country's economy as well as customers and employees. The hospitality industry generates revenue for local economies directly when tourists spend money on hotels, restaurants, and entertainment venues. It also helps economies of countries indirectly due to tourist spending on retail goods, foods, souvenirs, and crafts. Besides, tourism can stimulate the demand for services such as tour guides, translators, and operators, restaurants and entertainment services, and infrastructures such as airports, roads, and public transportation. Also relevant are the jobs

DOI: 10.4018/978-1-7998-2799-3.ch012

created by the industry. World Travel and Tourism Council recently published an article detailing the economic impact of the tourism industry. In 2017 the hospitality industry accounted for 313 million jobs worldwide. It translates to 9.9% of total employment and 20% of all global net jobs created in the past decade (World Travel & Tourism Council, 2018). Hospitality also supports jobs in the arts and culture industries, helping and supporting the growth of the entertainment industry.

The benefit of the hotel industry is covering to sustain the country's economy. In IHG Annual Report and Form 20-F 2018 mentioned that in 2018, the global hotel industry is an USD525bn industry. The 54% of rooms are affiliated with a global or regional chain, up from 50% in 2012, and 46% are unaffiliated or independent hotels. The top five hotel groups, namely IHG, Marriott, Hilton, Wyndham, and Accor, account for 25% of market share, rising from 19% in 2012, and 58% of the global development pipeline of hotels in planning or under construction. At a country level, in the United States, for example, there has been a significant rise in occupancy rates. Using statistics from Smith Travel Research (STR), Sheel (2017) reports that the year-to-date occupancy increased 0.6% to 67.4%, relative to 66.9% for the same period in 2016. The year-to-date average daily rate, or ADR (September), grew 2.0% to $127.14.

Zhukova (2018) stated about the challenges experienced by the hotel industry regarding the changing priorities of the modern consumer that:

....the use of mobile payments, the changing priorities of the modern consumer, the environmentally friendly thinking of the consumer, the individualization of services in hotels, the presence of chatbots, the creation of smart and wellness rooms with using voice activation and Internet commands, installation in hotels of digital kiosks, self-check-in and check-out of a guest using information technologies, digital equipment polar fund hotels, use of a mobile phone as a lock to the number, availability in the hotels of interactive social networking innovations, the use of SMS-notifications in the loyalty program, the active use of internet marketing. The hotel industry is already actively introducing and involving modern innovations in its business practice (Zhukova, 2018, p. 62).

There are several metrics for measuring hotel industry performance. RevPAR[1] is an essential indicator of the value guests ascribe to a given hotel, brand, or market and increases when guests stay more often or pay higher rates. Rooms supply is also significant because it is reflective of the attractiveness of investing in the hotel industry and is influenced by the assumed profitability of a brand or market. Driven by strong economic fundamentals, the global hotel industry has seen growth in both RevPAR and rooms supply over the past decade. It also plays a vital role economically, accounting for every 1 in 10 jobs around the world (IHG, 2018, p. 8).

The forces of digital technology are rapidly transforming all industries, and the hospitality industry is included. According to a Meltwater report, Asia's hospitality sector has a growth of 4 - 5% per annum. All related parties, customers, employees, and hotel investors increasingly expect services and operational processes to be available on-demand and fully connected. The hospitality industry is, therefore, rapidly adapting services to be more responsive, more flexible, and more efficient to deliver rich guest experience and long-term business earnings. Hotels that fail to transform digitally will more or less be at a significant disadvantage in terms of both operational efficiencies and meeting customer expectations (HospitalityBiz, 2018). In this chapter, the hotel sector is focused on a subset of the hospitality industry. Samples of Taiwan hotels are taken in this chapter. In-depth interviews with the executives of the hotels located in Taipei provides rich and insight information to understand the situation of Taiwan hotel industry and see the disruptive technology that might be both beneficial and harmful to the hotel industry in Taiwan.

BACKGROUND

With the changes in technology, the hospitality industry has been undergoing tremendous changes and disruptions. New challenges, as well as new opportunities, makes the hospitality industry attractive. New trends that have reshaped the hospitality industry, especially in the hotel business, are such as virtual communities by which social networks have had a profound impact on customers; sharing economy like Airbnb represents a significant disruption in the hotel industry; and online travel agents. These are a few new trends among several others that have an impact on the hotel industry, and they are making the competitive landscape even more robust than ever (Masset and Weisskopf, 2019). It contributes to the result that not all hotels have been successful in these new trends. Some hotels are successful even though in a highly competitive environment, while others are bankrupt and close down and make the end of their hospitality businesses. The questions rise to both practitioners and scholars what makes for a successful hotel? The research article by Jeffrey, Barden, Buckley, and Hubbard (2002) gives insights into hotel management. They had followed 15 years of hotel occupancy analysis in England. Using time-series analyses of daily and monthly occupancy rates in different samples of hotels in England for over 15 years, the result reveals consistent temporal components of occupancy performance. These differentiate hotels in terms of overall occupancy levels, seasonality, length of a season, trend, and within-week variations. They concluded that the components are related to the characteristics of hotels and their management (Jeffrey, Barden, Buckley and Hubbard, 2002).

The hospitality industry experienced dynamic growth and challenges, not only the global crisis but also market changes, consumer behavior, and technological trends. With an increasing role of the hospitality industry to the economy of the country, research on hospitality has been conducted worldwide. The aim of the studies in this field may differ depending on each research initiative; however, the ultimate goals of such research contribute to improve and sustain the hospitality industry. This will indirectly and relatively impact the economy of the countries or regions. Accounting, the language of business, is therefore required to keep up with changes made to each particular area or industry, hospitality is no exception. The more sophisticated the business environment is, the increasing the role of accounting, especially on management accounting, to meet the demand of management becomes.

In a situation where the competitiveness between hotels is increasing, hotel managers increase the focus on improving their performance as it can become their advantage. With competitive benchmarking, these improvements must be identified and made by the hotel management (Min & Min, 1997; Tsai, Song, and Wong, 2009). The vital managerial competencies required and the importance of hospitality experience of hotel financial controllers are studied. Data from a longitudinal study of the hotel financial controllers have been analyzed. The same authors gathered information using a content analysis of job advertisements, a survey of U.K. hotel financial controllers, and interviews with key financial managers in Burgess's (2017) research. The findings show that although there are many similarities with the generic accounting profession, the hotel finance role is unique. Their experience, combining with hospitality management competencies, can support operational managers. Their understanding of the complexity of the operation, in a dynamic and perishable environment, enables hotel financial controllers to act, on behalf of stakeholders, as business advisors to other managers, thereby enhancing profitability (Burgess, 2017).

The hospitality industry is growing worldwide. However, there are variations among countries. Several researchers who studied the hospitality industry focused on the country context. For example, Jeffrey, Barden, Buckley, and Hubbard (2002) studied the hotel industry in England; Min & Min, 1997; Tsai, Song, and Wong (2009) studied in U.K. hotel market; Briciu, Scorte and Mester (2013) investigated the hospitality industry in Bihor County, Romania; El-Shishini (2017) examined the relationship between contextual variables and the use of managerial accounting techniques in hotels industry in Bahrain; Pavlatos (2011) conducted research on 85 leading hotels in Greece; Makrigiannakis and Soteriades (2015) investigated 48 four and five-star hotels in Greek; Yu and Huimin (2007) analyzes and compares Chinese hotel financial performances; Shieh (2012) using 68 international tourist hotels in Taiwan from 1997 to 2006 as samples of the studies, are among many others.

In this chapter, the focus is on the Taiwan hotel industry. Taiwan has a dynamic capitalist-type economy with a semi-presidential government. The country enjoys strong growth sustained by the dynamism of the entire region, a population with high purchasing power, and an economy focused on technology. From the statistics, Taiwan has one of the lowest fertility rates in the world, which portends a looming labor shortage soon. In an attempt to counter this, new Taiwan regulations provide preferential tax incentives to foreign professionals employed in Taiwan. The goal is aimed at improving the overall environment for recruiting and attracting professionals from other countries.

There are significant drawbacks, however, for conducting business in Taiwan, the main one being the lack of direct communication with China, and another is a minimal domestic market. Nonetheless, Taiwan is ranked very high because of the ease of doing business and its strategic geographical location, which serves as an entryway to China and the ASEAN markets. Despite these drawbacks, with modern infrastructure and innovation capabilities, Taiwan is an attractive destination for investors (Market Entry, 2018).

Despite the conflict between China and Taiwan, the Peoples' Republic of China (PRC) is still the largest market for Taiwan businesses. However, in 2018, the number of Chinese visitors has dropped dramatically by 37.9%: from 974,000 to 605,000. Based on Market Entry (2018), the number represents the lowest number since 2012. Decreasing numbers of PRC visitors and tour groups have caused the industry to focus on regional source markets to partially offset this trend. In the long-term, Taiwan needs more innovation in its tourism sector in order to capture higher-rated foreign independent travelers (Voellm, 2017). Fulcoon was somewhat pessimistic when he concluded that the hospitality industry in Taiwan might not be expected to return to normal or rise again soon (2018).

MAIN FOCUS OF THE CHAPTER

This chapter discusses the Taiwan hotel industry, the research method used for data collection, the definition of critical terms regarding complexity, and findings of this research and conclusion and suggestion in this order. The questions addressed in this present chapter are i) how are management tools implemented in the Taiwan hotel industry?; ii) how advanced of technological-based implemented in Taiwan hotel industry, especially on complex hotels; iii) what are the revenue and cost structures of the Taiwan hotels; iv) what management tools are being used by the hotel executives in Taiwan? In this chapter, the impact of technology on the sales revenue of the hotel, the implementation of a technological business model, revenue and cost structures, the implementation of management accounting practices, financial and non-financial performance are discussed.

LITERATURE REVIEW

Hospitality has been widely interested and studied by several researchers. Briciu, Scorte and Mester (2013) investigated the hospitality industry in Bihor County, Romania, regarding the impact of accounting information on managerial decision. They conclude that 65.93% of their surveyed managers believe that the accounting information is very useful in decision making, compared to 62.63% who believe in financial accounting. The most important factor influencing the quality of management decisions is the lack of cost calculation system (88%) and the subjectivity of the accountants (78%).

El-Shishini (2017) examined the relationship between contextual variables and the use of management accounting techniques in the hotels industry in Bahrain. He believes that the size of the hotel, the intensity of competition, and the quality level of the hotel have a positive effect on the use of management accounting techniques. He explored these assumptions through the distribution of the questionnaires to the 37 hotels in Bahrain. He found that benchmarking, absorption costing, budgeting for organizational activities, in ranking order, are the most use by the hotel managers, while activity-based costing (ABC), activity-based management (ABM), and long-range forecasting are the least used by the hotel managers. He concluded that traditional management accounting techniques were widely used at hotels more than the recent management accounting techniques. Also, his result indicated that the intensity of competition has a significant positive relationship with the use of management accounting techniques, while the size and quality level of the hotel does not have any relation with the use of management accounting techniques. Part of this finding contradicts to the conclusion made by Fathi, Dozdahiri (2015), which the authors stated that advanced technologies are implemented in hotels to lower their expenses. The same authors used ABC implementation as a focus of their study and concluded that organizational, technological, individual, and environmental factors influence on ABC implementation in the hotel industry. Pavlatos (2011) researched 85 leading hotels in Greece regarding the impact of strategic management accounting and cost structure on the ABC system. The same author found that the adoption of the ABC system is positively associated with the extent of use of strategic management accounting techniques and with the cost structure while the size of the hotel has no relation to the adoption of the ABC system. In other words, no matter what size the hotel is, the ABC system can be adopted. This finding, again, contradicts a result of Poldrugovac, Tekavcic and Jankovic (2016) as they found that there is a significant relationship between sizes and hotel efficiency.

The majority of Greek hotels operate on a seasonal basis for about six to seven months per year. As a result, prices are heavily discounted, and contribution margins are low (Makrigiannakis and Soteriades, 2015). The authors investigated 48 four and five-star hotels in Greek. They found that size and sales mix structure have a positive impact on the number of operational departments' analyses of contribution margins. In other words, hotels with a high sales volume are more sensitive to calculation accuracy of profit sensitivity analyses and thus make extensive use of cost information for pricing.

Not only the studies conducted in developed countries, a developing country such as China is also in the pace of changes. The current development of the hotel industry in China, particularly the rapid growth of Chinese hotel brands and international hotel companies, is also concerned. The challenges facing hotel financial management in China and the rapid globalization of hotel development and operation and financial performance are the concern of researchers. The article by Yu and Huimin (2007) analyzes and compares hotel financial performances managed by international hotel companies, Chinese hotel management companies, and Chinese independently-managed hotels. They provided recommendations for improving hotel financial reporting and management for China's hotel industry (Yu & Huimin, 2007). The link between cost efficiency and financial performance, as it pertains to the hotel industry, is studied by Shieh (2012). This study employs the data envelopment analysis approach to estimate cost efficiency and uses three traditional financial indicators—the ratio of net operating profit before taxes, the ratio of earnings before taxes, and return on assets before taxes—to measure financial performance. Using 68 international tourist hotels in Taiwan from 1997 to 2006 as samples of the study, the significant finding indicates that cost efficiency is insignificantly associated with the financial performance regardless of the financial performance variable used (Shieh, 2012).

The optimal occupancy rate, operational, and profitability efficiency of Taiwan's international tourist hotels are studied by Chiu and Huang (2011). They stated in their research that high performance in operational effectiveness does not necessarily ensure high profitability. The same study shows that increasing sales is not the best way to improve performance, actually better to decrease the occupancy rate to enhance operational and profitability efficiencies. Evidence from the same authors concluded that the inconsistent occupancy rate targets could be remedied through an empirical model (Chiu & Huang, 2011).

Of course, both financial and non-financial performance is of the management concerns. Recent research on non-financial performance measurement in Cape Town, South Africa hotel industry, is studied by Mjongwana and Kamala (2018). They found that their 100 responding hotels use non-financial performance measurements to improve the hotel operation. Customer-oriented measures were the most frequently used by the small and medium-sized hotels in Cape Town – sales growth percentage, and occupancy levels, guest satisfaction (with the results of means of 4.77, 4.73, and 4.67 out of 5, respectively). These non-financial performance measurements provide businesses with feed-forward information and are used for improving profitability, productivity, and effectiveness as well as improving decision making (Mjongwana and Kamala, 2018).

The past literature gave heed on how the hotel managements used managerial tools to support their management decision. Due to a different environment, different economic context, competitive environment, technological advancement, country development, and culture, a country context is commonly used as a subject of the study. In this study, an in-depth investigation of the use of management accounting tools in the Taiwan hotel industry is studied. Like other hospitality industries in other countries, Taiwan is facing significant competitive challenges. Taiwan tourism is an essential component of Taiwan's economy, as well as a significant source of foreign exchange revenues, contributing 9.3% of the country's GDP (Mordor Intelligence, 2018). The political issue between Taiwan and mainland China plays a vital role in Taiwan's hotel industry development. This article does not touch on this topic instead of the management accounting tools being used by the management of Taiwan hotels.

The argument makes in this chapter is that the level of management tools used by the hotel management depends on the *"complexity"* of the hotels. The more complex hotels either on the type of hotels as chain hotels, the multiple locations of the hotels, and the luxuries of the hotels which generally classified by the number of stars; the more sophisticated management tools would have been used by the hotel management. The hotels with such complexity will use more often and sophisticated or advanced levels of management tools to help in decision making. The level of complexity of the hotels that the more complex they are, the full range of management tools will be sought for by management of the hotels. The proposition makes in this paper does not directly relate to the size of the hotels. However, there is a tendency that the larger hotels will possibly use more often and more advanced in management tools due to the rising level of issues to solve, and at the same time, larger hotels would have a financial capacity to invest in management systems to assist in their businesses.

RESEARCH METHODOLOGY

Much of the previous research surveyed in the previous section did not use in-depth interviews, while most of the researchers used questionnaires. Thus, the findings of those earlier studies maybe, to some extent, less reliable due to such obstacles such as questionnaire bias, including the way individual questions were designed, and in the way, the questionnaires were administered. Additionally, there may have been unanticipated communication barriers between the investigator and the respondents. Another perceived flaw in the previous research related to the use of a predetermined list of responses, rather than the use of open-ended approaches (Choi and Pak, 2005). To overcome these issues, this present study uses an in-depth interview method to investigate and to understand more deeply regarding management practices in the hotel industry and its digital technology transformation.

As the study aimed to gain knowledge from those directly involved in the digital transformation and management tools, this present study focused on five leading hotels in Taipei, the capital city of Taiwan. The five hotels were selected because of their unique qualities and were representative of 3 – 5-star hotels. Permission to carry out this study was negotiated beforehand by an exchange of written and verbal communication. In each case, the Chief Financial Officers (CFOs) agreed to collaborate.

This research is a qualitative exploratory study with phenomenological design to explore the hotel industry in Taiwan. Intensive individual in-depth interviews of the CFOs of the five hotels in Taipei were conducted during November – December 2018. The interviews were carried out mostly in English, while specific questions and responses needed an interpreter. The interpreter worked in parallel with the researcher to ensure the understanding and accuracy of the exchanges. Each of the in-depth interviews took place at the respective hotel venues and were between 1.5 – 2.0 hours in length.

An interview protocol was used to ensure consistency among the interviews, and thus increase the reliability of the findings. Most of the questions were open-ended. After each interview, the recorded data were then transcribed and reviewed. Patterns or themes among the participants were identified and grouped. The data from the interview was labeled 'Case 1', 'Case 2', and so on. The finding will be presented according to size – the number of hotel rooms – beginning with the smallest and ending with the largest.

FINDINGS

The Characteristics of the Selected Hotels in Taiwan

To classify the hotel industry, a degree of complexity is considered. Awang, Ishak, Radzi and Taha (2008) define the degree of environmental complexity of the hotel industry according to 6 dimensions: 1) geographic concentration of competitors; 2) geographic concentration of industry sales; 3) geographic concentration of labor availability; 4) level of products/services differentiation; 5) geographic concen-

Figure 1. Case studies used in this research
Source: Compiled by author

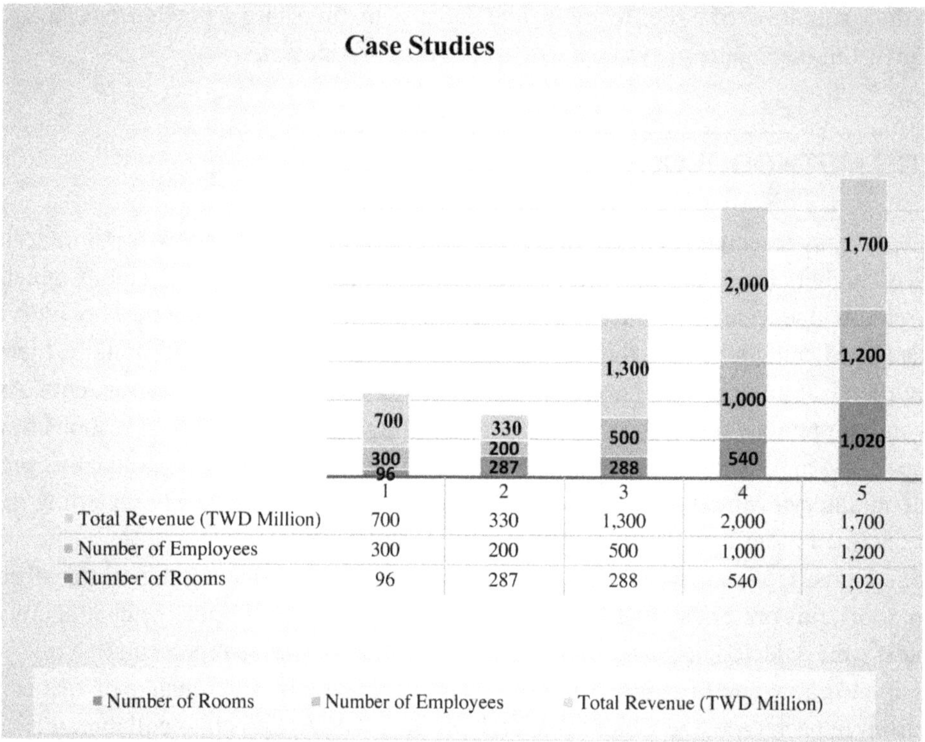

tration of customers; and 6) technological diversity used in the industry. The level of complexity of the hotel industry affects hotel classification. Other classification may base on size, star, location/clientele, ownership, level of services. The Institute of Hotel Management Bhubaneswar, Ministry of Tourism, Government of India (2017), for example, uses a classification system based on sizes (small, medium, large, very large); on Stars (1 star up to 5 stars); on locations (downtown hotel, sub-urban hotel, resort hotel, airport hotel, motel, floatels and hotels); on clientele (business or commercial hotels, transient hotels, boutique hotels, residential hotels, suite hotels, bed & breakfast hotels, casino hotels, conference centers, and green hotel); on ownership formality (independent hotels, chains, management contracts, franchise, referral groups, time-share hotels, condominium); on levels of services (upper-market which are luxury, world-class services hotels; mid-market; budget/economy hotels).

In this study, the levels of complexity are: highly complex, complex, and less complex[2]. The level of complexity is determined by the combination of at least three of the following: 1) number of stars, 2) size according to the number of rooms, 3) the number of employees, and 4) ownership.

The Impact of Technology on the Sales Revenue of the Hotel

As changing technology affects the hotel industry as a whole, the environment of the hotel industry has changed dramatically. For example, the competition between traditional hotels and Airbnb is intensifying every single day. Early January 2018, Airbnb announced it is expanding its experiences offerings to an additional 1,000 cities. Meanwhile, the lodging industry is not only adding its offerings but stepping up

Table 1. Descriptive information of cases 1 to 5 of the study (Number of stars is in the parenthesis)

Case 1 (****)	**Case 2** (***)
The "Case 1" hotel is a four-star hotel, established in 1998. Being a small luxury four-star hotel, it can provide high-quality service. It features a famous restaurant - the steakhouse, which was recommended by Michelin. Even though the hotel is quite small, it has a high standard of service quality.	The "Case 2" hotel has been operating for 39 years (since 1979) as a family business. The ownership has now been transferred to the second generation. The current Chief Executive Officer (CEO), who is an owner of the hotel, has an overseas experience over 20 years.
Case 3 (*****)	**Case 4** (*****)
The Case 3 hotel was established in 1999. It is a large five-star hotel, under the Tourism Group of Taiwan, and is located in a business area of Taipei. The Group also operates other hotels as well as a theme park and property residences. This hotel has been financially supported by its group.	The Case 4 hotel is a franchised luxury hotel under the IHG (InterContinental Hotels Group PLC) group. The IHG business model predominantly franchises its brands and manages hotels on behalf of third-party hotel owners. The hotel in Taipei is a five-star hotel and has been in operation for more than 20 years.
Case 5 (****)	
The 'Case 5" hotel has incorporated since1962. It was Taiwan's first international-standard luxury hotel in Taipei's city center, and in 1982 it was listed on the Taiwan Stock Exchange. It is generally regarded as one of Taiwan's most respected and successful companies. The company is owned by a group of families. The owner of the Case 5 hotel is also the CEO of the hotel. The Case 5 hotel in Taipei is especially well-known for its Michelin chef and restaurant, a steak house, and a five-star bakery.	

Source: Compiled by author

lobbying efforts in local and federal circles for stricter regulations governing Airbnb (Source: Forbes.com, 2018). With Airbnb's steady growth, it affects the revenues of the hotel industry. The question regarding the impact of technology such as Airbnb, robotic hotels, and the use of technology by customers was in the research concern.

Case 1 hotel which is a small hotel to some extent mentioned during the interview that:

....There is a small effect on the hotel due to the point that the hotel even though a small hotel. However, it is a luxurious hotel by which premium at all things applies such as room service premium, restaurant menu at premium grade, business function at premium quality, for instance. However, in the long run, perhaps the hotel would have been affected by such technology. (Excerpt from the interview on 20 December 2018)

Case 2 hotel stated regarding the customer choice with the aid of technology that:

.....Once a customer has made their choice of destination, the customer usually searches for the best hotel deals with just a matter of minutes. With the advancement of technology, it shapes how lodgings and restaurants are found and booked by the customer. Once the customer arrives at the hotel, they can check-in using electronic kiosks, or they can make online or mobile check-ins. No need to wait in line for the room keycard. In a luxurious hotel, customers can control all the functions of the room from a single device. This advanced technology has increased customer experience. (Excerpt from the interview on 7 November 2018 with the respondent Case 2 Hotel)

Case 3 hotel stated regarding the benefit of technology and cost reduction:

.....Most industries, the hotel industry, is no exception, agree to emergence of cost-saving technologies that help with reducing fixed cost and overhead cost. Technology can also help with guest loyalty. For example, reducing paper consumption, newspapers are normally provided to the customer as a complimentary. The newspaper consumption is challenging for hotels as it can be seen as a waste. However, not to offer the newspapers seems unusual, especially for luxurious hotels. With the use of technology, the hotel could offer a choice to the customer to download the newspapers and magazines of their preferences. This option has added value to the service, customize the choice for the individual customer and at the same time safeguard the environment, and increase customer satisfaction.

Besides, using technology can help the hotel to save the cost of energy, to predict the maintenance of the hotel facilities, to enhance the guest experiences, to remote checking-in and checking-out, to room services, for example. (Excerpt from the interview on 18 December 2018 with the respondent Case 3 Hotel)

The security issue is developed both positive and negative from the use of technology. The Case 5 Hotel talked about the paramount importance of security in the digital age.

....Hotel can use technology to improve security for the guests. For example, biometric authentication offers a secure method of identification proof, and such increases the level of safety to the guests. Some of the hotels use fingerprint activated room entry, and others use facial recognition. This technology is still in early stages. Some of the guests may have adverse reaction to this new technology. However, many agree that it is important, and having such a system will enhance security measures. Most of the customer appreciate hotels that paying attention to the security system regardless of their adverse interaction with such technology. (Excerpt from the interview on 6 December 2018 with the respondent Case 5 Hotel)

Implementation of a Technological Business Model

It is perhaps easy to assume customers prefer the "human" touch more than technology. However, business management should be careful about making such assumptions. Previous research shows an undoubtedly positive response to technology; there are other benefits to be gained from advanced technology implementation. For example, hotels can use of the various types of technology generates enormous quantities of data, which when properly analyzed and managed, enables the hotel to personalize their services in a way customer have never experienced before. Another sample is an analysis of voided checks in hotel restaurants and bars that can help management identify the cause of problems and take appropriate steps to improve and prevent future customer dissatisfaction and losses. Technology can help to track how often individual guests book hotels, the average length of a stay, their typical room service orders, and the movies they watch give hotels the opportunity to personalize offers from meal choices to entertainment. This makes the customer feel doubly welcome and valued.

Sustainability in a highly competitive hotel business environment requires effective business strategies and tools. Technology, such as an online booking system, including checking-in and check-out are typical in the modern-day hotel industry. However, small hotels may still be reluctant to adopt such time-saving and cost-saving tools. Larger hotels are more likely to have such supporting tools in place, while small hotels may not. Data from the interview with the CFO of the Case 2 hotel revealed adopting such technology would hurt an important customer base, namely Japan. Such a tool would be unfamiliar to that segment. Hence the old system which relied on long-term relationships with travel agencies continued.

However, the new CEO dramatically changed the business model, as revealed in the following excerpt from the interview.

Under the modern management of the new CEO, instead of relying more on a long-term relationship with travel agencies, the new CEO has started a business via an online travel agency (OTA) such as Agoda, booking.com, TripAdvisor, Priceline, Liberty Holidays, Expedia and Hotel.com. With the new media platform, the hotel attracted a broader range of customers from all over the world. Even though the primary customers are still from Japan (50%), customer diversification has varied since the new media platform was adopted. The second group is Mainland China customers, and the third is local Taiwanese customers. (Excerpt from the interview on 7 November 2018 with the respondent Case 2 Hotel)

Most of the hotels, either small or large adopt some sorts of technology to help them to reach customers easily. The Case 2 Hotel mentioned about the implementation of OTA that:

After the implementation of OTA, sales revenue increased by about 15%. While this practice appeared to be rather new for small hotels, it was quite usual in larger hotels where major management decisions in terms of technology adoption and implementation usually are a head office decision. The hotel management needs to train and retrain the staff to understand the system and to use the information generating from the system effectively. (Excerpt from the interview on 7 November 2018 with the respondent Case 2 Hotel)

Due to customer behavior has changed regarding technology, the business model of the hotel needs to give heed and change accordingly. The question regarding the generation and the hotel business model was mentioned by the Case 2 Hotel:

Customers in different generations has their technology preferences. For example, younger customers or customers in g[3] preferences are different from customers in other generations. Most Generation Z people equip with advanced technology at their hands. The use of mobile phones is sky-rocking in this generation. People at this generation book their hotel room, restaurant, and other entertainment services via their mobile phone. The customer behavior changes and differ from the customer in generation X[4]Y[5] or baby boomer[6]. As hotel management, it is therefore suggested to study the customer habits and preferences of this new generation customer and keep an eye on that moving forward. The trends and tendencies of how people of different generations use technology are vital to the hotel industry. Furthermore, the customers and/or prospective employees are a powerful force as they are driving purchases. In our hotel, we are thinking of implementing the technology to motivate the customer and keep the customer's royalty. I think it will help in sustaining the hotel revenue in the long run. (Excerpt from the interview on 7 November 2018 with the respondent Case 2 Hotel)

Form the interviews; it is found that the business model of hotels has been transformed to some extent regarding the implementation of advanced technology. For small hotels reaching out to customers may be a lot easier than the past with the aid of technology. Adopting new technology for larger hotels may not be a problem. However, the larger size of the hotel makes it sometimes slow and difficult to change the staff's familiarity. Training is vital and is a key to train the staff to get the most benefit out of technology.

Revenue and Cost Structures

The hotel industry is seeing strong revenue growth, mainly driven by room services and restaurants. Cost control is, therefore, essential, especially for a business that involves labor services. Cost control is a practice of identifying and reducing operational expenses to increase the margin for businesses. In the hotel industry, RevPAR and labor cost breakdown, ratio to sales, for example, are standard performance metrics. They provide important financial information within a specific period. Management can assess such data as room occupancy and the price of the average room. However, RevPAR alone may not the only metric which needs to be used. For example, it could be the case that a hotel may have a low RevPAR but at the same time having many rooms that would yield higher revenues. Therefore, a hotel manager often considers it together with average daily rate and occupancy rate as performance measures. All the hotels, of course, seek to increase the average daily rate by focusing on pricing strategies, including upselling, cross-sale promotions, and complimentary offers such as free spa and dining service in the hotel, while an overall economy is an additional significant factor in price setting.

Cost classifications involve the separation of a group of expenses of a business into different categories. Cost classifications for hotels can be classified as fixed costs and variable costs. They could be labeled as direct costs and indirect costs. The cost structure of hotels varies. For example, in the Case 2 hotel, the direct cost was the cost of goods/services sold with an average of 56% of the hotel revenue, while the indirect cost was about 20%. However, the most profitable department was in the rental department, which earned the highest contribution margin, while the restaurant department did not do so well as the rental department. In Case 5, the hotel's restaurant earned a major international award. Some 60% of the total hotel revenue came from that restaurant compared to 40% of the room rental division.

The RevPAR of Case 5 was set at NTD 4,500 per day (room rate of NTD 5,000 multiplied by 90% of occupancy rate). The restaurant revenue last year (2017) was NTD 1,200 million, where the fixed costs were 30%, and the variable costs were 40% of revenue. As a result, the margin for the restaurant was about 30% of its income. What follows are statements from the respondent:

For the room revenue, the hotel reported its revenue last year of NTD 500 million, with a fixed cost of 30% and a variable cost of 10%, leaving the hotel margin from this revenue at about 60% of its revenue. The profit margin of room revenue was 60%, while that of the restaurant was 30%. The profit margin of room revenue was double of the restaurant revenue. This is due to the cost structures of the Hotel, where room revenue the cost is more on fixed, while the restaurants are more on both variables (40%) and fixed (30%). (Excerpt from the interview on 6 December 2018 with the respondent Case 5 Hotel)

Implementing Management Accounting Practices

The accounting department regularly prepares financial statements – basically the monthly income statement. While other reports from each profit department are prepared on a fortnight basis. The contribution margin income statements are prepared by each profit department. They are monitored by the CFO and CEO in the meeting normally held twice a month for planning and decision-making purposes. The contribution margin income statement shows all variable expenses deducted from sales and arrives at a contribution margin. Fixed expenses are then subtracted to arrive at the net profit or loss for the period. The amounts remaining can be used to cover fixed costs and contribute to operating profit. Variance reports, budgeting reports, break-even-point reports, together with RevPAR, are typically reported on a regular basis.

Despite the highly competitive environment for the hospitality industry, complex accounting tools are not commonly used as in the case of Case 2 hotel management. Basic and less complicated management accounting reports are more commonly used. Complex accounting reports such as balanced scorecard, activity-based costing, total quality management report, lean, or just in time report are not practiced in that hotel.

There are other types of reports which are less complicated and which seem to be more practical for a small specialized hotel such as the Case 2 hotel. For example, total quality management (TQM) is commonly used in many industries, including, but not limited to, manufacturing, banking and finance, and medicine. It focuses on developing a work culture that emphasizes customer satisfaction. All members of the organization participate in improving processes, products, and services. The goal is to achieve continual improvement of business operations.

The Case 2 hotel strives to ensure all employees work toward common goals of improving service quality, as well as improving the procedures that are already in place. Particular emphasis is put on fact-based decision making and using performance metrics to monitor progress. High levels of organizational communication are encouraged to maintaining employee involvement and morale. The CFO did not recognize the term 'Total Quality Management (TQM)', however from the interview, all the process that the hotel practiced are those embodied in the TQM concept, but in a simplified version. Moreover, the CFO's office requires less time consuming and less complicated reports. More information comes from the interview.

For instance, the hotel has adopted the system known as Hotel Housekeeping Standard Procedures and Standard Operating Procedure (SOP)[7]. The housekeepers need to execute cleaning and maintenance tasks at various places in the hotel based on the set standards. The housekeeping staff can contribute to retaining guest satisfaction as well as to generate new guests willing to repeat their visits to the hotel. This will brings in more revenue to the hotel. What the hotel performs towards guest satisfaction and works productivity together, is an integral part of total quality control. The second example is on supply chain management. Due to the restaurant and cafeteria departments, ingredients supplying to the Hotel are from contracted suppliers. The hotel's goal when it comes to food and beverage services is to provide the top quality food and beverages. All cuisines and bakery are in-house produced. The hotel keeps a 10-day delivery contract with its food suppliers. With the software assisting in the control stock and inventory, the urgent purchase can be made. The hotel has at least three potential suppliers at a time. If one does not qualify either due to price, quality, quantity, or delivery schedule, the next potential supplier can replace such purchasing orders. (Excerpt from the interview on 7 November 2018 with the respondent Case 2 Hotel)

The concept of 'just in time (JIT) and lean' has often been employed in the manufacturing industry; however, to the hotel industry, perhaps some justification may be needed. The Hotel has implemented the JIT and lean system by having a lot size of ingredients such as food, to be delivered, and lead time (10 days in this case) that is adjusted to consumption. These practices are similar to those used in the manufacturing industry but less complicated.

As a standard practice, the accounting department prepares the financial statements – basically the income statement on a monthly basis. In the Case 5 hotel, accounting software is used. The operational

Table 2. Cost driver for cost allocation

Expenses	Cost Driver for Cost Allocation Purposes
1. CEO expense	Budget revenue (TWD)
2. Human Resource expense	Number of employees
3. MIS expense	Equipment quantity
4. Purchase expense	Percentage of working hours
5. Banquet expense	Banquet revenue (TWD)

Source: Compiled by author

accounting software – Oracle, Infrasys (*Infrasys Gourmate software for all Food and Beverage Sectors in the hotel*), and the Taiwan Accounting System (ALL NEW GL) are being used. As an example, Infrasys software can integrate with external systems such as hotel property management systems, membership systems, catering software, as well as accounting and inventory management systems. The software can switch ordering modes (fine dining / fast food) in real-time. With the aid of accounting software, it usually takes three business days to prepare and complete the financial statements — the segment reports prepared using the contribution income statement format. Variance reports, budgeting reports, break-even-point reports, together with RevPAR, are reported regularly.

The CEO of the Case 5 hotel initiated the activity-based costing (ABC)[8] system some three years ago. Each year, in the executive meeting, the topic of the cost drives is the topic of discussion. Information on the cost drivers is collected and presented for discussion. Every department is represented at this meeting, and a decision is reached as to which are the best cost drivers for their cost allocation. There were 28 cost drivers in 2018. Examples of the main cost drivers are illustrated below:

The cost allocation using ABC was about NTD$ 300.00 million, which was 70% of total expenses. The problem that the Case 2 Hotel experienced in implementing ABC was an argument about the cost drivers being used and allocated to the expenses of related departments. Several departments tried to negotiate about which costs would be allocated to them.

Oracle Hospitality (OPERA Cloud Services) is an enterprise platform for hotel operations and distribution. It offers comprehensive, next-generation capabilities hotels can use to enhance guest experiences and improve operating efficiency. It creates a platform for a hotel property-management system and a point-of-sale system. It manages the hotel, food, and beverage operations. Integrated Oracle Hospitality hotel solutions increase operating efficiency and enable better guest experiences for independent hotels and hotel chains (ORACLE, 2019).

The top accounting programs used in the hotel industry today are i) FreshBooks, ii) Sage Intact, iii) Inn-Flow, iv) Hotelogix, and v) M3AS software. To supply quality materials to the hotel, the hotel involves over 1,000 suppliers throughout the world. The contracts with suppliers are renegotiated every three months, based on quality and quantity considerations.

Financial and Non-Financial Performance of the Hotels

Hotel management personnel routinely concern themselves with financial performance measurements. Most use financial ratios such as return on assets (ROA), the return on equity (ROE), percentage of the cost of goods sold to total revenue, profit margin ratio, and net operating profit ratio. Unlike financial performance measures, non-financial performance measurements (NFPMs) provide the hotel with feed-forward information, a future-oriented measure. These are more relevant for planning purposes and day-to-day operations.

Hotels have used non-financial measurements to provide a closer link to long-term organizational strategies, while financial performance measures generally focus on annual or short-term performance against accounting standards. NFPMs are also progressive concerning meeting and exceeding customers' expectations as well as gaining and maintaining a competitive advantage over competitors. As mentioned by Micheli and Manzoni (2010), non-financial performance measurements are more critical to decisions that affect profitability and other long-term strategic goals. Hotels also have used several measures based on customer perspectives, such as:

- Guest complaints on employee helpfulness, on facilities, on services standards and other types of complaints
- Guest satisfaction surveys via online travel agencies
- Percentage of sales growth on a monthly period, quarterly period, and annually
- Room occupancy percentages
- Number of repeat customers
- Market share percentage

For the internal business operation, hotels also use non-financial performance measurements such as

- Ability to adjust, rectify, and respond to guest's complaints/requests
- Response time to guest requests
- Hotel supplies meeting the hotel purchasing standards (time, quality, quantity, price)
- Training hours of the staff (for example, training staff on work attitudes, how to deal with customer complaints, restaurant management, hospitality administration, and catering)

The CFO of the Case 2 hotel discussed the issue of customer satisfaction:

Customer satisfaction is one of the primary non-financial indicators that hotel management uses to monitor performance. For example, the hotel has set the customer satisfaction score at least 8.5 from 10 point-scale, which the survey was done via the online travel agencies. Furthermore, the hotel will take consideration of the customer complaints thoughtfully. The hotel has strategically tied the employee's benefit or bonuses to customer satisfaction. The hotel has differentiated its strategy by building customer experience as the culture of the hotel. The bonuses to employees who receive complaints and who do not show improvement will be deducted according to the hotel's criteria. (Excerpt from the interview on 7 November 2018 with the respondent Case 2 Hotel)

The Case 2 interviewee stressed that the effort to personalize the services to its customer would build, retain, and increase a healthy relationship. Other hotel financial managers also emphasized customer relationships. This is especially true in the case of the leading luxury hotel with its long history in the hospitality industry. Valuing long-term customers is essential.

SOLUTIONS AND RECOMMENDATIONS

In this chapter, the impact of the hotel industry on the economy is discussed. The direct impact would be the employment that hospitality creates to society. An indirect impact is a further employment that is created as a result of hospitality, such as jobs created in the supply and delivery industry, marketing, tourism, entertainment business, or even in the farming industry. The hotel industry contributes to a nation's GDP, the tax it contributes, and the employment it creates. As technologies continue to mature, the interplay of AI, IoT, blockchain, and other advanced technologies has been expanding and mutually reinforcing in wide practical operations and poses crucial questions to practitioners and policymakers. The hotel industry is nevertheless facing the challenge of reducing its operational costs significantly, and to structure an efficient internal process of it is to compete successfully in a competitive market. Taiwanese hotels are no exception. Having the right tools and right systems are therefore, of great importance to hotel managers on a day-to-day and long-term basis.

Concluding from the interviews, the size of the organization is positively associated with the sophistication used of tools and systems by the hotels. Small hotels tend to be less advantage in terms of having sophisticated technology. However, with the cost of technology is becoming cheaper and cheaper, small hotels can benefit from advanced technology. While larger hotels may benefit from the economies of scale, the flexibility to equip to the advanced technology is less than small hotels. However, with the use of AI, larger hotels are more ready to invest and implement across the board.

Secondly, the use of advanced technologies requires highly skilled personnel. The hotels need to train and retrain the staff to understand how to use the technologies, to learn a new operational way when advanced technology is interplayed. The implementations of advanced technology due to the complication when implementing require to have strong support from the management of the hotel.

And, thirdly, complexity. The hospitality industry, specifically hotels, to able to survive in today's highly competitive market, must adopt and make use of management tools for making routine decisions, and for the continuous management of many critical operations. This research learned that some hotels are still not generally using very sophisticated management system; most of the hotels used tools relate to cost determination and for controlling purposes.

The basic principles that the hotel must adhere to in order to retain the consumer and ensure their repeated visits to the hotel should be in practices; for example: addressing individual guest by name; collecting contact information from guests and e-mail in order to have further contact for recognition and next invitation; encouraging guests to write reviews on social media; studying the preferences and requirements of guests; providing quality service in every service area of the hotel; preferably direct booking. Furthermore, the recommendations for hotel management is perhaps to create programs to increase consumer, as well as a set of technical innovations to meet the changing consumer behavior. Hotels should be proactive regarding the service quality and see the changing environment as a challenge.

FUTURE RESEARCH DIRECTIONS

It is hoped that this study fills a gap in the literature by using an in-depth research methodology that focused on how management tools and practices are being accepted and being implemented. Of course, other aspects should be further explored, too. The interaction of Information Technology (IT) and contemporary management systems is an emerging research area. In order to sustain the hotel industry in the long run, financial issues regarding hotel performance could be further explored. Another important topic for research could be identifying factors that affect the improvement of hotel financial performance, for example, exploring the relationship between Accounting Information Systems (AIS) user satisfaction. As this research is a qualitative one, future research may consider covering the quantitative research using questionnaires in order to gather more extensive information. A mixed-method between qualitative (using interview and/or focus group) and quantitative (using questionnaire or secondary data for statistical analysis) can enhance a deeper understanding of the hotel industry. The next research could increase the number of participants. If the interview is to be conducted, more interviewees should be sought to cover hotels at different characteristics. The research may be conducted frequently due to the fast-changing of technology, which may affect the performance of the hotel industry. The rapid change could be formed, and hotels may incorporate advanced technology to use in their organizations easily.

CONCLUSION

In this chapter, Taiwanese hotel executives were interviewed. The questions wrapped around the advanced technology being used, adopted, and implemented in their hotels. The hotels in this research are highly complex, complex, and less complex. The hotel responses were well aware of the advanced technology that might affect them in the short and long run. Technological software was implemented in a highly complex hotel, but not all. Even though advanced technology should be employed in their opinion, they do not take a big step to change the hotels dramatically. The use of robots and machines may be easily implemented in a less complex hotel but not a highly complex or a five-star one.

In order to sustain the hotel industry, hotel management may consider the use of advanced technologies to help in both operation and management. The interview data revealed that many sophisticated practices were not being used. Among these are the following: the lean method in warehouse management, reduction of movement of housekeeping, supply chain management software, just-in-time, and real-time information reports. These tools were rarely mentioned by interviewees. When compared to the manufacturing industry, for example, where such sophisticated tools are familiar, hotels continue using rather simple management accounting practices. Even though the hotel business is highly competitive and becoming more so, there seems to be a reluctance to consider implementing practices that would improve the precision of key decisions and increase sustainability. Particularly owners of smaller hotels are urged to consider adding more sophisticated management tools to their operations. Consumers are changing, too, especially the Millennials who are typically and increasingly looking for unique and authentic experiences. They have money and time to travel, and they increasingly expect technology to aid, inform, and enrich their travels. Several new 'smart' applications now available include intuitive booking apps, chatbots, and mobile check-in/check-out. Today's and future hotel customers expect technology to be integrated into many areas of the travel experience. To meet this trend, the ability of hotel companies to work in partnership with technology providers has become increasingly important. Hotel owners and investors should take note.

Also, another feature of today's world has to do with data generation, storage, and uses. Cloud storage has further changed the game, giving accommodation providers, such as hotels, easy access to diverse real-time data. Such data enables a more personalized and efficient service. Operationally it allows providers to use data to tailor guest experiences faster and create a more personalized relationship with them. With this trend comes a growing responsibility to handle data responsibly, respecting consumer preferences, and rights. With the equip technology at hand, digital transformation in the hospitality industry could smoothly offer good experiences to their guests and could assist the hotels regarding cost-saving and revenue opportunities and enabling hotel owners/investors to reach new levels of profitability. Many hotels quickly define their specific goals of digital transformation, the enabler that accomplishes it is always connecting and reaching customers to the hotel services, while the unable ones may lose the opportunity to earn profits. Both cannot deny the role of digital technology that plays a significant role in the emerging hospitality industry.

ACKNOWLEDGMENT

The author would like to thank the Ministry of Foreign Affairs of the Republic of China (Taiwan) for the support through the 2018 Taiwan Fellowship Program (MOFATF201800090). Special thanks to the CFOs and senior financial managers of the five hotels for their cooperation and for providing vibrant and insightful information.

REFERENCES

Awang, K. W., Ishak, N. K., Radzi, S. M., & Taha, A. Z. (2008). Environmental variables and performance: Evidence from the Hotel Industry in Malaysia. *International Journal of Economics and Management*, *2*(1), 59–79.

Briciu, S., Scorte, C., & Mester, I. (2013). The impact of accounting information on managerial decision – Empirical study conducted in the hospitality industry entities in Romania. *Theoretical and Applied Economics*, *9*(586), 27–38.

Burgess, C. (2017). Managerial Competencies for U.K. Hotel Financial Controllers: Are They Hospitality Managers or Accountants? *The Journal of Hospitality Financial Management*, *25*(1), 27–43. doi:10.1080/10913211.2017.1313612

Chiu, Y.-H., & Huang, C.-W. (2011). Evaluating the optimal occupancy rate, operational efficiency, and profitability efficiency of Taiwan's international tourist hotels. *Service Industries Journal*, *31*(13), 2145–2162. doi:10.1080/02642069.2010.503889

Choi, B. C., & Pak, A. W. (2005). A catalog of biases in questionnaires. *Preventing Chronic Disease*, *2*(1), A13. PMID:15670466

CIMA. (2008). Activity Based Costing Topic Gateway Series No. 1. Available at http://www.cimaglobal.com/Documents/ImportedDocuments/cid_tg_activity_based_costing_nov08.pdf.pdf (Retrieved on 24 August 2019).

El-Shishini, H. (2017). The Use of Management Accounting Techniques at Hotels in Bahrain. *Review of Integrative Business and Economics Research*, *6*, 64–77.

Fathi, Z., & Dozdahiri, E. S. M. (2015). A survey of activity-based costing in hotel industry. *Management Science Letters*, *5*, 855–860. doi:10.5267/j.msl.2015.6.012

Fulcoon, M. (2018). Taiwan's Hotels Grapple with Oversupply. Taiwan Business Topic. Available at https://topics.amcham.com.tw/2018/12/taiwans-hotels-grapple-with-oversupply/ (Retrieved on 3 August 2019)

Gerdeman, D. (2018). The Airbnb Effect: Cheaper Rooms For Travelers, Less Revenue For Hotels. Forbes.com Feb 27, 2018. Retrieved from https://www.forbes.com/sites/hbsworkingknowledge/2018/02/27/the-airbnb-effect-cheaper-rooms-for-travelers-less-revenue-for-hotels/#7612dd3d672f

HospitalityBiz. (2018). The Digital hotel has Arrived. Retrieved from https://www.hospitalitynet.org/news/4090589.html (Retrieved on 1 December 2019).

IHG. (2018). IHG Annual Report. Available at https://www.ihg.com/hotels/gb/en/ (Retrieved on 12 March 2019).

Institute of Hotel management Bhubaneswar, Ministry of Tourism, Government of India. (2017). Available at http://www.ihmbbs.org/upload/2)Classifi%20of%20Hotels.pdf (Retrieved on 5 August 2019).

Jeffrey, D., Barden, R. R. D., Buckley, P. J., & Hubbard, N. J. (2002). What Makes for a Successful Hotel? Insights on Hotel Management Following 15 Years of Hotel Occupancy Analysis in England. *Service Industries Journal*, *22*(2), 73–88. doi:10.1080/714005078

Makrigiannakis, G., & Soteriades, M. (2015). Management Accounting in the Hotel Business. *International Journal of Hospitality & Tourism Administration*, *8*(4), 47–76. doi:10.1300/J149v08n04_03

Market Entry. (2018). Tourism and Hotel Industry in Taiwan: Analysis of Growth, Trends and Progress (2015-2020). Available at https://www.mordorintelligence.com/industry-reports/market-entry-tourism-and-hotel-industry-in-taiwan (Retrieved on 3 August 2019).

Masset, P., & Weisskopf, J. P. (2019). Top 10 Trends Reshaping. *Hospitals (Rockville)*, *2019*. Available at https://hospitalityinsights.ehl.edu/2019-top-hospitality-trends Retrieved 25 August 2019

Micheli, P., & Manzoni, J. F. (2010). Strategic Performance Measurement: Benefits, Limitations and Paradoxes. *Long Range Planning*, *43*(4), 465–476. doi:10.1016/j.lrp.2009.12.004

Min, H., & Min, H. (1997). Benchmarking the quality of hotel services: Managerial perspectives. *International Journal of Quality & Reliability Management*, *14*(6), 582–597. doi:10.1108/02656719710186209

Mjongwana, A., & Kamala, P. N. (2018). Non-financial performance measurement by small and medium sized enterprises operating in the hotel industry in the city of Cape Town. *African Journal of Hospitality, Tourism, and Leisure*, *7*(1), 1–26.

Mordor Intelligence. (2018). Industry Report. Available at https://www.mordorintelligence.com/industry-reports (Retrieved on 8 March 2019).

ORACLE. Oracle Data Sheet, Oracle Hospitality OPERA Cloud Services, Available at http://www.oracle.com/us/industries/opera-cloud-services-2870466.pdf (Retrieved on 20 January 2019).

Pavlatos, O. (2011). The impact of strategic management accounting and cost structure on ABC systems in hotels. *The Journal of Hospitality Financial Management*, *19*(2), 37–55. doi:10.1080/10913211.2011.10653912

Poldrugovac, K., Tekavcic, M., & Jankovic, S. (2016). Efficiency in the hotel industry: An empirical examination of the most influential factors. *Economic Research-Ekonomska Istraživanja, 29*(1), 583–597. doi:10.1080/1331677X.2016.1177464

Sheel, A. (2017). Hotel Industry Performance in 2016–2017 and the JHFM Index. *The Journal of Hospitality Financial Management, 25*(2), 75–76. doi:10.1080/10913211.2017.1406743

Shieh, H.-S. (2012). Does Cost Efficiency Lead to Better Financial Performance? A Study on Taiwan International Tourist Hotels. *The Journal of Hospitality Financial Management, 20*(1), 17–30. doi:10.1080/10913211.2012.10721889

Tsai, H., Song, H., & Wong, K. K. F. (2009). Tourism and Hotel Competitiveness Research. *Journal of Travel & Tourism Marketing, 26*(5–6), 522–546. doi:10.1080/10548400903163079

Voellm, D. J. (2017). Taiwan Hotel Market Update. HVS Asia-Pacific. Available at https://www.hvs.com/article/8094-taiwan-hotel-market-update (Retrieved on 3 August 2019).

World Travel & Tourism Council. (2018). Travel and Tourism Economic Impact 2018: World. Available at https://www.wttc.org/ (Retrieved on 15 March 2019).

Yu, L., & Huimin, G. (2007). An Analysis of Hotel Financial Management in China. *The Journal of Hospitality Financial Management, 15*(2), 39–48. doi:10.1080/10913211.2007.10653841

Zhukova, M. A. (2018). Development of the hotel industry based on the use of information technologies. Управление, (4), 62. doi:10.26425/2309-3633-2018-4-62-66

ENDNOTES

1 Revenue per available room (RevPAR) is a performance metric used in the hotel industry. It is calculated by multiplying a hotel's average daily room rate (ADR) by its occupancy rate. It may also be calculated by dividing a hotel's total room revenue by the total number of available rooms in the period being measured.

2 A highly complex hotel is a 4-5 star hotel with a number of rooms over 400 and a number of employees over 500. The ownership of this highly complex hotel can be a chain, management contract, or franchise hotel. The target customer of this type is an upper market.

Complex hotel is a 3-4 star hotel with a number of rooms between 100-400 and a number of employees between 300-500. The ownership can be either chain, management contract, or franchise and focuses on upper to mid-market.

Less complex hotel is a 1-2 star hotel with less than 100 rooms and has staff less than 300 staff. It is an independent hotel which aims for mid-market or budgeted customer.

3 eneration Z, or simply Gen Z, is the cohort of people born after the Millennials. Demographers and researchers typically use the mid- to late-1990s as starting birth years. Source: Dimock, Michael. "Defining generations: Where Millennials end and Generation Z begins". *Pew Research Center.* Retrieved 26 March 2019.

4 Generation X, commonly abbreviated to *Gen X*, is the generation following the baby boomers. Demographers and researchers typically use starting birth years ranging from the early-to-mid 1960s

and ending birth years in the early 1980s. Source: Horovitz, Bruce (4 May 2012). "After Gen X, Millennials, what should the next generation be?". *USA Today*. Retrieved 24 November 2012.

5 Millennials, also known as *Generation Y*, are the cohort of people following Generation X. Researchers and popular media use the early 1980s as starting birth years and the mid-1990s to early 2000s as ending birth years, with 1981 to 1996 a widely accepted definition. Source: Horovitz, Bruce (4 May 2012). "After Gen X, Millennials, what should next generation be?". *USA Today*. Retrieved 24 November 2012.

6 Baby boomers, also known as *Generation W*, the Me Generation or Boomers, are the generation that was born mostly following World War II, typically born from 1946 to 1964. Increased birth rates were observed during the post–World War II baby boom making them a relatively large demographic cohort. Source: Constance Ridley-Smith (21 January 2019). *Beyond Bricks and Mortar: Sym-Q: 5 Factors That Predict Intercultural Success*. ISBN 9781491859537

7 A standard operating procedure (SOP) is a set of step-by-step instructions compiled by an organization to help workers carry out complex routine operations. SOPs aim to achieve efficiency, quality output, and uniformity of performance while reducing miscommunication and failure to comply with industry regulations.

8 It is a costing method that identifies activities in an organization and assigns the cost of each activity to all products and services according to the actual consumption by each. The latter utilizes cost drivers to attach activity costs to outputs. CIMA (Chartered Institute of Management Accountants) defines ABC as an approach to the costing and monitoring of activities, which involves tracing resource consumption and costing final outputs. Resources are assigned to activities, and activities to cost objects based on consumption estimates. The latter utilizes cost drivers to attach activity costs to outputs (CIMA, 2008).

Chapter 13 Digital Government Competences for Digital Public Administration Transformation

Fatma Bouaziz
Faculty of Economics and Management, University of Sfax, Tunisia

ABSTRACT

The development of digital government leads to the digital transformation of public administration. This requires a mix of competences to design, implement, and manage digital government endeavors and the resulting transformations and to use new digital tools. However, researchers agree that digital government competences are under-studied. Moreover, the rare existing studies focus on singular aspects of digital government competences such as technical skills. Thus, this chapter proposes a synthesis of the extant literature about the digital government competences and how they are approached by researchers. The supply-side perspective and the demand-side perspective are proposed to organize the extant literature. Competences are categorized according to each perspective. An empirical investigation is needed to determine the most important competences and to propose the necessary training programs for the lacking ones.

INTRODUCTION

Governments around the world have been intensively adopting information and communication technologies (ICTs) and more recently digital technologies, such as mobile technologies, social media, the Internet of things, wireless technology, sensor networks, artificial intelligence and Big Data (Gil-Garcia et al., 2018). Digital technologies lead to the development of digital government. It is "the use of digital technologies, as an integrated part of governments' modernization strategies, to create public value" (OCDE, 2014, p.6).

The UN reports (2002 - 2018) demonstrate a continuous and heavy use of digital technologies by public administration to deliver public services and interact with citizens, businesses and other stakeholders. Promoting efficiency, efficacy and transparency, reducing corruption and enhancing citizen's

DOI: 10.4018/978-1-7998-2799-3.ch013

empowerment are among the expected benefits. They are going to be achieved thanks to the potential of digital technologies as a driver of changes and transformations or what is called digital transformation (Hess et al., 2016; Nwankpa and Roumani, 2016; Heilig et al., 2017). Digital transformation is concerned with the changes that digital technologies can bring about in a company's business model which result in changed products, organizational structures or in automated processes (Hess et al., 2016). The same case is in the public sector that can be reorganized by making use of ICTs, which have the potential to rethink public services; how they are produced and which actors are involved in the process (Schulz and Schuppan, 2012).

However, it seems that the expected benefits in terms of digital transformation are not always achieved. The failure rate of e-government projects in developing countries is very high; 35% have been classified as a total failure and 50% as a partial failure (Heeks, 2003). Moreover, many examples of organizations have been unable to keep pace with new digital reality (Hess et al., 2016), while the success of digital transformation lies especially in the adoption of organizational aspects and not only in using advanced technologies (Heilig et al., 2017). For governments, they need to change the way they work and organize themselves, work collaboratively and engage with citizens and businesses (OECD, 2019). This will require having the needed skills and abilities to design, implement and manage e-government projects and use new digital tools (Bertot and Jaeger, 2008; OECD, 2019). Without the required competences, projects risk being hindered. This may be an issue in developing countries where a chronic lack of qualified staff and inadequate training of the human resources are common features of the public administration (Bannister, 2003; Ndou, 2004, etc.).

Griffin and Schappun (2010, p.6) defined e-government competencies as "the skills, abilities and knowledge required of a person to carry out a task or a set of tasks to a degree of competence (i.e. a standard of performance, usually specifically defined)". However, e-government competences are rarely discussed in the academic debate (Schulz and Schuppan, 2012; Hofmann and Ogonek, 2018). It is also difficult to find an established understanding of these competences, especially with the fact that the rare existing studies have focused on singular aspects of digital competences such as technical skills (Hofmann and Ogonek, 2018). Moreover, a comprehensive empirical analysis of the required digital competences does not exist in both public and private sectors (Hofmann and Ogonek, 2018).

Since scientific literature focusing on digital government competences hardly exists, this chapter intends to deal with this theoretical gap. A literature review will be made based on the scarce previous researches on e-government competences. The main objective is to propose a synthesis about the digital government competences and how they are approached in the extant literature. To achieve this objective, three sections are going to be developed. The first section will deal with the digital government and the digital transformation of the public administration. The second section will discuss the criticality of competences for digital government success. The third section will deal with competency requirements for digital government.

BACKGROUND

Digitization, Digitalization, and Digital Transformation

Several definitions of digital transformation have been proposed in the extant literature. Some of these definitions are selected in table 1.

Table 1. Definitions of digital transformation

Authors	Definitions
Liu (2011, p.1730)	Digital transformation is "an organizational transformation that integrates digital technologies and business processes in a digital economy"
Hess et al. (2016, p.124)	"Digital transformation is concerned with the changes digital technologies can bring about in a company's business model, which result in changed products or organizational structures or in the automation of processes. These changes can be observed in the rising demand for Internet-based media, which has lead to changes of entire business models (for example in the music industry"
Nwankpa and Roumani (2016, p.2)	"digital transformation is defined as an organizational shift to big data, analytics, cloud, mobile and social media platforms"
Heilig et al. (2017, p.3).	"We use the term digital transformation to exclusively refer to the required transformations driving the digitalization in and between organizations based on a digital strategy. More specifically, a digital transformation refers to a broader process of transforming an organization or a network of organizations on different levels (e.g., strategy, governance, leadership, culture, people, technology, etc.) by making use of digital technologies and concepts, to which we refer to as enablers".
Warner and Wäger (2018)	"The use of new digital technologies, such as mobile, artificial intelligence, cloud, blockchain, and Internet of things (IoT) technologies, to enable major business improvements to augment customer experience, streamline operations, or create new business models".
Barthel and Hess (2019, p.4)	"Digital transformation describes the use of digital technologies to enable or trigger major business change. […] The transformation goes beyond simply digitizing resources, and can involve the transformation of products, services, processes, communication channels, or complete business models. In that regard, digital transformation can be seen as a specific type of organizational transformation based on digital innovation. In our understanding, digital innovations are innovations that integrate a (digital) technological solution, as well as a (digital) business solution"
Dobrica (2019, p.38).	"Digital transformation is about doing things differently-creating a completely new business model by using modern information and computer technologies. Digital transformation leverages existing knowledge to profoundly change the essence of the organization-its culture, management strategy, technological mix, and operational setup. It places the customer at the center of all its decisions and actions"

Reviewing the definitions mentioned in table 1 shows that they are structured around four key concepts. These are *digitization*, *digitalization*, *digital technologies*, and *transformation*. In fact, Dobrica (2019) advanced that digital transformation assumes an umbrella role, encompassing digitization and digitalization as its constituting components and regarding them as necessary steps in the big frame of an organization's digital transformation. These concepts are successively defined in the following.

Digitization is the "encoding of analog information into digital format. It makes physical products programmable, addressable, sensible, communicable, memorable, traceable and associable" (Yoo et al., 2010, p.725, in Warner and Wager, 2018, p.45).

Digitalization is a socio technical process of using advances in digital infrastructures to analyze, interpret, and shape changes in social, institutional, and increasingly cognitive contexts (Warner and Wager, 2018, p.8). It creates potent digital affordances that likely have a transformative effect upon the organization of the economic activity by supporting radical business model innovation (Autio et al., 2018, p.76).

Digital technologies are "ICTs, including the Internet, mobile technologies and devices, as well as data analytics used to improve the generation, collection, exchange, aggregation, combination, analysis, access, searchability and presentation of digital content, including for the development of services and apps" (OECD, 2014, p.6). Heilig et al. (2017) designate digital technologies by enablers which involve novel IT delivery models (e.g., cloud/fog computing), pervasive computing (e.g., internet of things, cyber-physical systems), mobile computing, block chain, and data science supporting tools (e.g., big data, machine learning) and decision-making. Viewed as combinations of information, computing, communication, and connectivity technologies digital technologies are fundamentally transforming business strategies, business processes, firm capabilities, products and services, and key inter firms relationships in extended business networks (Bharadwaj et al., 2013: 471, in Warner and Wager, 2018, p.45).

Transformation in business English is the process of completely changing the character or appearance of something in order to improve it. When associated to digital technologies, digital transformation is perceived as an ongoing process of using new digital technologies in everyday organizational life, which recognizes agility as the core mechanism for the strategic renewal of an organization's business model, collaborative approach, and eventually the culture (Warner and Wäger, 2018). Furthermore, digital transformation is much more than a mere process redesign. It is about structuring new business operations to facilitate and fully leverage firms' core competence through digital technology in order to attain competitive advantage (Liu, 2011, p.1730). Relatedly, Singh and Hess (2017, p.124) suggest the term "transformation" rather than "change" to emphasize that an organization's digital transformation goes far beyond functional thinking and holistically considers the "comprehensiveness of actions" that must be taken to exploit the opportunities or avoid the threats that stem from digital technologies.

However, if digital technologies are a key concept to talk about digital transformation, they should only be seen as a means. Digital transformation is fundamentally not about technology. It goes beyond using advanced technologies and methods to optimize business to imply the adaption of organizational aspects (Heilig et al., 2017; Dobrica, 2019). This means that senior leadership teams must find ways and strategies to capitalize on new and unexpected business model innovations that optimize customer needs and experiences. Accordingly, digital business transformation occurs when companies "pursue new revenue streams, products and services, and business models" (Dobrica, 2019).

For more clarity, Dobrica (2019) has reviewed the terms digitization, digitalization and digital transformation through five facets: focus, goal, activity, tools, and challenges. The table 2 is reproduced to give a summary.

Table 2. Digitization, Digitalization and Digital Transformation facets

	Digitization	Digitalization	Digital Transformation
Focus	Data conversion	Information processing	Knowledge leveraging
Goal	Change analog to digital format	Automate existing business operations and processes	Change company's culture, the way it works and thinks
Activity	Convert paper documents, photos, microfilms, LPs, films and VHS tapes to digital format	Creation of completely digital work processes	Creation of a new digital company or transformation to a digital one
Tools	Computers and conversion/ encoding equipment	IT systems and computer applications	Matrix of new (currently disruptive) digital technologies
Challenge	Volume Material	Price Financial	Resistance to change Human resource
Example	Scanning paper-based registration forms	Completely electronic registration process	Everything electronic, from registration to content delivery

Source. Dobrica, S. (2019). From Digitization, through Digitalization, to Digital Transformation. Online Searcher, p. 37.

Thus, the digital technologies are drivers of changes and transformations, leading to digital transformation. The latter is more about the changes that digital technologies can induce in a company's business model than digital technologies. It results in changed products, organizational structures or in automated processes (Hess et al., 2016). However, this will be a challenging endeavor and digital transformation attempts can easily fail since managerial, cultural and societal challenges may arise (Heilig et al., 2017, p.3). Particularly, societal challenge is related to humans need to be trained for having the ability to use new digital solutions and to (re-)solve problems which are considered to be even more complex as more data is coming in and system-wide implications are more demanding (Heilig et al., 2017, p.3).

Nowadays, digital transformation has become imperative not only for organizations seeking to survive and attain competitive advantages in a digital economy, but also for government and public organizations (Liu, 2011; Dobrica, 2019).

Digital Government for Digital Transformation in the Public Administration

Governments around the world are extensively using digital technologies at the different levels and branches of the public administration. UN reports (2002-2018) show a great interest for these technologies introduced into government processes to support internal functioning and operations and external relationships. Gil-Garcia et al. (2018) have noted that it becomes hard to imagine any government function or governance process that does not involve extensive use of information and technology. This extensive use of digital technologies refers to digital government.

Digital government is "government facilitated by the internet-web and allied information and telecommunication technologies" (Zwass, 2006). It is "the use of digital technologies, as an integrated part of governments' modernization strategies, to create public value. It relies on a digital government ecosystem that is comprised of government actors, non-governmental organizations, businesses, citizens' associations and individuals; which supports the production of and access to data, services and content through interactions with the government" (OCDE, 2014, p.6). Digital government is "a concept that has broadened in scope from an early focus on the use of ICT for government administration to the more recent notion that information and technology influence administration, management, and governance" (Gil-Garcia et al., 2018, p.634). By this, it encompasses two principal domains, which are e-government and e-democracy (Zwass, 2006).

E-government refers to "the use by the governments of ICTs, and particularly the Internet, as a tool to achieve better government" (OECD, 2014, p.6). It refers to public administration services provision and information dissemination through ICTs. E-government addresses the needs of all levels of government as service providers (Zwass, 2006). For that purpose, e-government is based on electronic interactions among public administration, citizens, businesses, employees, and actors in civil society. All of these interactions rely on the use of ICTs to provide information and services and to realize internal administration activities, management and decision-making.

E-democracy refers to "the use of ICTs to support the process of broad participation of citizens and of those who represent others, in debates, political discourses, surfacing and elaboration of societal issues, emergence of representation, polling, and voting" (Zwass, 2006). It involves e-voting and political participation online.

Like digital transformation, digital government initiatives are perceived as complex change efforts using digital technologies to transform the operation and effectiveness of government (Pardo, 2000). Accordingly, digital government is interpreted as "a phenomenon" that "involves new styles of leadership, new decision-making processes, different ways of organizing and delivering services, and new concepts of citizenship" (Gil-Garcia et al., 2018, p.634). Schulz and Schuppan (2012) declare that the public sector as a whole can be reorganized by making use of ICTs and that ICTs have the potential to rethink which public services are produced and how and which actors are involved in the process. Pardo (2000) added that digital government is not about using computers or building web sites for information access, it is rather about transforming the relationship between government and the public. Maintaining a primary focus on the business of government and *not* on the technologies is one of the challenges to these efforts (Pardo, 2000).

Thanks to digital technologies, the public administration can better respond to stakeholders' requests and preferences. This requires the development of a digital government strategy that is congruent with a modernization strategy focused on the public value creation and based on the abilities of digital technologies (Bouaziz, 2020, forthcoming). In fact, according to the OECD, digital technologies should not just be used to digitize existing government processes and to offer public services online. Governments should prioritize using digital technologies and data to rethink the design and implementation processes of public services and policies and that's for the purpose of achieving more citizen-driven approaches. Ultimately, transformed public governance should lead to outcomes that best meet user needs (OECD, 2014, OECD, 2019). Moreover, to become digital, governments need to adopt and use digital technologies and data as strategic components of their efforts to modernize the public sector. Digital technologies and data reuse need to be integrated in core processes and activities in order to establish new ways of working and promote greater openness and collaboration (OECD, 2014; OECD, 2019).

Thus, the evolution towards the digital government fits under a paradigm shift of PA modernization and change of management more than an automatic application of ICTs to processes, structures, and operational modes (Bouaziz, forthcoming, 2020). In fact, these technologies are disruptive in organizations and nations, changing traditional forms and altering the status quo (Valle-Cruz, 2019). They are capable of changing the ways of communication-and interaction between users and organizations, facilitating communication and interactions, as well as reducing time and costs. Therefore, the governments' use of ICTs should be combined with organizational changes to improve the structures and operations of PA and to be able to create public value. This requires new governance and institutional frameworks and new capabilities and skills development to sustain a digital culture in the public sector (OECD, 2014; OECD, 2019).

Competency Challenge in the Case of Digital Government

With the development of digital technologies, authors have studied their impacts on the workforce and the necessary competences to employability. The conclusion is that they induce greater flexibility in working time and change in both job profiles and competences and leadership dimensions (Ganz et al., 2019). Furthermore, new digital competences and various skills beyond the technical ones are needed (Hofmann and Ogonek, 2018).

However, the need for new competences may be more critical in the case of digital government initiatives. Grundén (2012, pp.146-147) ascertained "it is important to have competent personnel with requisite ICT skills and other organization management attributes to run back-end e-Government applications for it to succeed. Also, there is need to have citizens and businesses who have the requisite ICT skills and understand the benefits of engaging in e-Government for them to continue consuming e-Government services". The author added that "the success of e-Government depends largely on human skills". E-government initiatives imply several stakeholders such as politicians, enterprises and citizens, public officers, e-government systems users and project managers (Bouaziz, 2011, p.150). The lack of willingness, competences and commitment of these stakeholders may impediment the e-government projects success (Bouaziz, 2011, p.150). The lack of competences and ICT skills is also commonly mentioned as a failure factor or a barrier of e-government projects implementation (Ndou, 2004; Ebrahim and Irani, 2005; Twizeyimana and Andersson, 2019). This is particularly in developing countries, where the public administration suffers from a chronic lack of qualified staff and inadequate human resources training (Bannister, 2003; Ndou, 2004). If suitable solutions such as training, competence development and lifelong learning are not initiated this risks to be deepened as the rate of changes increases and digital technologies continue to evolve.

In fact, the OECD (2018, 2019) insists on the association of the public sector digital transformation with a transformed and empowered civil service workforce, enriched by a new mindset, new capabilities and, above all, adequate leadership and digital skills to successfully lead to user-driven services. Hofmann and Ogonek (2018) recognized that the individuals need a diversified skill set in times of digitalization. They use the definition of digital competences proposed by Vieru et al. (2015, 4683). Accordingly, "digital competence is an individual capacity to use and combine one's knowledge (i.e. know-what), skill (i.e. know-how), and attitude (i.e. know-why) associated with three related competence areas, technological, cognitive and social, to use new or existing ICT to analyze, select and critically evaluate information in order to investigate and solve work-related problems and develop a collaborative knowledge base while engaging in organizational practices within a specific organizational context." This definition offers a framework of digital competence that encompasses three learning domains, knowledge (know-what), skills (know-how) and attitude (know-why) and three competence areas, namely technological competence, cognitive competence and social competence.

According to Vieru et al. (2015) and Hofmann and Ogonek (2018), the technological area is underpinned by the knowledge, skills and attitude needed to explore new technological contexts and face technological problem in a flexible way. The cognitive area is underpinned by the knowledge, skills and attitude needed to "read, select, interpret and evaluate data and information taking into account their pertinence and reliability". The social area is underpinned by the knowledge, skills and attitude needed to interact with other individuals collaboratively using available ICT along the line of the existing organizational work norms and values. In addition, the social area is concerned with issues related to responsible, ethical and safe use of ICT as well as to risks awareness, privacy, property rights and social media exposure.

Hofmann and Ogonek (2018) added a new dimension, namely "impact awareness". It refers to "the ability to evaluate the impact of digitalization on processes and activities outside of the digital world" (Hofmann and Ogonek, 2018, p.131). Awareness of digitalization and its impacts by the managers is crucial since they are seen as role models and provide the basis for a digital-friendly organization culture.

Although, the extant literature highlights the criticality of competences to digital government success, studies on the topic seems to be scarce. Only few scientific articles about competency requirements for e-government exist (Griffin and Schuppan, 2010). Moreover, the e-government competences are rarely discussed in the academic debate and even an established understanding of e-government competences does not exist (Schulz and Schuppan, 2012). This lack of literature has recently been highlighted by Hofmann and Ogonek (2018). They stated that "research has not yet presented a comprehensive empirical analysis of the required digital competences in the public sector".

RESEARCH METHODOLOGY

The aim of this chapter is to propose a literature review about the required competences for digital government and digital public administration transformation. To attain this objective, a research methodology based on the following steps is pursued:

- **The first step** is to inventory the previous works dealing with competency, skills, public administration and e-government.
- **The second step** is based on a content analysis of the selected works. This is done by identifying the unit of analysis retained in these works, i.e. citizens, employees in private organizations, public administration's staff and managers, and ICT specialists, and leads to classifying them under two perspectives of analysis, which are the supply-side perspective and the demand-side perspective.
- **The third step** consists to identify the competences categories discussed in the retained works and to categorize them around these categories.
- **The fourth step** is to make a synthesis by proposing a list of competences and skills according to each perspective of analysis.

Results about competency requirements for digital government are shown in the next section.

RESULTS

Digital transformation induces modifications in organizational structures, processes and skill sets that are necessary to cope with and exploit digital technologies (Hess et al., 2016). To successfully adopt digital technologies and conduct digital transformation, human resources skills, abilities and competences are of paramount importance. In particular, the OECD (2003) considers that all the staff in the public administration should hold e-government skills. Both ICT specialists and civil servants need to be armed with the required competences to successfully lead digital government initiatives. Hybrid skills are required. They encompass a large variety of skills and competences (i.e., mixed competences) which go far beyond a limited set of IT-related competences (Schulz and Schuppan, 2012).

The rare extant literature on the topic has proposed different frameworks and sets of the required e-government competences. This may add difficulties in understanding the required competences of e-government. Furthermore, choosing the framework to apply becomes fuzzy. For a better understanding of digital government competences, an analysis of these frameworks and associated works has been made. This analysis shows that they can be classified based on two major perspectives. These perspectives are designated as follows: the supply side perspective and the demand side perspective.

The supply-side perspective deals with the required competences for the public personnel involved in digital government initiatives. The public personnel may belong to three communities of public servants, which are the information technology community, the information management community and the service community (OECD, 2003; Parrado, 2005; Ojo et al., 2007; Sali, 2011). It may have different roles: staff, mid-level management and senior-level management (Schappun, 2008; Schulz and Schappun, 2012; Hunnuins and Schappun, 2013). Thereby, different sets of competences are proposed in the literature according to the roles of public servants and the communities to which they belong. However, it is worth noting that these sets of competences overlap. A clear frontier can not be traced since the

Table 3. Competences according to the roles of public personnel

Staff	Mid-Level Managers	Senior Level Managers
IT Literacy Information Processing IT specialist Process Management Organizational Design Cooperation Competence Communicative Competence Self-Control	Project Management Process Management Organizational Design Risk Management IT Strategy Skills Communicative Competence Time Management Cooperation Competence Leadership	IT Strategy Organizational Design Project Management Risk Management Change Management Communicative Competence Cooperation Competence Leadership

public servants accomplish their different roles as part of a community to which they belong and at the same time members in the digital government project team.

According to the roles of public personnel, Schappun (2008), Schulz and Schappun (2012) and Hunnuins and Schappun (2013) have found that, in addition to the IT-related competences, a large variety of skills and competences are considered important in the context of e-government projects. They are differentiated between generic government skills and competences and newly arising core e-government skills and competences. The generic government skills and competences contain personal competences in terms of creativity, self-control and motivation and self-management. They furthermore encompass policy and legal skills (policy process, administrative law and cultures, specialized law) and change-related skills (project and change management skills and implementation competence). The latter categories are also more or less generic competences that are required in the public administration, but which are necessary in order to implement the transformational changes.

The core e-government skills and competences include the e-government management skills and competences (risk management, quality management, performance management and contract management). In addition, they encompass e-government design competences (organizational design, process design, IS design, IT specialist, and marketing skills), e-Competences (IT literacy, information processing, and media skills) and e-Policy competences (e-Strategies and e-Policies, models and concepts and information processing law). These comprise rather new competences that arise in the context of e-government. Table 3 gives a summary of the most important competences according to the roles of public personnel.

According to the public servants communities, competences vary with the community to which public personnel belong. Three communities are identified. These are the information technology community, the information management community and the service community (OECD, 2003; Parrado, 2005; Ojo et al., 2007; Sali, 2011):

- **Information technology community:** is formed by technical staff that works primarily in fields such as information technology supplies and services, telecommunications, IT consultancy, multimedia and Internet-based products and services.
- **Information management community:** includes specialists and managers responsible for the provision of information management services. The members of this community have the role of shaping the content management to be produced for different output media. It encompasses professions like librarians, archivists, and access to information and privacy specialists, communication managers and records managers among others.

Table 4. E-government skills identification in four skill sets

Information Technology Skills	Information Management Skills	Information Society Skills	Updated Management Skills
Basic IT literacy Specialist IT skills	Information Management Privacy and security protection Information user service skills	ICT vision abilities ICT strategy and planning Organizational change	Risk management Communication and coordination Financing arrangements skills Public-private partnerships

Source: Sali, M. (2011). Exploring e-government skills and competencies required in public sectors : A literature study. 1-29 https://pdfs.semanticscholar.org/fe54/2fb75dd5234007e39de3ef25484482421005.pdf

- **Service community:** is formed by general managers and front line staff providing services to citizens and other stakeholders by traditional or electronic means. This community's actors develop the whole life cycle of a service. All legal, financial, managerial and other professional experts typically belong to this community.

Each community of public servants should hold a mix of competences among four sets according to its needs. These sets of competences are the information society skills, the information technology skills, the information management skills and the business skills/ updated management skills (OECD, 2003; Parrado, 2005; Ojo et al., 2007; Sali, 2011).

The information society competences constitute the contribution of ICT to the Service Community (e.g. the ability to use ICT resources to implement an organization's e-government strategy in accordance with its overall strategy). The information technology competences are skills necessary to implement e-governance such as IT literacy, hardware, software, and communication. The information management competences are about the deployment of knowledge resources within the public administration and the sharing of knowledge. The updated management competences represent the contribution of the service community to the IT and IM communities (e.g. traditional managerial skills are insufficient for new organizational ICT needs; hence, managers need skills to manage organizational changes resulting from e-government). Table 4 resumes these skills.

Unlike the supply side perspective, the demand side perspective is interested in the required digital government competences for citizens and business employee as the users of e-government systems, services and information. Although authors (Khan et al., 2010; Vrabie, 2012) have included public administration employee when analyzing citizens and business competences, they will not be considered under the demand side perspective in our analysis since they are taken into account under the first perspective. From a demand side, Van Deursen and Van Dijk (2008; 2009), Khan et al. (2010) and Vrabie (2012) have been interested in the required skills for e-government services and information consumption. Different categories are proposed, but their content is almost identical. Khan et al. (2010) proposed and validated a user centric approach for e-skills identification and development. For these authors, e-skills are "the set of skills, knowledge, and concepts that are needed for effective consumption-access locate, operate, manage, understand, and evaluate-of e-services provided in different stages of e-government." (Khan et al., 2010, p.3). The authors classified the e-skills needed by citizens and private sector employees for e-service consumption as technical (PC usage skills, standard program skills, and applied e-skills) and Internet skills (operational, formal, strategic, and informational skills). Van Deursen and Van Dijk (2008; 2009) have identified digital skills. These are the same as Internet skills of Khan et al. (2010), i.e.

operational, formal, strategic, and informational skills. Vrabie (2012) discussed computer skills needed by citizens. These are basic aptitudes (computers hardware, files and folders, using simple programs, web browsing basics, e-mail basics), seeking information skills (searching for information, security / precaution) and e-participation skills. They also interfere with the e-skills of Khan et al. (2010) and the digital skills of Van Deursen and Van Dijk (2008; 2009).

Unlike previous authors, Fielder et al. (2016, p.23) has offered a more recent framework of digital competences. It is comprised of 21 competences grouped under five areas: information and data literacy; communication and collaboration; digital content creation; safety; and problem solving. The sets of competences proposed by Fielder et al. (2016) are retained in the present work (figure 1). Several reasons motivated this choice. First, the framework offers a good starting point for developing a derivative framework for consumer competence in the digital context and reveals gaps where new competences need to be defined (Fielder et al., 2016). Second, it underlines the safety competences, in particular "protecting health and well-being" and "protecting the environment", which are among the public values that the digital government initiatives seek to realize (Deng et al., 2018; Karunasena and Deng, 2012). Third, it includes current competences that are not involved in the previous works of Van Deursen and Van Dijk (2008; 2009); Khan et al. (2010) and Vrabie (2012).

SOLUTIONS AND RECOMMENDATIONS

Figure 1 resumes the competences required to conceive, lead and use digital government initiatives. Two branches in this figure are elaborated. The first branch deals with the skills analyzed under the demand-side perspective. The second one deals with those studied under the supply-side perspective. These skills refer to the competences according to the public servants' roles and the communities to which they belong.

From a demand-side perspective, five categories of competences are retained which are information and data literacy skills, communication and collaboration skills, digital content creation skills, safety skills and problem-solving skills. These skills are a prerequisite for citizens and business employees to be able to use the digital government systems that are implemented by public administrations. If they do not have the prerequisite level of competences, they cannot adequately use these systems and benefit from their advantages. These competences are technical to use digital applications and tools. They also encompass abilities to treat information, resolve problems, exchange with each other, and collaborate in a secure context. The lack of educating and informing citizens as well as business employees about the value of e-government systems and the lack of competences to profit from them may cause the failure of these initiatives. In this case, the government should play the main role to overcome this issue. In fact, it is necessary to ensure that users have the abilities and the necessary skills to use the developed systems. Whether giving computers and Internet connections, implementing online services or creating pages on social networks is not enough; instead, the targeted users should be equipped with different sets of skills so as to be able to exploit the full potential of these initiatives. This can be achieved by introducing adequate training programs into the teaching curriculum and by planning the necessary communication and training actions in favor of the business employees concerned by the e-government systems. A full strategy to enhance digital literacy should thereby be elaborated in parallel to the development of the digital government and the digital technologies. This strategy may lie on e-learning possibilities and MOOCs.

Figure 1. Digital government competences

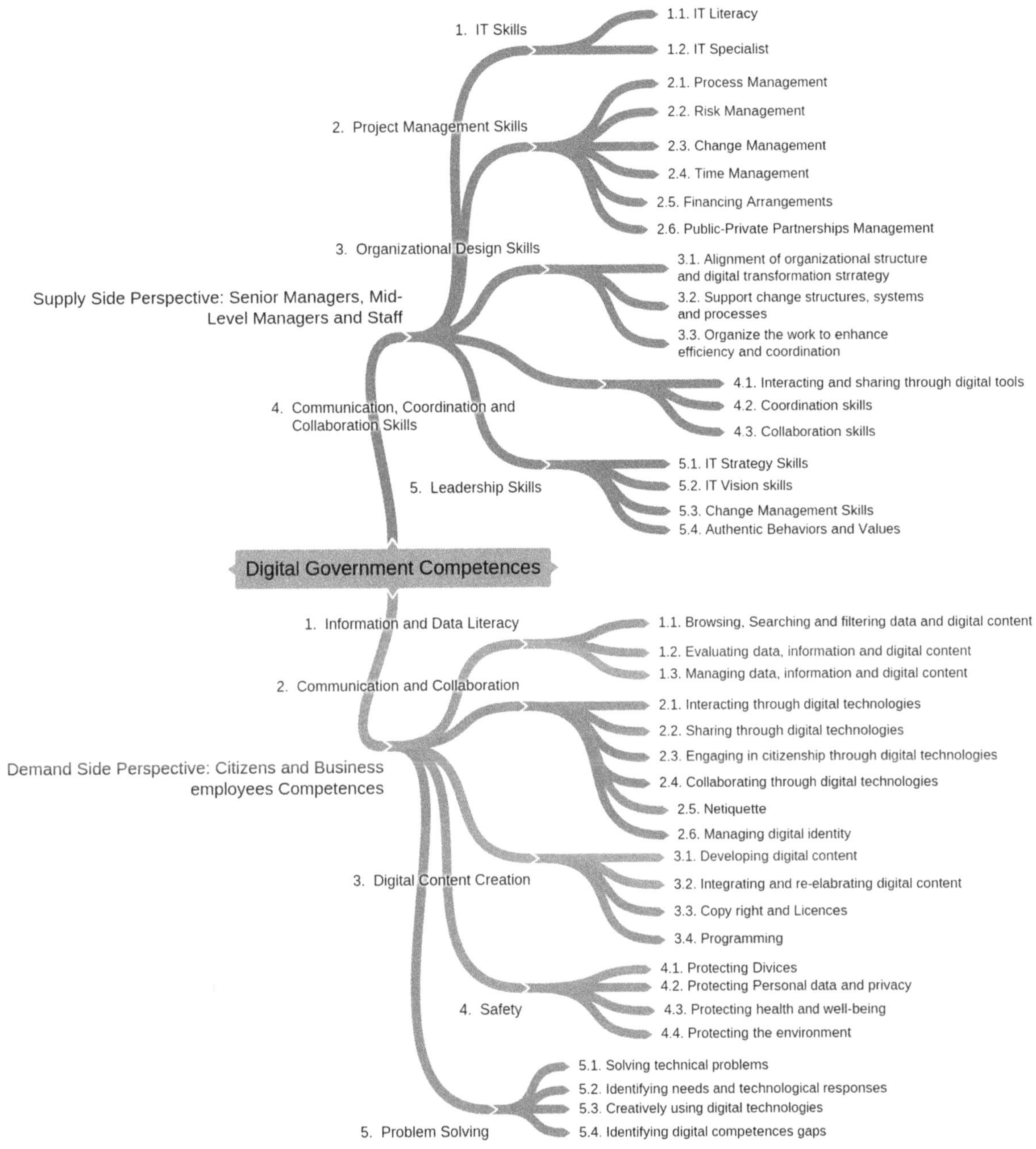

From a supply-side perspective, competences are identified according to the public servants roles and the public servants communities. These skills are varied according to the different levels of employees and managers and their different posts and ranks. However, they present commonalities and redundancies, which is understandable since the public servants accomplish their tasks while at the same time belong to a given community. Focusing on these skills, it follows that digital government requires up-to-date skills that are not simply technical since public servants need both basic technical understanding of digital tools (IT literacy) and broad skills so that they can make decisions about digital government projects and lead digital transformation of the public administration. Based on the literature review, five categories of competences are proposed, which are IT skills, project management skills, organizational design skills, communication, coordination and collaboration skills and leadership skills (IT strategy skills; IT vision skills; change management skills).

- **IT skills:** cover IT literacy skills and IT specialist skills. They concern the ones-necessary to public employees to enable them to use digital tools and applications to accomplish their tasks and understand the specificities of the online environments when navigating and interacting with citizens. They are also related to technical and specialized skills necessary to IT specialists so as to design and implement the technical features of a digital government project. These skills must be continuously updated so that IT specialists can follow the technological evolution.
- **Project management skills:** according to their roles in a digital government project, public servants need competences in risk management, process management, change management, time management, financial arrangements, public-private partnerships management, etc.
- **Organizational design skills:** digital government leads to the digital transformation of the public administration, which undergoes transformations in its design characteristics. This means that changes in structure, processes, roles, relationships, value creation mechanisms and business models should occur to take advantage of digital technologies. The choices in organizational design affect the likelihood of the innovation process success (Grimpe et al., 2019). Thus, conducting these changes requires organizational design skills to adequately define organizational changes, assess their impacts, evaluate the needed change efforts and prepare the change plans to implement them.
- **Communication, coordination and collaboration skills:** cover those skills necessary to use digital tools (wiki, web conferencing, survey tools, etc.) to transmit available information and ideas to others, to be sure that the meaning is effectively expressed and to create content collaboratively (Agyapong et al., 2018). Coordination and collaboration are a prerequisite to ensure information sharing within the organization and across departments. Several digital government initiatives transcend traditional organizations boundaries that require at the same time coordination and collaboration frameworks so as to have the ability to closely collaborate with other organizations. Digital technologies may be helpful in communication, coordination and collaboration within and across organizations.
- **Leadership skills (IT strategy skills; IT vision skills; change management skills):** are critical to ensure that a common understanding and a sense of mission are shared across all levels of government. Leadership skills are needed in all stages of digital government projects to gain acceptance, put implementation frameworks in place, manage change and sustain support for the project (OECD, 2003). Moreover, leadership skills are necessary at the political and administrative levels to develop a shared vision on the criticality of digital government to the modernization of the public administration. On the one hand, political leaders show political willingness and support to digital government and digital transformation strategies. This helps inscribe digital government initiatives in broad strategic objectives and modernization strategy. On the other hand, administrative leaders define action plans to concretize political vision. However, digital government brings digital transformation of public administrations and the expected benefits may take time to be achieved, which risks leading to resistances. This requires that leaders show strong commitment to deal with changes, to contribute positively in resolving problems and to define adequate visions

and plans. In this case, authentic leaders (Crawford et al., 2019; Lloyd-Walker and Walker, 2011) play a key role to positively influence other public servants' behaviors and adequately manage changes and resistances so that they can lead to success. Lloyd-Walker and Walker (2011) noted that authentic leaders have "confidence, optimism, hope, self-efficacy and resilience. They are clear about their own values and moral perspectives, knowledge and strengths and are equally aware of these attributes within others". Authentic leaders show authentic behaviors according to ethical behavioral practices and consequently affect followers' comportments. Unfortunately, it seems that many unethical practices have induced ethical behaviors deterioration among public servants, while improvement in ethical behavioral practices will increase the reputation of the public sector governance, not only over a short term but also over the long term (Abdullah et al., 2019).

FUTURE RESEARCH DIRECTIONS

This work inventories the required skills for digital government and public administration's digital transformation. The proposed list of competences helps managers elaborate an assessment tool for evaluating skill gaps and equipping public servants, citizens and enterprises employees by the needed ones. Digital tools may be used for conducting this assessment and acquiring the lacking competences. In fact, assessment procedures must be defined a priori to identify the level of available skills and those that are lacking and to plan solutions so as to upgrade the required skills, such as training and public-private partnerships. Taking steps to identify and provide the needed competences for public administrations to efficiently implement digital government initiatives as well as to maximize their benefits is a prerequisite. From a human resources management perspective, this proposes to adopt competency models and a competency-based approach in public administration. At this level, empirical research to recognize digital government competences at practical contexts is needed. Moreover, this work attire attention to the importance of the authentic leadership competency. The authentic leader's ability to establish and maintain relationships and to lead based on values contribute to the projects' success. Therefore, developing authentic behaviors practices among public servants and encouraging authentic leaders reinforces digital government projects. For future research, a focus is to make on the identification of the most important competences for each public servant category and the competency models to develop these skills in order to continuously improve and adapt human capital. Among these competences, leadership skills and authentic leadership deserve more investigation since authentic leaders affect followers' comportments while ethical behaviors are not usually evident in organizations.

CONCLUSION

The main objective of this chapter was to propose a literature review about the required competences for digital government and digital public administration transformation. Digital government is a challenging endeavor because public administration lacks necessary resources and skills to handle such initiatives and resulted transformations. Especially, with the permanent evolution of ICT tools and the development of digital technologies, the public administration should continuously upgrade the capacities of its human resources. The required competences for e-government have retained the attention of researchers.

However, they have adopted different perspectives of analysis, which make it difficult to find a common framework of these competences in the literature. Thereby, the required competences for digital government remain fuzzy. In this chapter, for a better understanding, a literature review was made and digital government competences were classified under two perspectives, namely the supply-side perspective and the demand side perspective (figure 1). The first one takes into account the competences of public servants according to the roles they assume and the community to which they belong. The second shows the citizens and business employees' competences required to use e-government systems, information and services. The analyzed literature shows that competences vary according to each perspective and its constituents. This will facilitate the comprehension of the different needs and prepare the necessary actions for formation, training plans and lifelong learning. Figure 1 gives a summary of the digital government competences cited by the literature. These competences deserve further investigations to identify the most important ones, assess the lacking ones and engage the suitable and necessary actions of formation and competences development. In fact, the availability of digital government skills will greatly affect an organization's ability to adopt digital government strategy and to conduct digital transformation. Finding public servants, citizens and business employees with the needed skills is critical to the success of digital government.

REFERENCES

Abdullah, A., Sulong, Z., & Said, R. M. (2019). An analysis on ethical climate and ethical judgment among public sector employees in Malaysia. *American Journal of Business*, *10*(2).

Agyapong, D., Brautlacht, R., & Owino, J. (2018). University Collaboration-Research Collaboration and Teaching Collaboration: Competency-Based Learning Using a Cross-Cultural Students' Project. Handbook of Applied Teaching and Learning.

Barthel, P., & Hess, T. (2019). Are Digital Transformation Projects Special? *Twenty-Third Pacific Asia Conference on Information Systems, China.*15. Retrieved from https://www.researchgate.net/publication/33445083

Bouaziz, F. (2011). E-Government Projects Risk Management: Taking Stakeholders in Perspective. In A. Thyab, A. Ajeeli, & Y. A. L. Al-Bastaki (Eds.), Handbook of Research on E-Services in the Public Sector: E-Government Strategies and Advancements, Chapter 13.

Bouaziz, F. (2020, forthcoming). From E-government to Digital Government: the Public Value Quest in the Tunisian Public Administration. In V. Ponnusamy, K. Rafique, & N. Zaman (Eds.), *Employing Recent Technologies for Improved Digital Governance*. IGI Global.

Crawford, J. A., Dawkins, S., Martin, A., & Lewis, G. (2019). Putting the leader back into authentic leadership: Reconceptualising and rethinking leaders. *Australian Journal of Management*; Advanced Online Publication. doi:10.1177/0312896219836460

Deng, H., Karunasena, K., & Xu, W. (2018). Evaluating the Performance of e-Government in Developing Countries: A Public Value Perspective. *Internet Research*, *28*(1), 169–190. doi:10.1108/IntR-10-2016-0296

Dobrica, S. (2019). From Digitization, through Digitalization, to Digital Transformation. *Online Searcher*, 35-39. Retrieved from https://www.researchgate.net/publication/332111919

Fielder, A., Vuorikari, R., Rodríguez-Priego, N., & Punie, Y. (2016). European Commission, & Joint Research Centre. Background review for developing the digital competence framework for consumers a snapshot of hot-button issues and recent literature. Luxembourg: Publications Office.

Ganz, W., Dworschak, B., & Schnalzer, K. (2019). Competences and Competence Development in a Digitalized World of Work. In I. L. Nunes (Ed.), *Advances in Human Factors and Systems Interaction* (Vol. 781, pp. 312–320)., doi:10.1007/978-3-319-94334-3_31

Gil-Garcia, J. R., Dawes, S. S., & Pardo, T. A. (2018). Digital government and public management research: Finding the crossroads. *Public Management Review*, *20*(5), 633–646. doi:10.1080/14719037.2017.1327181

Griffin, J., & Schuppan, T. (2010). E-Government Competencies: Looking Beyond technology, WSIS Forum 2010, E-Government Expert Group Meeting, May 14, 2010. Retrieved from http://unpan1.un.org/intradoc/groups/public/documents/un-dpadm/unpan039415.pdf

Grimpe, C., Murmann, M., & Sofka, W. (2019). Organizational design choices of high-tech startups: How middle management drives innovation performance. *Strategic Entrepreneurship Journal*, *13*(3), 359–378. doi:10.1002ej.1330

Grundén, K. (2012). Competence Development of E-Government: A Study Circle Approach. In Digital Democracy: Concepts, Methodologies, Tools, and Applications. . (Abstract) doi:10.4018/978-1-4666-1740-7

Heeks, R. B. (2003). E-Government for development causes of e-government success and failure: Factor Model. Retrieved from http://www.egov4dev.org/causegap.htm

Heilig, L., Lalla-Ruiz, E., & Voß, S. (2017). Digital transformation in maritime ports: Analysis and a game theoretic framework. *NETNOMICS: Economic Research and Electronic Networking*, *18*(2-3), 227–254. doi:10.100711066-017-9122-x

Hess, T., Benlian, A., Matt, C., & Wiesböck, F. (2016). Options for Formulating a Digital Transformation Strategy. *MIS Quarterly Executive*, *15*(2), 123–139.

Hofmann, S., & Ogonek, N. (2018). Different But Still The Same? How Public and Private Sector Organisations Deal with New Digital Competences. *The Electronic Journal of E-Government*, *16*(2), 127–135.

Hunnius, S., & Schuppan, T. (2013). Competency Requirements for Transformational E-Government. In *Proceedings 2013 46th Hawaii International Conference on System Sciences*, 1664-1673. 10.1109/HICSS.2013.139

Karunasena, K., & Deng, H. (2012). Critical factors for evaluating the public value of e-government in Sri Lanka. *Government Information Quarterly*, *29*(1), 76–84. doi:10.1016/j.giq.2011.04.005

Khan, G. F., Moon, J., Rhee, C., & Rho, J. J. (2010). E-government Skills Identification and Development : Toward a Staged-Based User-Centric Approach for Developing Countries. *Asia Pacific Journal of Information Systems.*, *20*(1), 1–31.

Liu, D., Chen, S., & Chou, T. (2011). Resource fit in digital transformation : Lessons learned from the CBC Bank global e-banking project. *Management Decision, 49*(10), 1728–1742. doi:10.1108/00251741111183852

Lloyd-Walker, B., & Walker, D. (2011). Authentic leadership for 21st century project delivery. *International Journal of Project Management, 29*(4), 383–395. doi:10.1016/j.ijproman.2011.02.004

Ndou, V. D. (2004). E - Government for Developing Countries : Opportunities and Challenges. *The Electronic Journal on Information Systems in Developing Countries, 18*(1), 1–24. doi:10.1002/j.1681-4835.2004.tb00117.x

Nwankpa, J. K., & Roumani, Y. (2016). IT Capability and Digital Transformation: A Firm Performance Perspective. In *Proceedings of the Thirty Seventh International Conference on Information Systems*, 1-16. Academic Press.

OECD. (2003). The e-government imperative. OECD e-government studies. Retrieved from www1.worlbank.org/publicsector/egov/E-governmentImperative.pdf.

OECD. (2014). Recommendation of the Council on Digital Government Strategies. Retrieved from http://www.oecd.org/gov/digital-government/Recommendation-digital-government-strategies.pdf

OECD. (2018). *Digital Government Review of Brazil: Towards the Digital Transformation of the Public Sector, OECD Digital Government Studies*. Paris, France: OECD Publishing; doi:10.1787/9789264307636-

OECD. (2019). Strengthening Digital Government. Retrieved from https://www.oecd.org/going-digital/strengthening-digital-government.pdf

Ojo, A., Janowski, T., Estevez, E., & Khan, I. K. (2007). Human Capacity Development for e-Government. *UNU-IIST-Report* (362). Retrieved from https://s3.amazonaws.com/academia.edu.documents/30890268/report362.pdf?

Pardo, T. A. (2000). Realizing the Promise of Digital Government: It's More than Building a Web Site. *Center for Technology in Government*, 12. Retrieved from http://www.cisp.org/imp/october_2000/10_00pardo.htm

Parrado, S. (2005). Skills for Electronic Service Delivery in Public Agencies. In M. *Khosrow-Pour (Ed.), Practicing E-Government: A Global Perspective* (pp. 310–328). IGI Global; doi:10.4018/978-1-59140-637-2.ch014

Sali, M. (2011). *Exploring e-government skills and competencies required in public sectors: A literature study*. 1-29. Retrieved from https://pdfs.semanticscholar.org/fe54/2fb75dd5234007e39de3ef25484482421005.pdf

Schulz, S., & Schuppan, T. (2012). Development of a European Framework for e-Government Competences, 14. Jörn von Lucke et al. (Hrsg.): Auf dem Weg zu einer offenen, smarten und vernetzten Verwaltungskultur Lecture Notes in Informatics (LNI), Gesellschaft für Informatik, Bonn. 47-60.

Schuppan, T. (2008). *Skill Requirements for e-Government*. EGPA Conference, September 3-5 in Rotterdam (The Netherlands). 1-18. Retrieved from https://www.scss.tcd.ie/disciplines/information_systems/egpa/docs/2008/Schuppan%20EGPA%2008.pdf

Schuppan, T. (2010). E-Government Competencies : Looking beyond Technology. In C. Shea, & G. Garson (Eds.), *Public Administration and Public Policy* (Vol. 155)., doi:10.1201/EBK1439807569-c22

United Nations. (2002). Benchmarking e-government: A Global perspective, assessing the progress of the UN member states. Retrieved from www.golconferenceca/presentations/eGovernment_UN.pdf.

United Nations. (2003). World Public Sector Report 2003: E-Government at the Crossroads, New York.

United Nations. (2004). Global e-government readiness report 2004: Towards access for Opportunity, New York. Retrieved from http://unpan1.un.org/intradoc/groups/public/documents/un/unpan019207.pdf

United Nations E-Government Survey. (2008). From e-Government to Connected. Governance.

United Nations E-Government Survey. (2012). E-Government for the People.

United Nations E-government Survey. (2014). E-government for the future we want.

United Nations E-government Survey. (2016). E-government in support of sustainable development.

United Nations E-government Survey. (2018). Gearing e-government to support Transformation towards Sustainable and resilient societies.

Valle-Cruz, D. (2019). Public value of e-government services through emerging technologies. *International Journal of Public Sector Management*, *32*(5), 530–545. doi:10.1108/IJPSM-03-2018-0072

van Deursen, A., & van Dijk, J. (2008). Measuring digital skills. Performance tests of operational, formal, information and strategic internet skills among the Dutch population. In *Proceedings of the 58th Conference of the International Communication Association*, 1-25, Academic Press.

van Deursen, A. J. A. M., & van Dijk, J. A. G. M. (2009). Improving digital skills for the use of online public information and services. *Government Information Quarterly*, *26*(2), 333–340. doi:10.1016/j.giq.2008.11.002

Vieru, D., Bourdeau, S., Bernier, A., & Yapo, S. (2015). Digital Competence: A Multi-dimensional Conceptualization and a Typology in an SME Context. In *Proceedings 2015 48th Hawaii International Conference on System Sciences*, 4681-4690. 10.1109/HICSS.2015.557

Vrabie, C. (2012). Computer competencies necessary for an effective e-governance, 12.

Warner, K. S. R., & Wäger, M. (2019). Building dynamic capabilities for digital transformation: An ongoing process of strategic renewal. *Long Range Planning*, *52*(3), 326–349. doi:10.1016/j.lrp.2018.12.001

Zwass, V. (2006). The web-internet compound as the infrastructure of digital government. *Business Process Management Journal*, *12*(1), 7–12. doi:10.1108/14637150610643715

ADDITIONAL READING

Al Salmi, M. (2017). Skills and Factors of e-Government : Case Study of Sultanate of Oman. *International Journal of Innovation, Management and Technology*, ▪▪▪, 313–319. doi:10.18178/ijimt.2017.8.4.747

Bassellier, G., Reich, B. H., & Benbasat, I. (2001). Information Technology Competence of Business Managers : A Definition and Research Model. *Journal of Management Information Systems, 17*(4), 159–182. doi:10.1080/07421222.2001.11045660

Gross Lopes, K. M. Macadar, M.A., & Luciano, E.M. (2019). Key drivers for public value creation enhancing the adoption of electronic public services by citizens. *International Journal of Public Sector Management*. doi:10.1108/IJPSM-03-2018-0081

Hazlett, S. A., & Hill, F. (2003). E-government: The realities of using IT to transform the public sector, *Managing Service Quality. International Journal (Toronto, Ont.), 13*(6), 445–452. doi:10.1108/09604520310506504

Hoffmann, T. (1999). The meanings of competency. *Journal of European Industrial Training, 23*(6), 275–286. doi:10.1108/03090599910284650

Lam, W. (2005). Barriers to e-government integration. *Journal of Enterprise Information Management, 18*(5), 511–530. doi:10.1108/17410390510623981

Lategan, T., & Fore, S. (2015). The impact of leadership styles on project success: Case of a telecommunications company. *Journal of Governance and Regulation., 4*(3), 48–56. doi:10.22495/jgr_v4_i3_p4

Markus, L. H., Cooper-Thomas, H. D., & Allpress, K. N. (2005). Confounded by Competencies? *An Evaluation of the Evolution and Use of Competency Models., 34*(2), 10.

Mergel, I., Edelmann, N., & Haug, N. (2019). Defining digital transformation : Results from expert interviews. *Government Information Quarterly, 101385*. doi:10.1016/j.giq.2019.06.002

van der Waldt, G. (2016). Towards an E-Governance competency framework for public service managers. The South African experiment. *African Journal of Public Affairs., 9*(4), 114–129.

KEY TERMS AND DEFINITIONS

Ability: The effective delivery of knowledge and skills in a given context.

Authentic Leadership: A pattern of leader behavior that draws upon and promotes both positive psychological capacities and a positive ethical climate.

Competency: The application of knowledge, skills and abilities in order to achieve the desired results.

Digital Competence: The set of knowledge, skills, attitudes and awareness that are required when using ICTs.

Digital Government: The use of ICTs to support the government administration, management, and governance.

Digital Technologies: These are mobile technologies, social media, GPS, the Internet of things, wireless technology, sensor networks, artificial intelligence, Big Data, nanotechnology, biotechnology, cognitive systems, renewable energy technologies and biometric software.

E-Democracy: The use of ICTs to support the process of participation of citizens, and their representatives in the debate, political discourse, surfacing and elaboration of societal issues, emergence of representation, polling, and voting.

Knowledge: The collection of information and experience that an individual possesses.

Skills: specific technical capabilities that enable an individual to perform a task.

Transformation: The process of completely changing the character or appearance of something in order to improve it. When associated to digital technologies, digital transformation is an ongoing process of using new digital technologies in everyday organizational life, which recognizes agility as the core mechanism for the strategic renewal of an organization's business model, collaborative approach, and eventually the culture.

Chapter 14
What Business Leaders Should Know About Cloud Computing:
Cloud Computing for Digital Business

Tomislav Videc
Oman Arab Bank, Oman

Jovana Zoroja
Faculty of Economics and Business, University of Zagreb, Croatia

Mirjana Pejic-Bach
https://orcid.org/0000-0003-3899-6707
Faculty of Economics and Business, University of Zagreb, Croatia

ABSTRACT

Cloud computing represents a popular and frequent notion in media and scientific research. However, in post-transition countries, there remains a low offer of cloud computing solutions and users have not still accepted this concept besides its broad spectrum of benefits. This chapter examines the possible reasons why, despite high media attention, there are not many examples of using cloud-computing solutions by companies in post-transition countries, using Croatia as a case study. For that purpose, the survey among Croatian enterprises has been conducted regarding cloud-computing usage in business. Collected data is analyzed using knowledge discovery from databases aiming to define firms' readiness when using cloud-computing solutions. Results indicate that Croatian enterprises recognize the value of cloud computing as a software-as-a-service, but significant barriers are present in terms of slow recognition of its business value. Steps for overcoming this obstacle are proposed in the chapter.

INTRODUCTION

Productivity and complexity of doing business has been increasing every day because of development and usage of modern IT products and services. Companies that are using advanced IT solutions are leaders

DOI: 10.4018/978-1-7998-2799-3.ch014

on the market and have higher competitive advantage compared to those who are lagging behind regarding IT applications. Companies should be ready to implement new technologies and to adapt daily to new business concepts based on new and innovative IT solutions such as cloud computing. Researchers predict significant growth of new market segment, which provide usage of existing technology in new business models, which enable more effective resources usage and time, invest in developing new and complex systems. In other words, instead of buying new system, the existing systems has been rent and paid by exploitation.

Software as a service is a new business model of application usage that alters classic application procurement with leasing systems of the same or similar functionality. This step in evolution does not represent technological advancement, instead, it is about renting a service, and the charge is directly related to the intensity of use. The technology used in such solutions allows for a dynamic increase in the available resources who act as a substitute for investment in the case of software as a service and cloud computing. This is extremely important in times of crisis when every investment is a major effort for an organization. A new way to use IT resources is to make cloud-based solutions and software as a service available to small and medium-sized businesses and new businesses that lack enough financial resources. This enables the entry of high technology and the latest business solutions into the most fragile part of the market that makes best use of available resources through inventiveness and adaptability.

Cloud computing and software as a service presents evolution in using IT at new and innovative level. One of cloud computing definitions is that final user via Internet can get service of data processing and data saving while he is familiar with all its functions (Zhang et al, 2010). The main reason for new business models are high costs of IT usage and complicated process of implementing new technology solutions in business processes. Cloud computing is not new technology, but usage of existing technology solutions and principles in new business models where is important to use existing applicative and infrastructure resources.

Previous research has indicated a strong relationship between the utilization of technologies and competitiveness of countries, companies, public organizations and individuals (Zoroja et al., 2016; Ollo- López et al., 2012). The same refers to cloud competing, which has been found as a key-technology for increasing competitiveness, both at company and public-organizations levels (Murugesan, 2011). However, most of the research of about cloud computing has been focused to developed countries, such as Ireland (Carcary et al., 2014) and overall European countries (Zaharia-Rădulescu et al., 2017), while the research on cloud computing in developing countries is still scarce (Yeboah-Boateng et al., 2014). Previous research on cloud computing in developing countries can be grouped into two broad groups. Frist group of the authors have investigated different aspects of cloud computing, such as motivations (Sharma et al., 2016; Hachicha et al., 2018; Hailu, 2012), and determinants of adoption (Mohammed et al., 2016; Abubakar et al., 2014). Second group of authors focus to the various aspects of cloud computing in specific countries, such as Nigeria (Dahunsi et al., 2015), or regions, such as Africa (Shaw et al., 2014). Gital et al. (2011) investigated the usage of cloud computing in higher education in Nigeria. Tijkongo et al. (2013) investigated the viability of cloud computing for SMEs in Namibia. Utilization of cloud computing in post-transition European countries is to our best knowledge not available in previous research, although other researchers have focused to different technologies, such as information systems overall (Kokles et al., 2015; Piotrowicz, 2015; Keszey, 2017; Lech, 2019), and specific technologies, such as business process management (Gabryelczyk et al., 2018). In order to shed some light in the area of cloud computing in post-transition countries, we focus our research to utilization of cloud computing in Croatian companies. Croatia is a post-transition country that has recently been accepted as the

28th member of European Union, and is still struggling with competitiveness at the global level, which indicates that it is a good representative of this group of countries.

Cloud computing represents a popular and frequent notion in Croatian journals and articles. Despite high media attention, there are not many examples of using cloud-computing solutions in Croatian companies. One of possible reasons is that many managers and professionals do not understand the possibilities and advantages of using cloud computing in business. In this paper, we want to define and present application of cloud computing in Croatian enterprises, as one of markets of post-transition countries. Although, notion regarding cloud computing solutions is very popular in Croatian business journals, there is still low offer of Croatian cloud computing solutions and users have not still accepted this concept will all its benefits. First, there are no investments in infrastructure and development of software solution. In addition, standardization of business decrease costs and final user buy service, which will improve his business processes and needs.

The goal of the paper is twofold. Firstly, we want to examine if the software usage as service can reduce costs and risks and secondly if the software usage as service can improve agility of company. The data will be collected through survey among Croatian enterprises and we will present case study. The research will be conducted among Croatian enterprises aiming to find out the level of knowledge regarding cloud-computing usage in business. Research analysis will be focused on safety of using cloud-computing services. Collected data will be analysed using knowledge discovery from databases aiming to define firms' readiness when using cloud-computing solutions. Cluster analysis will detect business verticals, which are ready to apply new models and solutions. In this analysis k-means cluster, analysis will be used (Jain, 2010). We will also pay attention to top managers, who are responsible for strategic decision making in companies, and their opinion and knowledge regarding implementation and usage of new technologies in business activities. This will give us better insight and better explanation about positive and negative characteristics regarding cloud computing, especially focusing on software as service.

The chapter is into five sections. In the first section Introduction, goals and research questions are defined and short description of selected data and methodology is presented. The second section is defining the broad filed of cloud computing and SWOT analysis is conducted in order to present current trend in using cloud computing in business. Methodology is described in the third section, defining theoretical aspect of knowledge discovery in databases and classification of data using cluster analysis. Fourth section presents the results of the cluster analysis, and in the last, fifth section, research results will be presented and discussed focusing on goals of the chapter and research questions.

BACKGROUND

The Notion of Cloud Computing

Cloud computing and software as a service (SaaS) represent the evolution of using IT in a new and innovative way. The reason for thinking about new business models can be found in the large increase in the cost of using IT tools and the increasingly complicated process of introducing them into the business processes of companies. The definition of cloud computing can be cited (Zhang et al, 2010): Provider's ability to provide computer processing and data storage services to the end user over the Internet. The end user is not familiar with the exact location or the exact way in which the service he / she purchases is built only with its business operation. In this way, the end user avoids investing in the infrastructure and building the solution itself and pays a fee for its use.

The use of computing in the cloud brings a series of benefits that are immediately converted into very exact savings that can be expressed financially. Precisely this is a cause of very rapid expansion of such solutions. Below are listed the main characteristics that distinguish the use of computing in the cloud of classical projects (Rosenberg, Mateos, 2011): (i) Resource pool - all users have at their disposal common resources which they use as they need to (ii) virtualization - all IT components that users use in direct or indirect way are virtualized and security copies can be made. All existing resources are effectively used in this way; (iii) the elasticity - all business needs are satisfied elastically with no investment in infrastructure or any of the second dependent resource; (iv) automation - all actions on the system are automated which increases reliability, and in the same time reduces errors, cost of maintenance and speed of the reaction system, and (v) a dynamic charge for use- approach allows charge to specifically spent resources without any capital investment and lump payment of licenses and services at certain time intervals.

The environmental aspect is becoming more and more important and the attitude towards natural resources, energy consumption and pollution is now intertwined with everyday activities. According to a 2006 study (Zhang et al., 2010) a total of 53% of all data centre costs were for power and cooling purposes. Given that in the US, 1.5% of total energy consumption went to the needs of the IT sector (with an average annual growth of 18%), while in Europe this figure is slightly higher. If we consider carbon dioxide emissions, in Europe the IT sector generates about 2% of which 1.75% is spent on the use of IT products and services (Etro, 2009). The average server load is only 6% (Marston et al., 2011).

Applying cloud computing (when appropriate) according to its features can lead to major organizational changes. Processes managed from the organization by deploying cloud computing are moving to an external system, which entails multiple consequences. Some of these changes are: (i) The organization is changing the need for IT professionals - programming and production change with project management, monitoring, controlling and process monitoring. Individuals will need to have much better insight into modern technology, trends and tools; (ii) IT resources in the organization are changing - the need for computing resources in the organization is decreasing because instead of local servers and data repositories, dislocated resources are accessed over the network, which only requires a web browser; (iii) Cloud computing must also be viewed in the context of business development strategies as it represents new tools in improving processes, implementing new technologies, and acquiring new knowledge. The financial aspect is significantly different from classic projects; (iv) According to some research (Bourne, 2012), cloud computing will enable the IT unit to be innovative, as well as increase its strategic impact; move from the domain of IT manager to the domain of CFO (Bourne, 2012).

Systematization of the above-mentioned features is necessary in order to see all aspects of cloud computing and software as a service in one place. Important aspects are listed in the SWOT analysis (Marston et al., 2010), as represented in Table 1.

Cloud Computing in Developed Countries

In just a few years, the idea of cloud computing has developed a dynamic market (Western Europe, the US and the Far East), with new solutions emerging every day. Every day, new projects are announced that should bring a lot of technical and business novelties (Marks et al., 2010). Service providers are generally grouped according to the type of service they offer, but there are also those that offer complete

Table 1. SWOT analysis of cloud computing

Strengths	Weaknesses
• Resources available on demand • Resource Scaling • No Unused Cost-Generating Resources • Reducing Power Consumption • Maintenance Cost Reduction • Reducing the Cost of Managing Technology Resources • Better cost control and projection • Minimum Implementation Time • Automation and standardization • Process Control • Real-time Business Insights • Access to new technologies • Less dependence of employees on the office, the ability to work from any location	• Loss of physical control over data • Users must accept the terms of the cloud computing business on their terms • When moving a part of the business to the cloud, the process is customized, not the application solutions themselves
Opportunities	**Threats**
• Access to modern technical and business solutions without capital investment • Access to modern technical and business solutions for those who did not have the conditions for it • Integration of available solutions in new and original ways (mashup) • Working in accordance with the new environmental standards • Cloud solutions change according to market and market conditions without direct and direct customer engagement • Insights into industry standards and best practices • Indirect knowledge transfer • Customers do not need the technical capacity for the system	• IT professionals who will be compromised by cloud computing • Data security • Increasing the bargaining power of a service provider after its service has been integrated into the client's business processes • Client over-dependency on service provider • Lack of standards for cloud computing • Legal aspect

Source: Marston et al., 2010.

solutions that range from infrastructure to software. In the infrastructure-focused part of the market, most firms can be divided into those that offer ready-made IaaS solutions (such as Amazon) and those that enable customers to achieve their IaaS solutions (e.g. IBM, HP, Dell) (Marks et al., 2010).

With the platform as a service, we also have two categories: those that provide management, monitoring, and billing services for cloud-based applications (such as Rightscale, Enomaly, and Eucalyptus) and those that provide the development environment and resources required to develop cloud applications (on the example of Amazon and Google) (Marks et al., 2010).

The most intense part of the market is the one related to cloud software solutions. Here are widely used solutions such as web search engines and email (for example Google, Yahoo, Microsoft), social networks (Facebook, Twitter, LinkedIn…), specialized process support tools (for example Salesforce). delivery of multimedia content (YouTube, Vimeo…), productivity tools in the mobile environment (DropBox, EverNote, Gladinet…). The reason why there are so many of these applications is the fact that when a system is provided with many users to create a user instance over the web, it is often about software as a service (Marks et al., 2010).

Innovations that bring technological advancements are increasingly interesting to the broad market today, and innovators are rewarded with business success (Chorafas, 2010). Apple, which over the long term has been able to prove that being different in an inventive way, leads to business success and significant market share.

Table 2. Use of basic cloud computing services in enterprises, 2016 (% of enterprises)

Country	Buy Cloud Computing Services Used Over the Internet	Buy E-Mail (as a CC Service)	Buy Office Software (e.g. Word Processors, Spreadsheets, etc.) (as a CC service)	Buy hosting for the Enterprise's Database (as a CC Service)	Buy Storage of Files (as a CC Service)
EU28	21	14	9	9	13
Belgium	28	16	13	13	19
Bulgaria	7	4	3	4	4
Czechia	18	14	7	6	9
Denmark	42	28	20	22	29
Germany	16	8	5	5	10
Estonia	23	15	10	6	12
Ireland	36	25	17	16	27
Greece	9	6	5	4	5
Spain	18	13	7	11	13
France	17	11	6	9	12
Croatia	23	17	11	11	13
Italy	22	18	8	8	9
Cyprus	15	11	8	5	11
Latvia	8	6	3	4	4
Lithuania	17	12	6	9	9
Luxembourg	19	12	9	10	13
Hungary	12	8	6	5	7
Malta	28	21	14	11	18
Netherlands	35	20	17	23	22
Austria	17	9	6	6	10
Poland	8	6	3	4	5
Portugal	18	14	8	7	11
Romania	7	5	3	4	4
Slovenia	22	13	11	8	11
Slovakia	18	15	9	6	9
Finland	57	42	29	25	36
Sweden	48	32	21	21	33
United Kingdom	35	22	17	14	24
Norway	40	28	21	25	28
The FYR of Macedonia	7	5	3	4	4

Source: Eurostat, 2017

According to research conducted by Gartner, cloud computing is the model that has significantly changed the entire IT market in the last two years (Gartner, 2012). The Gartner analyst points out that the changes caused by cloud computing have entered all parts of IT, and with all the positive impacts, the market still has no answer to all the risks that cloud computing brings with it for users and providers. As a direction of development, analysts cite a hybrid cloud and infrastructure that has been adapted for cloud computing. Predictions (Gartner, 2012) are that the cloud computing market in 2014 will weigh $ 150 billion, while other consulting firms are somewhat redundant and estimate slightly lower growth, with 2020 estimated market value at $ 250- $ 500 billion, with the bulk of SaaS service.

According to some authors, the increasing input of cloud solutions will also have significant macroeconomic consequences, even though IT budgets average around 5% of the total budget (Etro, 2009). First and foremost, the new model also means a whole new market, the creation of new firms that will offer services (primarily the SME segment), significant employment of highly educated workforce and the impact on GDP growth. Given these positive effects and a positive impact on the whole economy, some authors (Etro, 2009) recommend that the new business model be subsidized in order to increase the agility of the whole market and accelerate the turnover of money while raising the quality of services, innovation and development of SMEs large firms.

Cloud Computing in Post-Transition Countries

In order to analyse cloud computing usage in post-transition countries, we have collected data from Eurostat for the year 2016 regarding usage of basic cloud computing services in enterprises (% of enterprises) (Table 2) and usage of advanced cloud computing services in enterprises (% of enterprises) (Table 3). In Table 1 we have analysed five cloud computing services: (i) Buy cloud computing services used over the internet; (ii) Buy e-mail (as a CC service); (iii) Buy office software (e.g. word processors, spreadsheets, etc.) (as a CC service); (iv) Buy hosting for the enterprise's database (as a CC service) and (v) Buy storage of files (as a CC service). All selected countries use the most service *Buy cloud computing services used over the Internet.* Enterprises from most developed north European countries are leaders regarding cloud computing services via Internet: Denmark: 42; Finland: 57; Sweden: 48 and Norway: 40. They are also leaders in other four categories, while they foster development and usage of innovative ICTs the most. Less than 10% of enterprises use following cloud computing services in all five categories in following countries: Greece; Bulgaria; Latvia; Poland; Romania; The FYR of Macedonia.

In Table 3 we have analysed four advanced cloud computing services: (i) Buy finance or accounting software applications (as a CC service) ; (ii) Buy Customer Relationship Management software (as a CC service); (iii) Buy computing power to run the enterprise's own software (as a CC service) and (iv) Buy high CC services (accounting software applications, CRM software, computing power). All selected countries use the most service *Buy high CC services (accounting software applications, CRM software, computing power).* Enterprises from most developed north European countries are leaders regarding usage of high cloud computing services: Norway: 28; Finland: 34; Sweden: 28; Netherlands: 23; Ireland: 20; Denmark: 26. However, the lowest usage regarding enterprises from all selected countries is for category *Buy computing power to run the enterprise's own software.* Comparing results from Table 2 and Table 3 it can be concluded that enterprises use more basic cloud computing services than advanced cloud computing services, in both, developed and developing countries.

There is no large market for cloud computing in Croatia, and potential users show a high degree of caution even though the cloud computing model is suitable for recession because it gives businesses and institutions the opportunity to get tools to work without capital investment. In this way, not only is the capital investment turned into an operating expense, the service provider also assumes part of the

Table 3. Use of advanced cloud computing services in enterprises, 2016 (% of enterprises)

Country	Buy Finance or Accounting Software Applications (as a CC Service)	Buy Customer Relationship Management Software (as a CC Service)	Buy computing Power to Run the Enterprise's Own Software (as a CC Service)	Buy High CC Services (Accounting Software Applications, CRM Software, Computing Power)
EU28	7	6	4	11
Belgium	9	9	8	16
Bulgaria	2	1	1	3
Czechia	6	4	4	9
Denmark	19	14	14	26
Germany	4	3	3	7
Estonia	12	4	6	15
Ireland	13	11	8	20
Greece	2	2	2	4
Spain	5	5	5	9
France	5	5	3	8
Croatia	10	3	5	13
Italy	4	4	2	7
Cyprus	4	4	3	7
Latvia	3	2	2	4
Lithuania	7	4	6	10
Luxembourg	6	5	5	9
Hungary	4	3	3	6
Malta	6	6	6	11
Netherlands	19	14	7	23
Austria	4	4	4	8
Poland	2	2	2	4
Portugal	6	5	6	10
Romania	3	1	2	4
Slovenia	7	5	5	12
Slovakia	8	3	4	10
Finland	26	20	8	34
Sweden	21	14	12	28
United Kingdom	12	11	8	19
Norway	21	15	12	28
The FYR of Macedonia, the	3	2	2	4

Source: Eurostat, 2017

risk of its clients 'business because the dynamic payment guarantees that the service provider will feel any disruption to its clients' operations on its revenues. According to IDC (Baniček, 2011), the Croatian market is very small (in 2011 only HRK 68 million, representing only about 1% of annual IT spending in Croatia). The cloud solution offering has been slightly increasing over the last few years, but customer response is still inadequate. The conservative nature of small and medium-sized businesses struggling to survive today is worrying, so any cost reductions that can make their business easier should be a welcome help. Considering that the difficult economic situation in Croatia, which is a big problem for start-ups, the IT sector offers a number of start-up services where there are no investment costs for IT, but all infrastructure and software can be rented in the cloud and thus the risks and costs at the beginning reduce to a minimum. Interestingly, the markets of Western Europe, the US and the Far East are largely considering using or already using the cloud solution (Bourne, 2012). Less developed countries such as Croatia or Turkey are only making the first steps towards the widespread use of cloud computing (Kuyucu, 2010).

PERCEPTION OF CLOUD COMPUTING IN CROATIA ENTERPRISES

Methodology

Conducted survey consists of questions which can be divided into two main categories: demographic characteristics about person who is answering and issue regarding cloud computing. The survey was conducted using Google docs, which is also cloud computing service. The survey was filling using Web browser and data were saved in Google base which allow their extraction in different formats.

Using cluster analysis approach, we aim to determine dependent relationship among level of knowledge and readiness for using cloud computing business model as well as demographic data regarding employment, education and age. Cluster analysis implicit data categorization in groups where variables have similar characteristics but groups differ between themselves (Williams G. et al., 2009). There are different categorization methods. In this paper we will use k-means method where number of cluster is defined using following equation:

$$k \approx \sqrt{\frac{n}{2}}$$

There is 214 collected and fulfilled questionnaires which have been used for analysis and the number of cluster is ten.

Cluster analysis has been conducted using the answers of companies indicating their level of ICT usage in companies, for the following research questions:

- Do you use mobile devices in business?
- Can you still do your work while you are away from your workspace?
- Would you agree to use a web application (in the cloud) instead of an internal IT system (for example, email or CRM)?
- Does your company have computerized business processes (ERP, CRM, documentation management system, job management system, inventory and warehouse management system ...)?

Following information has been collected about enterprises and managers:

- Enterprises' industry
- Managers' age, education and management level

Managers were also asked about possible positive impact of cloud computing in their enterprises:

- adaptability of the company to new technological trends
- adaptability of the company to new business trends
- the speed of launching new products and services on the market
- reducing energy consumption and adopting environmental guidelines
- IT resources available upon request
- maintenance cost reduction
- standardization and control of business processes

In addition, managers were also asked about possible negative impact of cloud computing in their enterprises:

- security compromised by the physical location of the data
- Increasing dependency on cloud solution vendors
- loss of competitive advantage by standardizing business processes
- endangering the internal IT organizational unit
- Increased business risk because systems are outside the firm
- legal issues around damage repair, contract termination, data ownership with a cloud service provider

RESULTS

Perception of Advantages and Disadvantages of Cloud Computing

In conducted survey there were several questions regarding understanding of advantages and disadvantages of using cloud computing services. All advantages and disadvantages of using cloud computing services are described and presented in theoretical part of the chapter. In the Table 4, the respondents' answers are presented. Respondents are grouped according to industry where they are working (Public administration, Education, Energy, Pharmacy, Finance, Construction, Graphic, ICT, Manufacturing, Logistic, Media, Food, Student, Trade and Tourism) and positive characteristics of cloud computing services.

The same data regarding understanding of positive characteristics of cloud computing services are presented in Table 5 but according to level of education. Respondents' answers are quite similar, and there are no significant deviations in all seven categories. There are only differences in the category of respondents who have the highest degree of education, but according to low level of respondents (2), data cannot be used as relevant.

In Table 6 data about understanding positive characteristics of cloud computing services are presented regarding level of education. Workers and the lowest level of management recognize positive character-

Table 4. Understanding of positive characteristics of cloud computing services according to industries

Industry	Adaptability to New Technological Trends	Adaptability to new Business Trends	Adaptability to New Products and Services on the Market	Decreasing Consumption of Energy and Accepting Ecological Directions	IT Resources Available on Request	Decreasing Maintenance Costs	Standardization and Control of Business Processes
Public administration	47%	29%	0%	35%	35%	47%	29%
Education	83%	67%	33%	67%	67%	50%	33%
Energy	100%	50%	0%	50%	50%	50%	50%
Pharmacy	75%	50%	25%	50%	50%	25%	0%
Finance	50%	57%	14%	57%	57%	64%	14%
Construction	86%	57%	29%	57%	57%	71%	29%
Graphic	100%	0%	0%	0%	100%	100%	0%
ICT	65%	51%	32%	65%	77%	82%	40%
Manufacturing	33%	0%	33%	67%	67%	67%	33%
Logistic	33%	67%	33%	0%	0%	67%	33%
Media	44%	33%	33%	33%	78%	67%	56%
Food	50%	0%	0%	50%	50%	50%	50%
Student	50%	50%	50%	75%	25%	50%	25%
Trade	63%	50%	19%	31%	69%	63%	31%
Tourism	100%	100%	100%	100%	100%	0%	0%
Total	133	103	58	121	146	154	75

Source: Authors' analysis, 2012

Table 5. Understanding of positive characteristics of cloud computing services according to level of education

Level of Education	Adaptability to New Technological Trends	Adaptability to New Business Trends	Adaptability to New Products and Services on the Market	Decreasing Consumption of Energy and Accepting Ecological Directions	IT Resources Available on Request	Decreasing Maintenance Costs	Standardization and Control of Business Processes
PhD	100%	50%	0%	100%	50%	50%	0%
MBA	60%	47%	20%	50%	60%	57%	40%
High school	56%	34%	22%	51%	66%	66%	32%
Faculty	65%	53%	30%	59%	72%	68%	34%
Bachelor	60%	52%	32%	56%	68%	72%	40%
Total	133	103	58	121	146	154	75

Source: Authors' analysis, 2012

istics of cloud computing services the most. It is interesting to highlight that higher percentage workers and the lowest managment level agree that using cloud computing services can foster enterprises to adapt to new business trends compared to lower percentage of top management. The same effect can be seen

Table 6. Understanding of positive characteristics of cloud computing services according to management level

Management Level	Adaptability to New Technological Trends	Adaptability to New Business Trends	Adaptability to New Products and Services on the Market	Decreasing Consumption of Energy and Accepting Ecological Directions	IT Resources Available on Request	Decreasing Maintenance Costs	Standardization and Control of Business Processes
Workers / The lowest level	64%	53%	33%	62%	68%	75%	38%
Low level	65%	52%	24%	55%	65%	70%	32%
Middle level	53%	44%	22%	58%	75%	67%	33%
Top level	62%	24%	19%	38%	67%	76%	33%
Total	133	103	58	121	146	154	75

Source: Authors' analysis, 2012

Table 7. Understanding of negative characteristics of cloud computing services according to industries

Industry	Security Threat Because of Physical Data Location	Increasing Dependent of Cloud Computer Providers	Loss of Competitive Advantage Because of Standardization of Processes	Jeopardise of Internal IT Infrastructure	Higher Business Risk Because of Outsourcing Services	Legal Issues Regarding Elimination of Damages, Contract Cancellation, Data Ownership
Public administration	71%	41%	0%	0%	41%	59%
Education	50%	50%	0%	0%	50%	33%
Energy	50%	50%	0%	0%	0%	50%
Pharmacy	50%	75%	25%	0%	25%	50%
Finance	64%	71%	7%	0%	57%	36%
Construction	29%	57%	43%	0%	29%	71%
Graphic	100%	0%	0%	0%	100%	100%
ICT	52%	71%	11%	26%	51%	56%
Manufacturing	67%	67%	33%	0%	33%	67%
Logistic	33%	33%	33%	33%	33%	33%
Media	33%	56%	0%	22%	44%	56%
Food	0%	0%	0%	0%	0%	50%
Student	25%	55%	0%	25%	0%	25%
Trade	38%	56%	13%	19%	19%	44%
Tourism	100%	0%	0%	100%	100%	0%
Total	109	135	23	41	96	114

Source: Authors' analysis, 2012

Table 8. Understanding of negative characteristics of cloud computing services according to management level

Management Level	Security Threat Because of Physical Data Location	Increasing Dependent of Cloud Computer Providers	Loss of Competitive Advantage Because of Standardization of Processes	Jeopardise of Internal IT Infrastructure	Higher Business Risk Because of Outsourcing Services	Legal Issues Regarding Elimination of Damages, Contract Cancellation, Data Ownership
Workers / The lowest level	54%	66%	10%	21%	49%	47%
Low level	48%	56%	11%	23%	44%	64%
Middle level	56%	64%	17%	17%	36%	47%
Top level	38%	71%	5%	5%	43%	57%
Total	109	135	23	41	96	114

Source: Authors' analysis, 2012

regarding following two characteristics: Adaptability to new products and services on the market and Decreasing consumption of energy and accepting ecological directions.

In Table 7, the data regarding understanding negative characteristics of cloud computing services according to industries are presented. None of the respondents from following industries: Public administration, Education and Energy agree that characteristics: Loss of competitive advantage because of standardization of processes and Jeopardise of internal IT infrastructure present negative characteristics of cloud computing services. None of the respondents employed in Food industry (0%) agree that all mentioned characteristics of cloud computing services have negative impact on business, expect 50% of them who agree that only characteristic Legal issues regarding elimination of damages, contract cancellation, data ownership has negative impact on business.

In Table 8 analysis of negative characteristics of cloud computing services according to management level has been presented. Respondents' answers are quite similar. However, it is important to highlight that top managers have the lowest understanding regarding risk that are consequence of the new business model.

Table 9 presents classification of respondents' answers according to level of education. It is important to highlight answers given by respondents with the highest level of education, PhD, who has high

Figure 1. Number of enterprises across clusters
Source: Authors' analysis, 2012

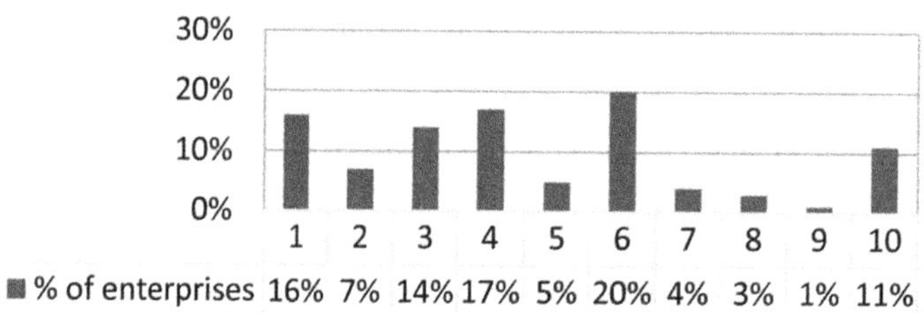

Table 9. Understanding of negative characteristics of cloud computing services according to level of education

Level of Education	Security Threat Because of Physical Data Location	Increasing Dependent of Cloud Computer Providers	Loss of Competitive Advantage Because of Standardization of Processes	Jeopardise of Internal IT Infrastructure	Higher Business Risk Because of Outsourcing Services	Legal Issues Regarding Elimination of Damages, Contract Cancellation, Data Ownership
PhD	100%	50%	50%	0%	100%	100%
MBA	47%	67%	17%	23%	53%	43%
High school	46%	56%	7%	17%	34%	41%
Faculty	53%	70%	10%	19%	46%	63%
Bachelor	48%	40%	8%	20%	44%	36%
Total	109	135	23	41	96	114

Source: Authors' analysis, 2012

understanding of most risk regarding usage of cloud computing services. However, there are only 2 respondents who has PhD degree so their answers are not relevant for the analysis.

Cluster Analysis Results

Figure 1 presents number of enterprises across clusters. The highest percentage of enterprises is in Cluster 1 (16%), in Cluster 4 (17%) and in Cluster 6 (20%). The lowest percentage of enterprises, less than 5%, is in following clusters: Cluster 5 (5%), in Cluster 7 (4%), in Cluster 8 (3%) and in Cluster 9 (1%).

Table 10 presents the mean values of research variables across clusters. In all clusters, except Cluster 8, respondents use mobile phones in everyday business activities. Furthermore, in all clusters, except Cluster 3, 7 and 10, respondents can conduct their business even if they are not at their work place. Enterprises from all 10 cluster, except Clusters 2, 5 and 9, use some of business information systems (ERP, CRM, DMS, …). Respondents from enterprises group in most of the clusters would like to use some Web applications instead of internal IT systems. It can be concluded that most of the enterprises are ready to use cloud computing services, but some changes and adaption have to be done. Results showed that Cluster 1 and Cluster 6 are convenient, according to all criteria, for implementation of cloud computing services. Cluster 3, 4, and 10 are almost ready for implementation, while Clusters 2, 5, 6, and 9 are not yet ready for implementation of cloud computing services. In Cluster 7 most of data are unknown or neutral, so data for Cluster 7 are not relevant in this case. Furthermore, 78 respondents are grouped into Clusters which satisfied all criteria important for implementation of cloud computing services and 91 of them are grouped into Clusters which are almost ready for implementation. Only 36 respondents are employ in enterprises that are not ready for implementation of cloud computing services.

Table 11 presents ANOVA analysis of cluster analysis regarding readiness for using cloud computing services. Anova analysis has been conducted on total sample of enterprises in order to determine statistical significant differences for selected variables. Research results showed that there is statistically significant difference for all four variables (p-value: 0,000).

Table 10. Usage of ICT services in enterprises across clusters

	Cluster									
	1+	2-	3*	4*	5-	6+	7-	8	9-	10*
Do you use mobile phones in business?	Yes	Yes	Yes	Yes	Yes	Yes	Yes	No	Yes	Yes
Can you conduct your business tasks even if you are not at your work place?	Yes	No	Yes and No	Yes	Yes	Yes	Yes and No	No	Yes	Yes and No
Would you agree to use some Web application (in cloud) instead internal IT system (e.g. email or RM)?	Yes, if Web application has better features than internal IT system	Yes, if Web application has better features than internal IT system	Yes, if Web application has better features than internal IT system	Yes	Yes	Yes	Do not know	Do not know	No	Yes
Does your enterprise use business information systems, e.g. ERP, CRM, DMS, ……)?	Yes, some of them	No	Yes, all of them	Yes, all of them	No	Yes, some of them	Just basic	Yes, all of them	No	Yes, some of them

Source: Authors' analysis, 2012

Note: (+) enterprise has adequate prerequisites for cloud computing implementation, (*) enterprise has partially fulfilled prerequisites for cloud computing implementation; (-) enterprise does not meet prerequisites for cloud computing implementation

Clusters' Characteristics According to Enterprises and Managers

Table 12 presents number of enterprises across clusters and industry. Most of the selected enterprises are from ICT industry (124) and most of them are grouped into Cluster 1 (23) and Cluster 6 (34) which means that they have adequate prerequisites for cloud computing services implementation. Respondents who are employ in Public administration are mostly unprepared to use cloud computing services. Most of the enterprises, from all industries, are grouped into Cluster 1 (35) and Cluster 6 (43), which means that they are ready for implementation. It can be concluded that respondents who are surrounded by technical advances are ready to use new and innovative technologies, such as cloud computing services.

Table 11. ANOVA analysis of cluster analysis regarding readiness for using cloud computing services

	Cluster		Error		F	Sig.
	Mean Square	df	Mean Square	df		
Do you use mobile phones in business?	0.348	9	0.070	204	5 .007	0.000***
Can you conduct your business tasks even if you are not at your work place?	17 .292	9	0.116	204	148 .596	0.000***
Would you agree to use some Web application (in cloud) instead internal IT system (e.g. email or RM)?	39 .996	9	0.129	204	310 .059	0.000***
Does your enterprise use business information systems, e.g. ERP, CRM, DMS, ……)?	16 .993	9	0.192	204	88 .569	0.000***

Source: Authors' analysis, 2012

Table 12. Number of enterprises across clusters and industry

Industry	Cluster										Total
	1+	2-	3*	4*	5-	6+	7-	8	9-	10*	
Public administration		4	2	2	1	1	4	2		1	17
Education		1			2		1			2	6
Energy		1	1								2
Pharmacy	2		1	1							4
Finance	3	2	1	3		2	1			2	14
Construction	1	1		2						3	7
Graphic					1						1
ICT	23	2	19	23	3	34		3	2	15	124
Manufacturing	1			1		1					3
Logistic		1	1				1				3
Unemployed						1					1
Media	1	3	1	1	1	1				1	9
Food				2							2
Student	1	1			1		1				4
Trade	3		3	2	2	3	1	2			14
Tourism			1								1
Total	35	16	30	37	11	43	9	7	2	24	214

Source: Authors' analysis, 2012

Note: (+) enterprise has adequate prerequisites for cloud computing implementation, (*) enterprise has partially fulfilled prerequisites for cloud computing implementation; (-) enterprise does not meet prerequisites for cloud computing implementation

The same distribution has been presented according to respondents' age (Table 13). Most of the respondents have 25 through 35 ages and who use in their everyday business activities new technical achievements, and most of them are ready to use cloud computing services. Very similar situation is with group where respondents have 35 through 45 ages. Elderly respondents are not prepared as the first two groups regarding accepting new technical services.

The same distribution has been presented according to respondents' educational level (Table 14). Most of the respondents finished faculty and are ready to use new technical achievements in their everyday business activities. Respondents who have a bachelor's degree are mostly employ in enterprises that have partially fulfilled prerequisites for cloud computing implementation or in enterprises that do not meet prerequisites for cloud computing implementation at all.

Table 15 presents understanding of cost effectiveness impact of cloud computing services implementation regarding managers' level. Top management support is crucial for new business model implementation. Research results showed that cloud computing services implementation has been supported by all managers' level.

Table 13. Managers' age for selected enterprises across clusters

Age Across Clusters	Cluster										Total
	1+	2-	3*	4*	5-	6+	7-	8	9-	10*	
18-25	1	1	2		1		1			1	7
25-35	18	6	16	21	7	25	4	4	1	17	119
35-45	15	8	9	16	3	17	4	3	1	6	82
45-55	1		2								3
More than 55		1	1			1					3
Total	35	16	30	37	11	43	9	7	2	24	214

Source: Authors' analysis, 2012

Note: (+) enterprise has adequate prerequisites for cloud computing implementation, (*) enterprise has partially fulfilled prerequisites for cloud computing implementation; (-) enterprise does not meet prerequisites for cloud computing implementation

Table 14. Distribution of respondents across clusters according to education level

Education Level	Cluster										Total
	1+	2-	3*	4*	5-	6+	7-	8	9-	10*	
PhD	1				1						2
MBA	6		4	5	1	7	3			4	30
High school	7	4	9	5		5	3	3	1	4	41
Faculty	18	9	15	25	5	26	2	3	1	12	116
Bachelor	3	3	2	2	4	5	1	1		4	25
Total	35	16	30	37	11	43	9	7	2	24	214

Source: Authors' analysis, 2012

Note: (+) enterprise has adequate prerequisites for cloud computing implementation, (*) enterprise has partially fulfilled prerequisites for cloud computing implementation; (-) enterprise does not meet prerequisites for cloud computing implementation

Usage of Cloud Computing in Enterprises

Table 15. Understanding cost effectiveness impact of cloud computing services implementation regarding managers' level

Managers' Level	Cluster										Total
	1+	2-	3-	4-	5-	6+	7-	8-	9+	10+	
Workers / The lowest level	29	15	1	3	2	5		18	8	10	91
Low level	24	10		2	3	6		12	6	3	66
Middle level	13	4	1	2	2	3	1	6	2	2	36
Top level	7	2			3	3		4		2	21
Total	73	31	2	7	10	17	1	40	16	17	214

Source: Authors' analysis, 2012

Note: (+) enterprise has adequate prerequisites for cloud computing implementation, (*) enterprise has partially fulfilled prerequisites for cloud computing implementation; (-) enterprise does not meet prerequisites for cloud computing implementation

Table 16. The level of using cloud computing services by enterprises

	Cluster									
	1+	2-	3*	4*	5-	6+	7-	8	9-	10*
Do you understand cloud computing services?	Yes	No	No	Yes	Yes	Yes	Yes	Yes	Yes	Yes
Do you use cloud computing services?	Yes	No	No	No	No	Yes	No	Yes	No	No
Which service is more affordable when starting a new enterprise: (1) cloud computing service, (2) own software and hardware?	Cloud computing service	Can not assess	Own software and hardware	Can not assess	Cloud computing service	Cloud computing service	Own software and hardware	Can not assess	Cloud computing service	Cloud computing service
Do you think that cloud computing services can decrease business costs?	Yes, but do not know how much	Can not assess	Yes, but do not know how much	Yes, but do not know how much	Can not assess	Yes, until 30%	Can not assess	Yes, but do not know how much	Yes, but do not know how much	Yes, until 50%

Source: Authors' analysis, 2012

Note: (+) enterprise has adequate prerequisites for cloud computing implementation, (*) enterprise has partially fulfilled prerequisites for cloud computing implementation; (-) enterprise does not meet prerequisites for cloud computing implementation

Table 16 presents level of using cloud-computing services by enterprises. Only respondents who are employ in enterprises grouped into Cluster 2 and Cluster 3 do not understand cloud computing services. However, cloud computing services are used only by enterprises grouped in Clusters 1, 6 and 8. Furthermore, most of the respondents agree that cloud computing services are more affordable than using own software or hardware (Clusters 1, 5, 6, 9 and 10) and that computing services can decrease business costs (Clusters 1, 3, 4, 6, 8, 9 and 10). It can be concluded that respondents mostly understand importance of cloud computing services but still some prerequisites have to be fulfilled to implement it.

Research results also showed that most of the respondents use some kind of cloud computing services for private and business purposes, although they are not aware of that, e.g. Facebook, Twitter, YouTube, DropBox, Evernote, Flickr. There are respondents who use cloud computing services although they are not aware that they are using it (19) and there are also respondents who use cloud computing services but they think that that are not cloud computing services (87). It can be concluded that most of the respondents do not understand concept of Internet services that they are using every day.

SOLUTIONS AND RECOMMENDATIONS FOR DIGITAL BUSINESS

Because of its particularity, cloud computing is not appropriate for every business situation. Data sensitivity, legal aspect, control, internal IT procedures impact on usage of cloud computing. The use of cloud computing should also be viewed from the perspective of the size of the business and its market position. Large companies are burdened with various protocols and business rules that are in place due to market or legal regulations, but also internal rules. Their evaluation of cloud computing must take all these aspects into account. Small and medium-sized businesses have far fewer constraints, so this new model is a great option given the crisis (Sultan, 2011).

Cloud computing and software as a service are exceptional for start-ups, that will benefit from a new business model as soon as possible, and the advent has opened new business opportunities that were impossible without this model (Williams, 2010). Agility and innovation in small businesses have a dominant influence and the value of a business idea is the judgment of their fate. New tools allow them to launch products and services very quickly and cheaply, and to test their business ideas very quickly and easily. There is no capital investment in cloud solutions, and costs are business dependent, so business risks are significantly reduced.

Reducing barriers to new firms will certainly have a significant impact on the entire market (Williams, 2010) and increases in the number of new firms will have a significant impact on medium and large established firms that can expect outflows and increased competition. An advantage over large firms is also the speed at which small new firms can implement new technological and business models such as cloud computing (Marks et al., 2010).

The use of cloud computing also has many financial implications, so a company moving to a cloud solution will have changes in balance, profit and loss, and some indicators of sizes will change too. Some of the economic indicators that will change digital business models are as following:

1. Short-term / Long-term assets - Reduction - no licenses converted to operating expenses and thus the investment is better distributed
2. Cash flow - even - instead of making significant payments to suppliers in the project, the cost is transformed into a monthly fee that monitors the use of the software, and thus the cost is evenly distributed over a time interval. This has the positive consequence and less need for borrowing due to the absence of large capital expenditure
3. Depreciation - as assets decrease, depreciation decreases as well. Classic projects show the values of licenses and systems in a firm's assets. As the entire cost is converted to operating cost here, there is no increase in assets and later no depreciation.
4. The self-financing factor - is growing - as assets are smaller
5. Indebtedness factor - declining - given that there are no major infrastructure expenditures and licenses that are a prerequisite for classic IT projects
6. Business Expenses - Growing - SaaS is paid as an operating expense by use, so capital investments are transformed into regular operating expenses
7. Overall economy - growing - no investment

8. Cost-effectiveness of business activity - first falling then rising
9. Cost-effectiveness of financing activities - growing - no significant investment in infrastructure, licenses and project, so there is no borrowing that is possible with a classic project
10. Turnover ratio - rising - no assets are used to increase revenue
11. Intensity of research and education - falling - it is not necessary to have individual IT knowledge within the company, but knowledge and product are rented as needed. It must be remembered that knowledge of the essential processes is necessary and that extracting knowledge from the firm can be disastrous for the future. Of course, the fall in funds invested in education does not have to happen, but it is possible to reduce the amounts. It will later be described how knowledge levels can be increased by deploying cloud computing.

FUTURE RESEARCH DIRECTIONS FOR DIGITAL BUSINESS

In the last few years, cloud computing and software as a service have not come off the front page, and market interest has grown. The media is very interested in the field and thousands of articles can be found on the Internet about the possible application, benefits and future of the new business model. Researchers agree that cloud computing has a bright future. Generating new ideas every day leads to an exponential expansion of new solutions that find their way to the end users very easily. In just a few years, all major software and hardware vendors have solutions that somehow have elements of the cloud solution. For example, they argue that modern technology and modern business models will have a huge impact on firms in the future that will lose the structure we know today, and their organizational structures will become much shallower. This will enable them to operate more agile, and the highest value, he believes, will be an ecosystem made up of employees, resources and users.

According to some authors (Rosenberg, Mateos, 2011), cloud computing and software as a service will have the greatest impact on start-ups. Investors in such businesses want to avoid any capital investment so cloud computing is a great business model for reducing risk and overall costs. The big impact on today's market will also result in the attempt of IT organizational units in large companies to integrate into cloud solutions without significantly losing their influence and importance. According to some theories, this will result in a whole host of new solutions that will represent only a step towards using the right cloud solutions (Rosenberg, Mateos, 2011). Private cloud has the potential to be imposed as a solution that will be appropriate for companies that handle sensitive data.

When analysing possible directions for the development of cloud computing, most authors agree that centralized controlled data storage is one of their top priorities. Security standards and issues must be addressed through an institutionalized protocol that will guarantee everyone a process over this data under controlled conditions. With advanced mobile devices accessing centralized data in the cloud according to certain rules, the implementation of a new generation of business solutions should be a tool for successful business for everyone. Research on cloud systems should be conducted in several of the following ways (Marston et al., 2010): economic models of cloud computing, IT strategy and cloud computing, technology required for cloud solutions, green IT and cloud computing, and prescribing regulations related to cloud computing. Given that this is a relatively new technological business model, the importance of research and standard setting is of great importance.

Cloud-based solutions will become less expensive, and deploying them will bring market advantage, primarily by looking at SaaS solutions. Cloud providers will become more efficient and effective, lowering service prices (Rosenberg, Mateos, 2011). The solutions that will be the foundation for cloud computing will be open source. Open source has a growing and stronger community, and a unique business model where classic licensing is lacking, has led to significant cost reductions and the popularization of new technology solutions. In a survey by Bourne (2012) for one of the most important cloud service providers, Google, the vast majority (94%) of Western European companies surveyed agree to use cloud solutions within the next year. As many as 96% of the surveyed think that such solutions can help them to do business and achieve better efficiency. Of these, 38% have already adopted some cloud solutions.

From all the above it can be concluded that cloud computing is one of the main directions of development of application of computing. In the future, we should expect an increasing exploitation of such services. The negative aspects, which are mainly related to the lack of standardization, will surely be the focus of several professionals trying to nullify them through standardization and methodology.

REFERENCES

Abubakar, A. D., Bass, J. M., & Allison, I. (2014). Cloud computing: Adoption issues for sub-saharan African SMEs. *The Electronic Journal on Information Systems in Developing Countries*, *62*(1), 1–17. doi:10.1002/j.1681-4835.2014.tb00439.x

Baniček, D. (2011) *Business.hr*. Retrieved August 27, 2019 from www.business.hr/b-it/cloud-trziste-u-hrvatskoj-tesko-tek-68-mil-kuna

Bourne, V. (2012). *Business impact of the Cloud, 2012*. Retrieved September 3, 2019 from lp.google-mkto.com/CFOCloudComputingResearch.html

Carcary, M., Doherty, E., Conway, G., & McLaughlin, S. (2014). Cloud computing adoption readiness and benefit realization in Irish SMEs—An exploratory study. *Information Systems Management*, *31*(4), 313–327. doi:10.1080/10580530.2014.958028

Chorafas, D. N. (2011). *Computing Strategies*. SAD: CRC Press Taylor & Francis Group.

Dahunsi, F. M., & Owoseni, T. M. (2015). Cloud computing in Nigeria: The cloud ecosystem perspective. *Nigerian Journal of Technology*, *34*(1), 209–216. doi:10.4314/njt.v34i1.26

Etro, F. (2009). The Economic Impact of Cloud Computing on Business Creation, Employment and Output in Europe. An application of the Endogenous Market Structures Approach to a GPT innovation. *Review of Business and Economic Literature, KU Leuven. Faculty of Economics and Business*, *0*(2), 179–208.

Gabryelczyk, R., & Roztocki, N. (2018). Business process management success framework for transition economies. *Information Systems Management*, *35*(3), 234–253. doi:10.1080/10580530.2018.1477299

Gartner. (2012). *Gartner press releases, 2012*. Gartner. Retrieved August 17, 2019 from www.gartner.com

Gital, A. Y. U., & Zambuk, F. U. (2011). Cloud computing: Solution to ICT in higher education in Nigeria. *Advances in Applied Science Research*, *2*(6), 364–369.

Hachicha, Z. S., & Mezghani, K. (2018). Understanding Intentions to Switch Toward Cloud Computing at Firms' Level: A Multiple Case Study in Tunisia. *Journal of Global Information Management*, *26*(1), 136–165. doi:10.4018/JGIM.2018010108

Hailu, A. (2012). Factors influencing cloud-computing technology adoption in developing countries. (Doctoral dissertation), Capella University.

Jain, A. K. (2010). Data clustering: 50 years beyond K-means. *Pattern Recognition Letters*, *31*(8), 651–666. doi:10.1016/j.patrec.2009.09.011

Keszey, T. (2017). Information systems in transition economies: Does ownership matter? *Information Systems Management*, *34*(1), 65–84. doi:10.1080/10580530.2017.1254456

Kokles, M., Romanová, A., & Hamranová, A. (2015). Information systems in the post-transition period in enterprises in Slovakia. *Journal of Global Information Technology Management*, *18*(2), 110–126. do i:10.1080/1097198X.2015.1052686

Kuyucu, A. D. H. (2010). *The Playground of Cloud Computing in Turkey*. Elsevier. Retrieved August 7, 2019 from www.elsevier.com/locate/procedia

Lech, P. (2019). Enterprise system implementations in transition and developed economies: Differences in project contracting and governance. *Information Technology for Development*, *25*(2), 357–380. doi: 10.1080/02681102.2018.1564726

Marks, E. A., & Lozano, B. (2010). *Executive's Guide to Cloud Computing. SAD.* John Wiley & Sons.

Marston, S., Li, Z., Bandyopadhyay, S., Zhang, J., & Ghalsasi, A. (2010). *Cloud computing — The business perspective*. Elsevier. Retrieved September 7, 2019 from www.elsevier.com/locate/dss

Mohammed, F., Ibrahim, O., & Ithnin, N. (2016). Factors influencing cloud computing adoption for e-government implementation in developing countries: Instrument development. *Journal of Systems and Information Technology*, *18*(3), 297–327. doi:10.1108/JSIT-01-2016-0001

Murugesan, S. (2011). Cloud computing gives emerging markets a lift. *IT Professional*, *13*(6), 60–62. doi:10.1109/MITP.2011.94

Ollo-López, A., & Aramendía-Muneta, M. E. (2012). ICT impact on competitiveness, innovation and environment. *Telematics and Informatics*, *29*(2), 204–210. doi:10.1016/j.tele.2011.08.002

Piotrowicz, W. (2015). Information technology and systems in the Visegrád group of countries (Czech Republic, Hungary, Poland, and Slovakia): A literature review. *Journal of Global Information Technology Management*, *18*(2), 77–93. doi:10.1080/1097198X.2015.1052684

Rosenberg, J., & Mateos, A. (2011). *The Cloud at Your Service. SAD.* Manning Publications.

Sharma, S. K., Al-Badi, A. H., Govindaluri, S. M., & Al-Kharusi, M. H. (2016). Predicting motivators of cloud computing adoption: A developing country perspective. *Computers in Human Behavior*, *62*, 61–69. doi:10.1016/j.chb.2016.03.073

Shaw, V., & Braa, J. (2014). Approaches to participatory design in Africa in the age of cloud computing. In *Proceedings of the 13th Participatory Design Conference: Short Papers, Industry Cases, Workshop Descriptions, Doctoral Consortium papers, and Keynote abstracts* – Volume 2, (pp. 45-48). ACM 10.1145/2662155.2662180

Sultan, N. A. (2011). *Reaching for the "cloud": How SMEs can manage*. Elsevier. Retrieved June 27, 2019 from www.elsevier.com/locate/ijinfomgt

Tjikongo, R., & Uys, W. (2013). *The viability of cloud computing adoption in SME's in Namibia. In 2013 Ist-Africa conference & exhibition* (pp. 1–11). IEEE.

Williams, M. I. (2010). *A Quick Start Guide to Cloud Computing*. United Kingdom: Kogan Page Limited.

Yeboah-Boateng, E. O., & Essandoh, K. A. (2014). Factors influencing the adoption of cloud computing by small and medium enterprises in developing economies. *International Journal of Emerging Science and Engineering*, *2*(4), 13–20.

Zaharia-Rădulescu, A. M., & Radu, I. (2017). Cloud computing and public administration: approaches in several European countries. In *Proceedings of the International Conference on Business Excellence* - Volume 11, No. 1, pp. 739-749. De Gruyter Open. 10.1515/picbe-2017-0078

Zhang, Q., Cheng, L., & Boutaba, R. (2010). *Cloud computing: state-of-the-art and research challenges*. Brazil: The Brazilian Computer Society.

Zoroja, J., & Pejic Bach, M. (2016). Impact of information and communication technology to the competitiveness of European countries-cluster analysis approach. *Journal of Theoretical and Applied Electronic Commerce Research*, *11*(1), i–x. doi:10.4067/S0718-18762016000100001

KEY TERMS AND DEFINITIONS

Cloud Computing: Represents a popular and frequent notion related to on-demand availability of data storage.

Cluster Analysis: Represent a method of grouping selected variables where objects in the same group are similar and differ to those from other groups.

Software as a Service (SaaS): Represent the evolution of using IT in a new and innovative way and this service is also knowen as Web-based software.

SWOT Analysis: Represent a planning method which can facilitate managers to identify strengths, weaknesses, opportunities and threats related to business processes.

Compilation of References

Abdullah, A., Sulong, Z., & Said, R. M. (2019). An analysis on ethical climate and ethical judgment among public sector employees in Malaysia. *American Journal of Business, 10*(2).

Abubakar, A. D., Bass, J. M., & Allison, I. (2014). Cloud computing: Adoption issues for sub-saharan African SMEs. *The Electronic Journal on Information Systems in Developing Countries, 62*(1), 1–17. doi:10.1002/j.1681-4835.2014.tb00439.x

ACARA. (2015). *National Assessment Program: ICT Literacy Years 6 & 10 Report 2014.* Australian Curriculum, Assessment, and Reporting Authority. Retrieved from http://www.nap.edu.au/verve/_resources/D15_8761__NAP-ICT_2014_Public_Report_Final.pdf

Ademola, O. E. (2017, October). *Reality of human decision-making: A leadership and management platform.* Paper presented at the meeting 9th iSTEAMS Multidisciplinary CrossBorder Conference, University of Ghana, Legon.

Adeyemi, O. A. (2017). *Organizational change and how it effects healthcare employees: A study on employee resistance to change in electronic medical record implementation (Doctoral dissertation).* (10602795 D.M.), Colorado Technical University, Ann Arbor. ProQuest Dissertations & Theses Global database.

Adjei-Bamfo, P., Maloreh-Nyamekye, T., & Ahenkan, A. (2019). The role of e-government in sustainable public procurement in developing countries: A systematic literature review. *Resources, Conservation and Recycling, 142*, 189–203. doi:10.1016/j.resconrec.2018.12.001

Afolabi, A., Ibem, E., Aduwo, E., Tunji-Olayeni, P., Oluwunmi, O., & Ayo-Vaughan, E. (2019). Gauging parameters for e-procurement acquisition in construction businesses in Nigeria. International Journal of Construction Management, 1-10.

Afutu-Kotey, R. L., Gough, K. V., & Owusu, G. (2017). Young Entrepreneurs in the Mobile Telephony Sector in Ghana: From Necessities to Aspirations. *Journal of African Business, 18*(4), 476–491. doi:10.1080/15228916.2017.1339252

Agarwal, R., DesRoches, C., & Jha, A. K. (2010). The digital transformation of healthcare: Current status and the road ahead. *Information Systems Research, 21*(4), 796–809. doi:10.1287/isre.1100.0327

Agarwal, R., Tanniru, M., & Wilemon, D. (1997). Assimilating Information Technology Innovations: Strategies and Moderating Influences. *IEEE Transactions on Engineering Management, 44*(4), 347–358. doi:10.1109/17.649864

Ageshin, E. A. (2001). E-procurement at work: A case study. *Production and Inventory Management Journal, 42*(1), 48–53.

Aghena W., De Smet, A. & Weerda, K. (2015). Agility: It rhymes with stability, *McKinsey Quarterly*, Dec.

Agyapong, D., Brautlacht, R., & Owino, J. (2018). University Collaboration-Research Collaboration and Teaching Collaboration: Competency-Based Learning Using a Cross-Cultural Students' Project. Handbook of Applied Teaching and Learning.

Ahmad, T., Aljafari, R., & Venkatesh, V. (2019). The Government of Jamaica's electronic procurement system: Experiences and lessons learned. *Internet Research, 29*(6), 1571–1588. doi:10.1108/INTR-02-2019-0044

Ahmed, F., As-Saber, S., Fry, S., & Smith, R. (2019). Bureaucracy and E-government: A study of e-procurement initiatives in Bangladesh.

Ali, W., Frynas, J. G., & Mahmood, Z. (2017). Determinants of corporate social responsibility (CSR) disclosure in developed and developing countries: A literature review. *Corporate Social Responsibility and Environmental Management, 24*(4), 273–294. doi:10.1002/csr.1410

Allman, S. (2009). Leadership vs. Management. *Successful Meetings, 58*(11), 12.

Altayyar, A., & Beaumont-Kerridge, J. (2016). External Factors Affecting the Adoption of E-procurement in Saudi Arabian's SMEs. *Procedia: Social and Behavioral Sciences, 229*, 363–375. doi:10.1016/j.sbspro.2016.07.147

Anderson, P. (1999). Perspective: Complexity Theory and Organization Science, *Organization science* (10:3), pp. 216-232.

Anderson, N., Potočnik, K., & Zhou, J. (2014). Innovation and creativity in organizations: A state-of-the-science review, prospective commentary, and guiding framework. *Journal of Management, 40*(5), 1297–1333. doi:10.1177/0149206314527128

Andries, P., & Debackere, K. 2007. Adaptation and Performance in New Businesses: Understanding the Moderating Effects of Independence and Industry, Small Business Economics (29:1-2), pp. 81–99.

Ansoff, H. I., Kipley, D., Lewis, A. O., Helm-Stevens, R., & Ansoff, R. (2019). *Implanting strategic management* (3rd ed.). Palgrave Macmillan / Springer. doi:10.1007/978-3-319-99599-1

Arbab Kash, B., Spaulding, A., Johnson, C. E., & Gamm, L. (2014). Success factors for strategic change initiatives: A qualitative study of healthcare administrators' perspectives. *Journal of Healthcare Management, 59*(1), 65–81. doi:10.1097/00115514-201401000-00011 PMID:24611428

Argyris, C. (1977). Double loop learning in organizations. *Harvard Business Review, No.77502.*

Arnold, J. A., Arad, S., Rhoades, J. A., & Drasgow, F. (2000). The Empowering Leadership Questionnaire: The Construction and Validation of a New Scale for Measuring Leader Behaviors, *Journal of Organizational Behavior* (21:3), pp. 249-269.

Askari, M. Y., & El-Refae, G. A. (2019). The rationality of irrational decisions: A new perspective of behavioral economics. *International Journal of E-Business Research, 17*(4), 388–401.

As-Saber, S., Rahman, S., Rahim, M., Cairns, G., Härtel, C., & Waheduzzaman, W. (2014). Promises and Pitfalls of Using Social Media in Public E-procurement: An Appraisal. *AJIS. Australasian Journal of Information Systems, 18*(3). doi:10.3127/ajis.v18i3.1097

Atchison, T. A. (1999). Managing change. *Frontiers of Health Services Management, 16*(1), 3–29. doi:10.1097/01974520-199907000-00002 PMID:10557432

Augier, M., & Teece, D. J. (2009). Dynamic Capabilities and the Role of Managers in Business Strategy and Economic Performance, *Organization Science, 20*(2), pp. 410-421.

Australian Commission on Safety and Quality in Health Care. (2011). *Patient-centred care: Improving quality and safety through partnerships with patients and consumers.* Retrieved from https://www.safetyandquality.gov.au/wpcontent/uploads/2012/03/PCC_Paper_August.pdf

Australian Government. (2015). *Embracing the digital age. National Innovation and Science Agenda.* Retrieved from http://www.innovation.gov.au/page/embracing-digital-age

Australian Health Minister's Conference. (2008). *National e-Health Strategy Summary*. Melbourne, Australia: Australian Government.

Avolio, B., & Gardner, W. (2005). Authentic leadership development: Getting to the root of positive forms of leadership. *The Leadership Quarterly*, *16*(3), 315–338. doi:10.1016/j.leaqua.2005.03.001

Avolio, B., Gardner, W., Walumbwa, F., Luthans, F., & May, D. (2004). Unlocking the mask: A look at the process by which authentic leaders impact follower attitudes and behaviors. *The Leadership Quarterly*, *15*(6), 801–823. doi:10.1016/j.leaqua.2004.09.003

Avolio, B., & Hannah, S. (2008). Developmental readiness: Accelerating leader development. *Consulting Psychology Journal: Practice and Research*, *60*(4), 331–347. doi:10.1037/1065-9293.60.4.331

Avolio, B., & Reichard, R. (2008). The rise of authentic followership. In R. Riggio, I. Chaleff, & J. Lipman-Blument (Eds.), *The Art of Followership: How Great Followers Create Great Leaders and Organizations* (pp. 325–337). San Francisco, CA: Jossey-Bass.

Awang, K. W., Ishak, N. K., Radzi, S. M., & Taha, A. Z. (2008). Environmental variables and performance: Evidence from the Hotel Industry in Malaysia. *International Journal of Economics and Management*, *2*(1), 59–79.

Axelrod, R., & Cohen, M. D. (1999). *Harnessing Complexity*. New York, NY: Basic Books.

Ayanso, A., Cho, D., & Lertwachara, K. (2010). The digital divide: Global and regional ICT leaders and followers. *Information Technology for Development*, *16*(4), 304–319. doi:10.1080/02681102.2010.504698

Azemi, A. (2018). Promoting innovation through systems thinking and systems design. In *Proceedings of the 2018 Annual IEEE International Systems Conference (SysCon)*. IEEE. 10.1109/SYSCON.2018.8369600

Aziz, L. (2007). *Managing change in healthcare information technology projects: the role of the project manager*. Project Management Institute.

Baker, C., Charlebois, M., Lopatka, H., Moineau, G., & Zelmer, J. (2016). Influencing Change: Preparing the Next Generation of Clinicians to Practice in the Digital Age. *Healthcare Quarterly*, *18*(4), 5–7. doi:10.12927/hcq.2016.24569 PMID:27009699

Banerjee, A. V., Cole, S., Duflo, E., & Linden, L. (2007). Remedying Education: Evidence from two randomized experiments in India'. *The Quarterly Journal of Economics*, *122*(3), 1235–1264. doi:10.1162/qjec.122.3.1235

Baniček, D. (2011) *Business.hr*. Retrieved August 27, 2019 from www.business.hr/b-it/cloud-trziste-u-hrvatskoj-tesko-tek-68-mil-kuna

Bansal, S., Khanna, M., & Jain, S. (2018). Corporate social responsibility rules in India: An Assessment. *Economic and Political Weekly*, *53*, 44–51.

Barasa, E. W., Molyneux, S., English, M., & Cleary, S. (2017). Hospitals as complex adaptive systems: A case study of factors influencing priority setting practices at the hospital level in Kenya. *Social Science & Medicine*, *174*(Feb), 104–112. doi:10.1016/j.socscimed.2016.12.026 PMID:28024239

Barbera, P., Bonneau, R., Egan, P. Jost, J., Nagler, J., Tucker, J. (2014). *Leaders or followers?* Measuring political responsiveness in the U.S. Congress using social media data. *110th American Political Science Association Annual Meeting.*

Barczak, G., Lassk, F., & Mulki, J. (2010). Antecedents of team creativity: An examination of team emotional intelligence, team trust and collaborative culture. *Creativity and Innovation Management*, *19*(4), 332–345. doi:10.1111/j.1467-8691.2010.00574.x

Barnea, A., & Rubin, A. (2010). Corporate social responsibility as a conflict between shareholders. *Journal of Business Ethics*, *97*(1), 71–86. doi:10.100710551-010-0496-z

Barnes, C. & Perpetual Guardian. (2019). White paper - The four-day week: Guidelines for an outcome-based trial - Raising productivity and engagement. Auckland, New Zealand: 4 Day Week Global Ltd.

Barnett, R., & Coate, K. (2005). *Engaging the Curriculum in Higher Education*. Maidenhead: Open University Press and McGraw Hill Education.

Barney, J. B. (1991). Firm resources and sustained competetitive advantage. *Journal of Management*, *17*(1), 99–120. doi:10.1177/014920639101700108

Barrett, M., Oborn, E., & Orlikowski, W. (2016). Creating Value in Online Communities: The Sociomaterial Configuring of Strategy, Platform, and Stakeholder Engagement. *Information Systems Research*, *27*(4), 704–723. doi:10.1287/isre.2016.0648

Barros, G. (2010). Herbert A. Simon and the concept of rationality: Boundaries and procedures. *Brazilian Journal of Political Economy*, *30*(3), 119, 455-472.

Bartezzaghi, E., & Ronchi, S. (2003). Internet supporting the procurement process: Lessons from four case studies. *Integrated Manufacturing Systems*, *14*(8), 632–641. doi:10.1108/09576060310503447

Barthel, P., & Hess, T. (2019). Are Digital Transformation Projects Special? *Twenty-Third Pacific Asia Conference on Information Systems, China.* 15. Retrieved from https://www.researchgate.net/publication/33445083

Bashshur, R. L., Reardon, T. G., & Shannon, G. W. (2000). Telemedicine: A new health care delivery system. *Annual Review of Public Health*, *21*(1), 613–637. doi:10.1146/annurev.publhealth.21.1.613 PMID:10884967

Bauer, J. C. (2002). Rural America and the digital transformation of health care: new perspectives on the future...third annual Southern Illinois Healthcare/Southern Illinois University Health Policy Institute, Rural Health Care: Current Issues and Future Directions, May 31 and June 1, 2001. *Journal of Legal Medicine*, *23*(1), 73–83. doi:10.1080/019476402317276678 PMID:11957332

Benabou, R., & Tirole, J. (2010). Individual corporate social responsibility. *Economica*, *77*, 1–19. doi:10.1111/j.1468-0335.2009.00843.x

Benbasat, I., Goldstein, D. K., & Mead, M. (1987). Strategy in Studies of Information Systems. *Management Information Systems Quarterly*, *11*(3), 369–386. doi:10.2307/248684

Benbya, H., & McKelvey, B. (2006). Toward a complexity theory of information systems development. *Information Technology & People*, *19*(1), 12–34. doi:10.1108/09593840610649952

Bennis, W., Sample, S. B., & Asghar, R. (2015). *The art and adventure of leadership*. San Francisco, CA: John Wiley & Sons.

Bensaou, M. (1999). Portfolios of buyer-supplier relationships. *MIT Sloan Management Review*, *40*(4), 35.

Billington, C. (2015). A change in mindset - building the digital future. *Public Sector*, *38*(2), 5–9.

Bilodeau, L. (2019). What is whole systems thinking? Retrieved from https://sustain.ok.ubc.ca/whole-systems-plan/what-is-it/

Blasé, J., & Blasé, J. (2002). The dark side of leadership: Teacher perspectives of principal mistreatment. *Educational Administration Quarterly*, *38*(5), 671–727. doi:10.1177/0013161X02239643

Bleich, M. R. (2014a). Developing leaders as systems thinkers--Part I. *Journal of Continuing Education in Nursing, 45*(4), 158–159. doi:10.3928/00220124-20140327-13 PMID:24702051

Bleich, M. R. (2014b). Developing leaders as systems thinkers--part II. *Journal of Continuing Education in Nursing, 45*(5), 201–208. doi:10.3928/00220124-20140424-13 PMID:24801681

Block, J. H., Kohn, K., Miller, D., & Ullrich, K. (2015). Necessity Entrepreneurship and Competitive Strategy, IZA Discussion Papers, No. 8219. *Small Business Economics, 44*(1).

Blyth, M. J. (2005). Learning from the future through scenario planning. Retrieved from http://www.fourscenes.com.au/LearningFromScenarios0305.pdf

Boddy, C. (2011). The corporate psychopaths theory of the global financial crisis. *Journal of Business Ethics, 102*(2), 255–259. doi:10.100710551-011-0810-4

Bogdanowicz, M. (2015). *Digital Entrepreneurship Barriers and Drivers*. In D. K. Boojihawon, & Z. M. Ngoasong (Eds.), Emerging digital business models in developing economies: The case of Cameroon. *Strategic Change-Briefings in Entrepreneurial Finance, 27*(2), 129–137. doi:10.1002/jsc.2188

Boisot, M., & MacMillan, I. (2004). Crossing epistemological boundaries: Managerial and entrepreneurial approaches to knowledge management. *Long Range Planning, 37*(6), 505–524. doi:10.1016/j.lrp.2004.09.002

Bondarouk, T., & Ruel, H. (2009). Electronic human resource management: Challenges in the digital era. *International Journal of Human Resource Management, 20*(3), 505–514. doi:10.1080/09585190802707235

Bonnet, D., Ferraris, P., Westerman, G., & McAfee, A. (2012). Talking 'bout a Revolution. *Digital Transformation Review, 2*(1), 17–33.

Bosilj Vukšić, V., Ivančić, L., & Suša Vugec, D. (2018). A Preliminary Literature Review of Digital Transformation Case Studies. In *ICMIT 2018: 20th International Conference on Managing Information Technology* (pp. 737–742). Retrieved from https://waset.org/Publication/a-preliminary-literature-review-of-digital-transformation-case-studies-/10009516

Bossert, O., Laartz, J., & Ramsey, T. J. (2014). Running your company at two speeds, *McKinsey Quarterly*, Dec.

Bouaziz, F. (2011). E-Government Projects Risk Management: Taking Stakeholders in Perspective. In A. Thyab, A. Ajeeli, & Y. A. L. Al-Bastaki (Eds.), Handbook of Research on E-Services in the Public Sector: E-Government Strategies and Advancements, Chapter 13.

Bouaziz, F. (2020, forthcoming). From E-government to Digital Government: the Public Value Quest in the Tunisian Public Administration. In V. Ponnusamy, K. Rafique, & N. Zaman (Eds.), *Employing Recent Technologies for Improved Digital Governance*. IGI Global.

Bourne, V. (2012). *Business impact of the Cloud, 2012*. Retrieved September 3, 2019 from lp.google-mkto.com/CFO-CloudComputingResearch.html

Bowersox, D. J., Closs, D. J., & Drayer, R. W. (2005). The digital transformation: Technology and beyond. *Supply Chain Management Review, 9*(1), 22–29.

Bowman, G. (2016). The practice of scenario planning: An analysis of inter- and intra-organizational strategizing. *British Journal of Management, 27*(1), 77–96. doi:10.1111/1467-8551.12098

Bralić, A., & Divjak, B. (2016). Use of MOOCs in Traditional Classroom: Blended Learning Approach. In *Proceedings of 9th EDEN Research Workshop*, Oldenburg. Retrieved from http://www.eurodl.org/index.php?p=special&sp=articles&inum=9&article=766

Brammer, S., Brooks, S., & Pavelin, S. (2006). Corporate social performance and stock returns: UK Evidence from disaggregate measures. *Financial Management*, *35*(3), 97–116. doi:10.1111/j.1755-053X.2006.tb00149.x

Brandon-Jones, A., & Kauppi, K. (2018). Examining the antecedents of the technology acceptance model within e-procurement. *International Journal of Operations & Production Management*, *38*(1), 22–42. doi:10.1108/IJOPM-06-2015-0346

Bratianu, C. (2017). *Strategic Thinking in Turbulent Times*. In C. Bratianu, A. M., Dima, & S. Hadad (Eds.), *Proceedings of the 11th International Conference on Business Excellence*, Bucharest University of Economic Studies, Bucharest ()10.1515/picbe-2017-0026

Briciu, S., Scorte, C., & Mester, I. (2013). The impact of accounting information on managerial decision – Empirical study conducted in the hospitality industry entities in Romania. *Theoretical and Applied Economics*, *9*(586), 27–38.

Briggs, C., & Makice, K. (2012). *Digital fluency: building success in the digital age*. Bloomington: SociaLens.

Brinker, S., & McLellan, L. (2014). The Rise of the Chief Marketing Technologist, *Harvard Business Review, 92*(7), pp. 82-85.

Briscoe, P. (2016). Global systems thinking in education to end poverty: Systems leaders with a concerted push. *International Studies in Educational Administration (Commonwealth Council for Educational Administration & Management (CCEAM)), 43*(3), 5-19.

Brockes, C., Grischott, T., Dutkiewicz, M., & Schmidt-Weitmann, S. (2017). Evaluation of the Education "Clinical Telemedicine/e-Health" in the Curriculum of Medical Students at the University of Zurich. *Telemedicine Journal and e-Health*, *23*(11), 899–904. doi:10.1089/tmj.2017.0011 PMID:28498779

Brown, M., & Mitchell, M. (2010). Ethical and unethical leadership: Exploring new avenues for future research. *Business Ethics Quarterly*, *20*(4), 583–616. doi:10.5840/beq201020439

Brown, S. L., & Eisenhardt, K. M. (1997). The Art of Continuous Change: Linking Complexity Theory and Time-Paced Evolution in Relentlessly Shifting Organizations. *Administrative Science Quarterly*, *42*(1), 1–34. doi:10.2307/2393807

Brunner, J. (1969). *Stand on Zanzibar*. New York: Ballantine.

Bryant, A. (2014). *An interpretative phenomenological analysis of student engagement in learning at the middle level.* (Education Doctoral Theses Paper), Northeastern University, USA. Retrieved from https://repository.library.northeastern.edu/files/neu:336553 (208)

Brzozowski, J., Cucculelli, M., & Surdej, A. (2014). Transnational ties and performance of immigrant entrepreneurs: The role of home-country conditions. *Entrepreneurship and Regional Development*, *26*(7-8), 546–573. doi:10.1080/08985626.2014.959068

Brzozowski, J., Cucculelli, M., & Surdej, A. (2017). The determinants of transnational entrepreneurship and transnational ties' dynamics among immigrant entrepreneurs in ICT sector in Italy. *International Migration (Geneva, Switzerland)*, *55*(3), 105–125. doi:10.1111/imig.12335

Burgess, C. (2017). Managerial Competencies for U.K. Hotel Financial Controllers: Are They Hospitality Managers or Accountants? *The Journal of Hospitality Financial Management*, *25*(1), 27–43. doi:10.1080/10913211.2017.1313612

Business Council of Australia. (2017, October). *Future-proof: Protecting Australians through education and skills.* Melbourne, Victoria. Retrieved from http://hdl.voced.edu.au/10707/440976

Butler-Henderson, K. (2017). *Health Information Management 2025: What is required to create a sustainable profession in the face of digital transformation?* Launceston, Australia: University of Tasmania.

Buvat, J., Crummener, C., Kar, K., Sengupta, A., Solis, B., Aboud, C., & Aoufi, E. H. (2018). The digital culture challenge: Closing the employee-leadership gap. Retrieved from Cambridge International Examination. (2016). *Global thinking*. Paper presented at the Cambridge Schools Conference, Kuala Lumpur. https://www.cambridgeinternational.org/Images/346617-global-thinking.pdf

Byrne, D. S. (1998). *Complexity theory and the social sciences: an introduction*. New York, NY: Routledge.

Cai, S., Jun, M., & Yang, Z. (2006). The impact of interorganizational internet communication on purchasing performance: A study of Chinese manufacturing firms. *The Journal of Supply Chain Management*, *42*(3), 16–29. doi:10.1111/j.1745-493X.2006.00014.x

Çakar, N., & Ertürk, A. (2010). Comparing innovation capability of small and medium-sized enterprises: Examining the effects of organizational culture and empowerment. *Journal of Small Business Management*, *48*(3), 325–359. doi:10.1111/j.1540-627X.2010.00297.x

Calabretta, G., Gemser, G., & Wijnberg, N. M. (2017). The interplay between intuition and rationality in strategic decision-making: A paradox perspective. *Organization Studies*, *38*(3-4), 365–401. doi:10.1177/0170840616655483

Campbell, D., Draper, R., & Huffington, C. (2019). *A Systemic Approach to Consultation. Systemic Thinking and Practice Collection*. London, UK: Routledge.

Cao, X., Zhang, X., & Xi, Y. (2011). Ambidextrous organization in harmony: A multi-case exploration of the value of HeXie management theory. *Chinese Management Studies*, *5*(2), 146–163. doi:10.1108/17506141111142799

Carcary, M., Doherty, E., Conway, G., & McLaughlin, S. (2014). Cloud computing adoption readiness and benefit realization in Irish SMEs—An exploratory study. *Information Systems Management*, *31*(4), 313–327. doi:10.1080/10580530.2014.958028

Carr, A. S., & Smeltzer, L. R. (2002). The relationship between information technology use and buyer-supplier relationships: An exploratory analysis of the buying firm's perspective. *IEEE Transactions on Engineering Management*, *49*(3), 293–304. doi:10.1109/TEM.2002.803389

Carsten, M., & Uhl-Bien, M. (2013). Ethical followership: An examination of followership beliefs and crimes of obedience. *Journal of Leadership & Organizational Studies*, *20*(1), 49–61. doi:10.1177/1548051812465890

Cash, P. J., Hartlev, C. G., & Durazo, C. B. (2017). Behavioural design: A process for integrating behaviour change and design. *Design Studies*, *48*, 96–128. doi:10.1016/j.destud.2016.10.001

Cavallo, A., Ghezzi, A., Dell'Era, C., & Pellizzoni, E. (2019). Fostering digital entrepreneurship from startup to scaleup: The role of venture capital funds and angel groups. *Technological Forecasting and Social Change*, *145*, 24–35. doi:10.1016/j.techfore.2019.04.022

Cerne, M., Bunjak, A., Wong, S., & Moh'd, S. (2019). Creative and deserving: Digital workers' transformation of subjective into objective creativity. Academy of Management Proceedings, 2019(1), 10070.

Černe, M., Jaklič, M., & Škerlavaj, M. (2013). Authentic leadership, creativity, and innovation: A multilevel perspective. *Leadership*, *9*(1), 63–85. doi:10.1177/1742715012455130

Chae, B. K. (2014). A Complexity Theory Approach to IT-Enabled Services (Iess) and Service Innovation: Business Analytics as an Illustration of Ies, *Decision Support Systems, 57*(1), pp. 1-10.

Chan, S. (2018). What the heck is digital transformation- Finally a definition anyone can understand? *Open Consulting Group*, May.

Chan, Y. E., Sabherwal, R., & Thatcher, J. B. (2006). Antecedents and Outcomes of Strategic IS Alignment: An Empirical Investigation, *IEEE Transactions on Engineering Management, 53*(1), pp. 27-47.

Chand, M. T., & Rosalie, L. (2011). Global competitiveness, consumer choice and 'country of origin' effect: An exploratory East-West study. *Asia Pacific Journal of Management*, *17*(3), 15.

Chand, M., & Ghorbani, M. (2011). National culture, networks and ethnic entrepreneurship: A comparison of the Indian and Chinese immigrants in the US. *International Business Review*, *20*(6), 13. doi:10.1016/j.ibusrev.2011.02.009

Chang, H.-L. (2019). Factors Affecting the Offshore Deployment of Interorganizational Systems in China: A Case Study Analysis. *Journal of Organizational Computing and Electronic Commerce*, *29*(3), 163–189. doi:10.1080/10919392.2019.1583811

Charlton, E. (2019, February 12). The results of Finland's basic income experiment are in. Is it working? *World Economic Forum*. Retrieved from https://www.weforum.org/agenda/2019/02/the-results-finlands-universal-basic-income-experiment-are-in-is-it-working/

Chase, Z., & Laufenberg, D. (2011). Embracing the squishiness of digital literacy. *Journal of Adolescent & Adult Literacy*, *54*(7), 535–537. doi:10.1598/JAAL.54.7.7

Chen, Y., Wang, Y., Nevo, S., Benitez-Amado, J., & Kou, G. (2015). IT Capabilities and Product Innovation Performance: The Roles of Corporate Entrepreneurship and Competitive Intensity, *Information & Management, 52*(6), pp. 643-657.

Cheong, M., Yammarino, F., Dionne, S., Spain, S., & Tsai, C. (2018). A review of the effectiveness of empowering leadership. *The Leadership Quarterly*, *30*(1), 34–58. doi:10.1016/j.leaqua.2018.08.005

Cherry, M. A. (2015). Beyond Misclassification: The Digital Transformation of Work. *Comparative Labor Law & Policy Journal*, (3): 577–602. Retrieved from https://heinonline.org/HOL/P?h=hein.journals/cllpj37&i=614

Chiu, Y.-H., & Huang, C.-W. (2011). Evaluating the optimal occupancy rate, operational efficiency, and profitability efficiency of Taiwan's international tourist hotels. *Service Industries Journal*, *31*(13), 2145–2162. doi:10.1080/02642069.2010.503889

Choi, B. C., & Pak, A. W. (2005). A catalog of biases in questionnaires. *Preventing Chronic Disease*, *2*(1), A13. PMID:15670466

Chorafas, D. N. (2011). *Computing Strategies*. SAD: CRC Press Taylor & Francis Group.

Christian Tanner, R. W. (2008). Current Trends and Challenges in Electronic Procurement: An Empirical Study. *Electronic Markets*, *18*(6), 6–18. doi:10.1080/10196780701797599

Chrysostome, E., & Arcand, S. (2009). Survival of Necessity Immigrant Entrepreneurs: An Exploratory Study. *Journal of Comparative International Management*, *12*(2).

Chung, F. (2019). 'Never been so disgusted': Coles customer's viral photo of baby formula 'express lane' causes outrage. *News.com.au*. Retrieved from https://www.news.com.au/finance/business/retail/never-been-so-disgusted-coles-customers-viral-photo-of-baby-formula-express-lane-causes-outrage/news-story/11f9a42ce383329864cd09365d48e1ce

CIMA. (2008). Activity Based Costing Topic Gateway Series No. 1. Available at http://www.cimaglobal.com/Documents/ImportedDocuments/cid_tg_activity_based_costing_nov08.pdf.pdf (Retrieved on 24 August 2019).

Clark, R. M. (2019). *Intelligence Analysis: A Target-Centric Approach* (6th ed.). Washington, DC: CQ Press / SAGE.

Clark, R. M., & Lowenthal, M. M. (2016). *The five disciplines of intelligence collection*. Thousand Oaks, CA: CQ Press / SAGE.

Cohn, D., & Caumont, A. (2016). *10 demographic trends that are shaping the U.S. and the world.* Retrieved from https://www.pewresearch.org/fact-tank/2016/03/31/10-demographic-trends-that-are-shaping-the-u-s-and-the-world/

Cohn, K. H. (2015). Training for physician engagement. *Talent Development, 69*(5), 58–62.

Coile, R. C. Jr. (2000). The digital transformation of health care. *Physician Executive, 26*(1), 8. PMID:10788126

Colbert, A., Yee, N., & George, G. (2016). The digital workforce and the workplace of the future. *Academy of Management Journal, 59*(3), 731–739. doi:10.5465/amj.2016.4003

Colbert, B. A., & Kurucz, E. C. (2007). Three Conceptions of Triple Bottom Line Business Sustainability and the Role for HRM. *Human Resource Planning, 30*(1), 21–29.

Coldwell-Neilson, J. (2016). Office for Learning and Teaching (OLT) National Teaching Fellowship: Unlocking the code to digital literacy, *Australian Learning and Teaching Fellows.* Retrieved from https://docs.education.gov.au/system/files/doc/other/fellows_2012-2016.pdf

Coldwell-Neilson, J. (2018). Digital Literacy Expectations in Higher Education. In *Proceedings of Open Oceans: learning without borders. 35th International Conference of innovation, practice and Research in the use of Education Technologies in Tertiary Education.* Deakin University, Australia.

Coldwell-Neilson, J. (2019). What is Digital Literacy [video file]. Retrieved from https://video.deakin.edu.au/media/t/0_80y0wiqq

Coldwell-Neilson, J., & Cooper, T. (2019). Digital literacy meets industry 4.0. In J. Higgs, W. Letts, & G. Crisp (Eds.), Education for employability (Volume 2): learning for future possibilities. Boston, MA: Brill Sense. pp. 37-50.

Coldwell-Neilson, J. (2017). Digital Literacy - A Driver for Curriculum Transformation. In *Proceedings of the Higher Education Research and Development Society of Australasia Conference, HERDSA '17*, Sydney, Australia.

Commins, J. (2014). Addressing physician engagement. *HealthLeaders, 17,* 8-8, 10.

Compaine, B. M. (Ed.). (2001). *The digital divide: Facing a crisis or creating a myth?* Cambridge, MA: MIT Press Sourcebooks. doi:10.7551/mitpress/2419.001.0001

Constant, A., & Shachmurove, Y. (2006). Entrepreneurial ventures and wage differentials between Germans and immigrants. *International Journal of Manpower, 27*(3), 208–229. doi:10.1108/01437720610672149

Cooper, T., Tierney, H., Brooker, M. R., & Sutcliffe, J. (2020, in review, for publication). Theorisation, Models of Youth Work and Digital Technologies: Opportunities and Challenges. In Z. Zaremohzzabieh, S. Ahrari, S. E. Krauss, A. A. Samah, & S. Z. Omar (Eds.), The Handbook of Research on Youth Work in a Digital Society. Hershey, PA: IGI Global.

Cooper, T. (2019). Calling out 'alternative facts': Curriculum to develop students' capacity to engage critically with contradictory sources. *Teaching in Higher Education, 24*(3), 444–459. doi:10.1080/13562517.2019.1566220

Cooper, T., & Scriven, R. (2017). Communities of inquiry in curriculum approach to online learning: Strengths and limitations in context. *Australasian Journal of Educational Technology, 33*(4), 22–37. doi:10.14742/ajet.3026

Cope, J. (2011). Entrepreneurial learning from failure: An interpretative phenomenological analysis. *Journal of Business Venturing, 26*(6), 604–623. doi:10.1016/j.jbusvent.2010.06.002

Corini, J. (2000). Integrating e-procurement and strategic sourcing. *Supply Chain Management Review, 2*(3), 70–75.

Crabtree, B. F., & Miller, W. L. (1999). *Doing qualitative research.* Thousand Oaks, CA: Sage.

Crawford, J., Dawkins, S., Martin, A., & Lewis, G. (2017). Understanding the organizational climate of unethical leadership in the Australian Football League. *The Journal of Leadership Studies*, *11*(2), 52–54. doi:10.1002/jls.21525

Crawford, J., Dawkins, S., Martin, A., & Lewis, G. (2018). Conceptualising authentic followers and developing a future research agenda. In D. Cotter-Lockard (Ed.), *Authentic Leadership and Followership: International Perspectives* (pp. 271–293). Cham, Switzerland: Palgrave Macmillan. doi:10.1007/978-3-319-65307-5_11

Crawford, J., Dawkins, S., Martin, A., & Lewis, G. (2019). Putting the leader back into authentic leadership: Reconceptualising and rethinking leaders. *Australian Journal of Management*; Advanced Online Publication. doi:10.1177/0312896219836460

Crawford, J., & Kelder, J.-A. (2019). Do we measure leadership effectively? Articulating and evaluating scale development psychometrics for best practice. *The Leadership Quarterly*, *30*(1), 133–144. doi:10.1016/j.leaqua.2018.07.001

Crawford, J., Kelder, J.-A., & Knox, M. (2020). What does it take to be a social entrepreneur? Authentic leaders and their effect on innovation. In Ç. Dogru (Ed.), *Leadership styles, innovation, and social entrepreneurship in the era of digitalization* (pp. 282–310). Hershey, PA: IGI Global.

Croom, S., & Brandon-Jones, A. (2007). Impact of e-procurement: Experiences from implementation in the UK public sector. *Journal of Purchasing and Supply Management*, *13*(4), 294–303. doi:10.1016/j.pursup.2007.09.015

Croom, S., & Johnston, R. (2003). E-service: Enhancing internal customer service through e-procurement. *International Journal of Service Industry Management*, *14*(5), 539–555. doi:10.1108/09564230310500219

Crossman, B., & Crossman, J. (2011). Conceptualising followership–a review of the literature. *Leadership*, *7*(4), 481–497. doi:10.1177/1742715011416891

Currell, R., Urquhart, C., Wainwright, P., & Lewis, R. (2000). Telemedicine versus face to face patient care: Effects on professional practice and health care outcomes. *Cochrane Database of Systematic Reviews*, (2). doi:10.1002/14651858.Cd002098 PMID:10796678

Dahunsi, F. M., & Owoseni, T. M. (2015). Cloud computing in Nigeria: The cloud ecosystem perspective. *Nigerian Journal of Technology*, *34*(1), 209–216. doi:10.4314/njt.v34i1.26

Daily News. (2019). Need for global thinking. *Daily News*.

Dale, C. (2003). The competitive networks of tourism e-mediaries: New strategies, new advantages. *Journal of Vacation Marketing*, *9*(2), 109–118. doi:10.1177/135676670300900201

Damanpour, F., & Evan, W. M. (1984). Organizational Innovation and Performance: The Problem of "Organizational Lag," *Administrative Science Quarterly, 29*(3), pp. 392–409.

Damanpour, F. (2017). *Organizational innovation*. Oxford University Press; doi:10.1093/acrefore/9780190224851.013.19

Dapp, T. (2014). Fintech – The digital (r)evolution in the financial sector. *Current Issues: Digital Economy and Structural Change*. Retrieved from www.dbresearch.com

Dattakumar, A., Gray, K., Henderson, K. B., Maeder, A., & Chenery, H. (2012). We are not educating the future clinical health professional workforce adequately for e-health.

Davidson, E., & Vaast, E. (2010). Digital Entrepreneurship and its Sociomaterial Enactment. In *43rd Hawaii International Conference on Systems Sciences Vols 1-5* (pp. 2978-+). Los Alamitos, CA: IEEE Computer Soc.

Davila, A., Gupta, M., & Palmer, R. (2003). Moving procurement systems to the internet: The adoption and use of e-procurement technology models. *European Management Journal*, *21*(1), 11–23. doi:10.1016/S0263-2373(02)00155-X

Davis, H. P. (2014). Managing black swan uncertainty: A grounded theory examination of portfolio managers' assessment of unknown risk. (3609144 Ph.D.), Capella University, Ann Arbor. ProQuest Central; ProQuest Dissertations & Theses Global database.

Davis, F. D. (1989). Perceived usefulness, perceived ease of use, and user acceptance of information technology. *Management Information Systems Quarterly*, *13*(3), 319–340. doi:10.2307/249008

Davis, F. D. (1989). Perceived usefulness, perceived ease of use, and user acceptance of information technology. *MIS quarterly*, 319-340. Competence: Findings of an Australian study. *Studies in Health Technology and Informatics*, *178*, 33–38.

Dawson, A., Hirt, M. & Scanlan, J. (2016). The economic essentials of digital strategy, *McKinsey Quarterly*, March.

Dawson, G. S., Watson, R. T., & Boudreau, M.-C. (2010). Information Asymmetry in Information Systems Consulting: Toward a Theory of Relationship Constraints, *Journal of Management Information Systems, 27*(3), pp. 143-177.

Day, G. S., & Hubbard, K. J. (2003). Customer relationships go digital. *Business Strategy Review*, *14*(1), 17–26. doi:10.1111/1467-8616.00240

De Velde, E. V., Vermier, W., & Corten, F. (2005). Finance and accounting: Corporate social responsibility and financial performance. *Corporate Governance*, *5*(3), 129–137. doi:10.1108/14720700510604760

de Zilwa, D. (2016). The strengths and capacities of authentic followership. *Leadership and Organization Development Journal*, *37*(3), 310–324. doi:10.1108/LODJ-01-2014-0010

de Zilwa, D., Lapierre, L., & Carsten, M. (2014). A new conceptual framework for authentic followership. In L. Lapierre, & M. Carsten (Eds.), *Followership: What is it, and why do people follow* (pp. 47–72). Bingley, UK: Emerald Group.

Deakin University. (2018). Professional Practice credentials. Retrieved from http://www.deakin.edu.au/ppc

Deloitte Access Economics. (2019). Australia's Digital Pulse: Developing the digital workforce to drive growth in the future. A report prepared for the Australian Computer Society. Retrieved from https://www.acs.org.au/content/dam/acs/acs-publications/Digital-Pulse-2019-FINAL-Web.pdf

Deng, H., Karunasena, K., & Xu, W. (2018). Evaluating the Performance of e-Government in Developing Countries: A Public Value Perspective. *Internet Research*, *28*(1), 169–190. doi:10.1108/IntR-10-2016-0296

Denison, D., & Mishra, A. (1995). Toward a theory of organizational culture and effectiveness. *Organization Science*, *6*(2), 204–223. doi:10.1287/orsc.6.2.204

Denti, L. (2011). *Leadership and innovation: How and when do leaders influence innovation in R&D teams?* Sweden: University of Gothenburg.

Dess, G., McNamara, G., Eisner, A., & Hyun, S. (2019). *Strategic management: Creating competitive advantages* (9th ed.). New York: McGraw-Hill Education.

Dhaliwal, D. S., Li, O. S., Tsang, A., & Yang, Y. G. (2011). Voluntary nonfinancial disclosure and cost of equity capital: The initiation of corporate social responsibility reporting. *The Accounting Review*, *86*(1), 59–100. doi:10.2308/accr.00000005

Dhanesh, G. S. (2015). Why corporate social responsibility? An analysis of drivers of CSR in India. *Management Communication Quarterly*, *29*(1), 114–129. doi:10.1177/0893318914545496

Dharmapala, D., & Khanna, V. (2016). The impact of mandated corporate social responsibility: Evidence from India's companies act of 2013, *Coase-Sandor Working Paper Series in Law and Economics*.

Dhillon, G. (2008). Organizational competence for harnessing IT: A case study. *Information & Management, 45*(5), 297–303. doi:10.1016/j.im.2008.01.008

Dobrica, S. (2019). From Digitization, through Digitalization, to Digital Transformation. *Online Searcher*, 35-39. Retrieved from https://www.researchgate.net/publication/332111919

Dods, J., Alem, L., & Nepal, S. (2012). *Caring for the last 3%: Telehealth potential and Broadband implications for remote Australia. CSIRO ICT Centre; 2012*. Retrieved from Canberra.

Domegan, C., McHugh, P., Devaney, M., Duane, S., Hogan, M., Broome, B. J., ... Piwowarczyk, J. (2016). Systems-thinking social marketing: Conceptual extensions and empirical investigations. *Journal of Marketing Management, 32*(11-12), 1123–1144. doi:10.1080/0267257X.2016.1183697

Domitriu, C., Timofti, L. C., Nechita, E., & Domitriu, G. (2014). The Influence of the Locus of Control and Decision-making Capacity upon the Leadership Style. *Social and Behavioral Sciences*, 141, 25 August, 494-499.

Drucker, P. E. (1984). The new meaning of corporate social responsibility. *California Management Review*, *26*(2), 55–63. doi:10.2307/41165066

Drucker, P. F. (1966). *The Effective executive*. New York: Harper & Row.

Drucker, P. F. (2002). *Managing in the Next Society*. New York: St. Martin S. Press.

Dubé, L., & Paré, G. (2003). Rigor in Information Systems Positivist Case Research: Current Practices, Trends, and Recommendations, *MIS Quarterly, 27*(4), pp. 597-636.

Dunleavy, P., Margetts, H., Bastow, S., & Tinkler, J. (2006). New public management is dead—Long live digital-era governance. *Journal of Public Administration: Research and Theory*, *16*(3), 467–494. doi:10.1093/jopart/mui057

Dutton, J. E. (1993). Interpretations on Automatic: A Different View of Strategic Issue Diagnosis, *Journal of Management Studies* (30:3), pp. 339-357.

Du, W. Y., & Li, W. J. (2018). *Digital Entrepreneurship Ecosystem as a New Form of Organizing: A Case of Zhongguancun*. Singapore: World Scientific Publ Co Pte Ltd.

Du, W. Y., Pan, S. L., Zhou, N., & Ouyang, T. H. (2018). From a marketplace of electronics to a digital entrepreneurial ecosystem (DEE): The emergence of a meta-organization in Zhongguancun, China. *Information Systems Journal*, *28*(6), 1158–1175. doi:10.1111/isj.12176

Eadie, R., Perera, S., & Heaney, G. (2010). A cross-discipline comparison of rankings for e-procurement drivers and barriers within UK construction organisations. *Journal of Information Technology in Construction,* 217-233.

Eadie, R., Perera, S., & Heaney, G. (2011). Analysis of the use of E-Procurement in the Public and Private Sectors of the UK construction Industry. *Journal of Information Technology in Construction*, *16*, 669–686.

Eadie, R., Perera, S., Heaney, G., & Carlisle, J. (2007). Drivers and barriers to public sector e-procurement within Northern Ireland's construction industry. *Journal of Information Technology in Construction*, *12*, 103–120.

Eason, C. C. (2016). Achieving Sustainable Effectiveness. *The Journal of Applied Management and Entrepreneurship*, *21*(2), 119–121. doi:10.9774/GLEAF.3709.2016.ap.00009

Ebel, P., Bretschneider, U., & Leimeister, J. M. (2016). Leveraging Virtual Business Model Innovation: A Framework for Designing Business Model Development Tools, *Information Systems Journal, 26*(5), pp. 519-550.

Economist, T. (2017). The everywhere stores: Online retail is booming in China. *The Economist, Oct. 26.*

Economist, T. (2019). Red Moon Rising: How China could dominate science. Should the world worry? *The Economist, Jan. 12.*

Economist, The. (2017, May 6). Data is giving rise to a new economy. Retrieved from https://www.economist.com/briefing/2017/05/06/data-is-giving-rise-to-a-new-economy

Economist. (2005). *The good company-Special Report*. Retrieved from https://www.economist.com/special-report/2005/01/22/the-good-company

Economist. (2017). *Donate, or else: Indian firms make the best of coerced do-goodery*. Retrieved from https://www.economist.com/business/2017/11/16/indian-firms-make-the-best-of-coerced-do-goodery

Economist. (2019). *I'm from a company, and I'm here to help*, August 24, 14-16.

Edejer, T. T.-T. (2000). Disseminating health information in developing countries: The role of the internet. *BMJ (Clinical Research Ed.), 321*(7264), 797–800. doi:10.1136/bmj.321.7264.797 PMID:11009519

Edirippulige, S., & Armfield, N. R. (2017). Education and training to support the use of clinical telehealth: A review of the literature. *Journal of Telemedicine and Telecare*, *23*(2), 273–282. doi:10.1177/1357633X16632968 PMID:26892005

Edirippulige, S., Armfield, N. R., & Smith, A. C. (2012). A qualitative study of the careers and professional practices of graduates from an e-health postgraduate programme. *Journal of Telemedicine and Telecare*, *18*(8), 455–459. doi:10.1258/jtt.2012.gth107 PMID:23209263

Edirippulige, S., Brooks, P., Carati, C., Wade, V. A., Smith, A. C., Wickramasinghe, S., & Armfield, N. R. (2018). It's important, but not important enough: eHealth as a curriculum priority in medical education in Australia. *Journal of Telemedicine and Telecare*, *24*(10), 697–702. doi:10.1177/1357633X18793282 PMID:30343657

Edirippulige, S., Samanta, M., & Armfield, N. R. (2018). Assessment of Self-Perceived Knowledge in e-Health Among Undergraduate Students. *Telemedicine Journal and e-Health*, *24*(2), 139–144. doi:10.1089/tmj.2017.0056 PMID:28708457

Edirippulige, S., Smith, A. C., Armfield, N. R., Bensink, M., & Wootton, R. (2012). Student perceptions of a hands-on practicum to supplement an online eHealth course. *Journal of Medical Internet Research*, *14*(6). doi:10.2196/jmir.2029 PMID:23246840

Edirippulige, S., Smith, A. C., Wickramasinghe, S., & Armfield, N. R. (2018). Examining the Influence of E-Health Education on Professional Practice. *Journal of Medical Systems*, *42*(11), 215. doi:10.100710916-018-1084-5 PMID:30276486

Eglene, O., Dawes, S., & Schneider, C. (2007). Authority and leadership patterns in public sector knowledge networks. *American Review of Public Administration*, *37*(1), 91–113. doi:10.1177/0275074006290799

Eisenhardt, K. M. (1989). Building Theories from Case Study Research. *Academy of Management Review*, *14*(4), 532–550. doi:10.5465/amr.1989.4308385

Eisenhardt, K. M. (1991). Better stories and better constructs: The case for rigor and comparative logic. *Academy of Management Review*, *18*(3), 620–627. doi:10.5465/amr.1991.4279496

Eisenhardt, K. M., & Graebner, M. E. (2007). Theory building from cases: Opportunities and challenges. *Academy of Management Journal*, *50*(1), 25–32. doi:10.5465/amj.2007.24160888

El-Shishini, H. (2017). The Use of Management Accounting Techniques at Hotels in Bahrain. *Review of Integrative Business and Economics Research*, *6*, 64–77.

Emory, C. W., & Cooper, D. (1991). *Business Research Methods*. Homewood, IL: R. D. Irwin.

Etro, F. (2009). The Economic Impact of Cloud Computing on Business Creation, Employment and Output in Europe. An application of the Endogenous Market Structures Approach to a GPT innovation. *Review of Business and Economic Literature, KU Leuven. Faculty of Economics and Business*, *0*(2), 179–208.

European Commission. (2004). E-Health, Making Healthcare Better for European Citizens: An Action Plan for a European E-health Area. Retrieved from https://ec.europa.eu/digital-single-market/en/news/e-health-making-healthcare-better-european-citizens-action-plan-european-e-health-area

European Commission. (2006). Recommendation of the European Parliament and the Council of 18 December 2006 on key competences for lifelong learning. *Official Journal of the European Union, L, 394*(10). Retrieved from https://eur-lex.europa.eu/legal-content/EN/TXT/HTML/?uri=CELEX:32006H0962&from=EN

Evans, C. P., & Gawer, A. (2016). *The Rise of the Platform Enterprise: A Global Survey*. Retrieved from https://www.thecge.net/app/uploads/2016/01/PDF-WEB-Platform-Survey_01_12.pdf

Faddis, A. (2018). The digital transformation of healthcare technology management. *Biomedical Instrumentation & Technology*, *52*(s2), 34–38. doi:10.2345/0899-8205-52.s2.34 PMID:29775387

Farani, A. Y., Karimi, S., & Motaghed, M. (2017). The role of entrepreneurial knowledge as a competence in shaping Iranian students' career intentions to start a new digital business. *European Journal of Training and Development*, *41*(1), 83–100. doi:10.1108/EJTD-07-2016-0054

Fathi, Z., & Dozdahiri, E. S. M. (2015). A survey of activity-based costing in hotel industry. *Management Science Letters*, *5*, 855–860. doi:10.5267/j.msl.2015.6.012

Ferber, J. (1999). *Multi-Agent Systems*. Harlow, UK: Addison-Wesley.

Fernández-Villacañas, M. A. (2016). The Global Strategic Intelligence Management; E-module multimedia for student's self-study. #Europehome Project, UE: European University Foundation.

Fernández-Villacañas, M. A. (2017). La inteligencia estratégica económica global: La capacidad anticipativa clave en la solución de los nuevos conflictos. In E. del Aire (Ed.), *La Nueva Economía de la Defensa en un Nuevo Orden Mundial. Reflexiones desde el Ámbito Aeroespacial* (pp. 137–163). Madrid, Spain: BBVA & Ministerio de Defensa de España.

Fielder, A., Vuorikari, R., Rodríguez-Priego, N., & Punie, Y. (2016). European Commission, & Joint Research Centre. Background review for developing the digital competence framework for consumers a snapshot of hot-button issues and recent literature. Luxembourg: Publications Office.

Filkins, B. L., Kim, J. Y., Roberts, B., Armstrong, W., Miller, M. A., Hultner, M. L., ... Steinhubl, S. R. (2016). Privacy and security in the era of digital health: What should translational researchers know and do about it? *American Journal of Translational Research*, *8*(3), 1560–1580. PMID:27186282

Flath, C. M., Friesike, S., Wirth, M., & Thiesse, F. (2017). Copy, Transform, Combine: Exploring the Remix as a Form of Innovation, *Journal of Information Technology, 32*(4), pp. 306-325.

Fleisher, C. S., & Blenkhorn, D. L. (ED). (2001). Managing frontiers in competitive intelligence. Westport, CT: Quorum Books.

Fleisher, C. S., & Bensoussan, B. (2007). *Business and Competitive Analysis: effective application of new and classic methods*. Upper Saddle River, NJ: FT Press.

Fleming, A., Wise, R. M., Hansen, H., & Sams, L. (2017). The sustainable development goals: A case study. *Marine Policy*, *86*, 94–103. doi:10.1016/j.marpol.2017.09.019

Forrester, J. W. (2007). System Dynamics: A personal view of the first fifty years. *System Dynamics Review*, *23*(2-3), 345–358. doi:10.1002dr.382

Foss, N. J., & Saebi, T. (2018). Business models and business model innovation: Between wicked and paradigmatic problems. *Long Range Planning*, *51*(1), 9–21. doi:10.1016/j.lrp.2017.07.006

Foundation for Young Australians. (2017). The new work smarts: thriving in the new work order. *Foundation for Young Australians (FYA)*. Retrieved from https://www.fya.org.au/our-research/#report-downloads

Fowler, B. L., Johns, J., Tanniru, M. R., Balijepally, V., Roumani, Y. F., Bobryk, D., & Mitchell, K. (2018). Engaging patients through Multi-Disciplinary Rounding -The case study at a Michigan hospital. *Journal of Hospital Administration*, *7*(5), 17–26. doi:10.5430/jha.v7n5p17

Fredrickson, B., & Joiner, T. (2002). Positive emotions trigger upward spirals toward emotional well-being. *Psychological Science*, *13*(2), 172–175. doi:10.1111/1467-9280.00431 PMID:11934003

Freeman, R. E. (1984). *Strategic management: A stakeholder approach*. Boston, MA: Pittman-Ballinger.

Friedman, M. (1970). The social responsibility of business is to increase its profits. *New York Times Magazine*, *32*, 122-126.

Friedman, T. L. (2007). *The world is flat, A brief history of the twenty-first century* (2nd rev ed.). New York: Farrar, Straus, and Giroux Publisher.

Fulcoon, M. (2018). Taiwan's Hotels Grapple with Oversupply. Taiwan Business Topic. Available at https://topics.amcham.com.tw/2018/12/taiwans-hotels-grapple-with-oversupply/ (Retrieved on 3 August 2019)

Gabryelczyk, R., & Roztocki, N. (2018). Business process management success framework for transition economies. *Information Systems Management*, *35*(3), 234–253. doi:10.1080/10580530.2018.1477299

Ganz, W., Dworschak, B., & Schnalzer, K. (2019). Competences and Competence Development in a Digitalized World of Work. In I. L. Nunes (Ed.), *Advances in Human Factors and Systems Interaction* (Vol. 781, pp. 312–320)., doi:10.1007/978-3-319-94334-3_31

Gardenswartz, L., Rowe, A., Digh, P., & Bennett, M. (2003). *The global diversity desk reference: Managing an international workforce*. Pfeiffer.

Gardner, P. J. (2009). Organizational change: All we want is better projects-why so difficult? *AACE International Transactions*, PM.03.01-PM.03.25.

Gardner, L., & Stough, C. (2002). Examining the relationship between leadership and emotional intelligence in senior level managers. *Leadership and Organization Development Journal*, *23*(2), 68–78. doi:10.1108/01437730210419198

Gardner, W., Cogliser, C., Davis, K., & Dickens, M. (2011). Authentic leadership: A review of the literature and research agenda. *The Leadership Quarterly*, *22*(6), 1120–1145. doi:10.1016/j.leaqua.2011.09.007

Garland, E., Fredrickson, B., Kring, A., Johnson, D., Meyer, P., & Penn, D. (2010). Upward spirals of positive emotions counter downward spirals of negativity: Insights from the broaden-and-build theory and affective neuroscience on the treatment of emotion dysfunctions and deficits in psychopathology. *Clinical Psychology Review*, *30*(7), 849–864. doi:10.1016/j.cpr.2010.03.002 PMID:20363063

Gartner. (2010). Market Overview: Consulting and Si Service Providers, Asia/Pacific 2010. Gartner.

Gartner. (2012). *Gartner press releases, 2012*. Gartner. Retrieved August 17, 2019 from www.gartner.com

Gartner. (2016). Gartner Survey Reveals Marketing Budgets Increased for Third Consecutive Year. In R. van der Meulen, & A. A. Forni (Eds.), *Newsroom,* Gartner.

Gartner. (2017). *Capitalizing on Your Business Ecosystems Economy: A Gartner Trend Insight Report.* Retrieved from https://www.gartner.com/doc/3759164/capitalizing-business-ecosystems-economy-gartner

Gates, B. (2000). Digital dividends. www.microsoft.com/billgates/speeches/2000/10-18digitaldividends

Gay, A. (2019). How digital literacy affects the modern workforce. Adobe Blog. Retrieved from https://theblog.adobe.com/how-digital-literacy-affects-the-modern-workforce/

Gekara, V., Snell, D., Molla, A., Karanasios, S., & Thomas, A. (2019). Skilling the Australian workforce for the digital economy. *National Centre for Vocational Education Research (NCVER)* Adelaide, SA. Retrieved from https://eric.ed.gov/?id=ED594704

Gerdeman, D. (2018). The Airbnb Effect: Cheaper Rooms For Travelers, Less Revenue For Hotels. Forbes.com Feb 27, 2018. Retrieved from https://www.forbes.com/sites/hbsworkingknowledge/2018/02/27/the-airbnb-effect-cheaper-rooms-for-travelers-less-revenue-for-hotels/#7612dd3d672f

Gerow, J. E., Grover, V., Thatcher, J. B., & Roth, P. L. (2014). Looking toward the Future of IT-Business Strategic Alignment through the Past: A Meta-Analysis, *MIS Quarterly, 38*(4), pp. 1059-1085.

Ghosh, S. (2018, July 18). Artificial intelligence will create as many jobs as it destroys, according to PwC. *Tech Insider.* Retrieved from https://www.businessinsider.com.au/pwc-artificial-intelligence-wont-destroy-all-jobs-2018-7

Gilad, B. (2004). *Early Warning: Using Competitive Intelligence to Anticipate Market Shifts, Control Risk, and Create Powerful Strategies.* New York: Amacom.

Gil-Garcia, J. R., Dawes, S. S., & Pardo, T. A. (2018). Digital government and public management research: Finding the crossroads. *Public Management Review, 20*(5), 633–646. doi:10.1080/14719037.2017.1327181

Gill, M. J. (2014). The Possibilities of Phenomenology for Organizational Research. *Organizational Research Methods, 17*(2), 118–137. doi:10.1177/1094428113518348

Gill, R. (2003). Change Management or Change Leadership. *Journal of Change Management, 3*(4), 307–318. doi:10.1080/714023845

Gilster, P. (1997). *Digital literacy.* New York, NY: Wiley Computer Pub.

Gilstrap, C., & Morris, A. (2015). Role fluidity and emergent followership: How nonprofit leaders communicate their relationship as/to followers. *Journal of Nonprofit Education and Leadership, 5*(3), 153–173.

Ginsberg, A. & Venkatraman, N. (1992). Investing in New Information Technology: The Role of Competitive Posture and Issue Diagnosis, *Strategic Management Journal* (13: S1), pp. 37-53.

Gioia, D., Schultz, M., & Corley, K. (2000). “Organizational identity, image, and adaptive instability,” *Academy of Management Review*, 25: 63–81.

Gital, A. Y. U., & Zambuk, F. U. (2011). Cloud computing: Solution to ICT in higher education in Nigeria. *Advances in Applied Science Research, 2*(6), 364–369.

GIZ. (2018). *Guide for Mapping the Entrepreneurial Ecosystem: Observe – Analyse – Visualise.* Retrieved from https://www.coursehero.com/file/41633904/GuideforMappingtheEntrepreneurialEcosystempdf/

Goffee, R., & Jones, G. (2006). *Why should anyone be led by you: What it takes to be an authentic leader*. Boston, MA: Harvard Business School Press.

Gould, D. (1994). Immigration Links to the Home Country: Empirical Implications for US Bilateral Trade Flow. *The Review of Economics and Statistics*, *76*(2), 302. doi:10.2307/2109884

Grant, M., & Booth, A. (2009). A typology of reviews: An analysis of 14 review types and associated methodologies. *Health Information and Libraries Journal*, *26*(2), 91–108. doi:10.1111/j.1471-1842.2009.00848.x PMID:19490148

Grant, R. M. (2016). *Contemporary Strategy Analysis: Text and Cases Edition*. Hoboken, NJ: John Wiley & Sons.

Gray, K., Dattakumar, A., Maeder, A., Butler-Henderson, K., & Chenery, H. (2014). *Advancing ehealth education for the clinical health professions: Final report 2014*.

Green, A. E. (2017). Implications of technological change and austerity for employability in urban labour markets. *Urban Studies, 54*(7), 1638. Sage.

Greenes, R. A., & Shortliffe, E. H. (1990). Medical informatics. An emerging academic discipline and institutional priority. *Journal of the American Medical Association*, *263*(8), 1114–1120. doi:10.1001/jama.263.8.1114

Greenhalgh, T., & Papoutsi, C. (2018). Studying complexity in health services research: Desperately seeking an overdue paradigm shift. *BMC Medicine*, *16*(1), 95. doi:10.118612916-018-1089-4 PMID:29921272

Gregg, M. (2011). *Work's intimacy*. Cambridge, UK: Polity.

Gregory, A., Tharyna, R., & Whittaker, J. (2014). Corporate social responsibility and firm value: Disaggregating the effects on cash flow, risk and growth. *Journal of Business Ethics*, *124*(4), 633–657. doi:10.100710551-013-1898-5

Griffin, J., & Schuppan, T. (2010). E-Government Competencies: Looking Beyond technology, WSIS Forum 2010, E-Government Expert Group Meeting, May 14, 2010. Retrieved from http://unpan1.un.org/intradoc/groups/public/documents/un-dpadm/unpan039415.pdf

Griffin, J. J., & Mahon, J. F. (1997). The corporate social performance and corporate financial performance debate: Twenty five years of incomparable research. *Business & Society*, *36*(1), 5–31. doi:10.1177/000765039703600102

Grimpe, C., Murmann, M., & Sofka, W. (2019). Organizational design choices of high-tech startups: How middle management drives innovation performance. *Strategic Entrepreneurship Journal*, *13*(3), 359–378. doi:10.1002ej.1330

Gronn, P. (2002). Distributed leadership as a unit of analysis, *Leadership Quarterly*, 13: 423–451.

Grundén, K. (2012). Competence Development of E-Government: A Study Circle Approach. In Digital Democracy: Concepts, Methodologies, Tools, and Applications. . (Abstract) doi:10.4018/978-1-4666-1740-7

Gunasekaran, A., McGaughey, R. E., Ngai, E. W., & Rai, B. K. (2009). E-Procurement adoption in the Southcoast SMEs. *International Journal of Production Economics*, *122*(1), 161–175. doi:10.1016/j.ijpe.2009.05.013

Gunasekaran, A., & Ngai, E. W. (2008). Adoption of e-procurement in Hong Kong: An empirical research. *International Journal of Production Economics*, *113*(1), 159–175. doi:10.1016/j.ijpe.2007.04.012

Gupta, M., & Narain, R. (2012). A survey on supplier relationship in e-procurement in Indian organisations. *International Journal of Logistics Systems and Management*, *12*(1), 89–121. doi:10.1504/IJLSM.2012.047060

Gupta, M., & Tripathi, S. (2018). A framework for buyer satisfaction in e-procurement in Indian scenario: An integrated ISM and SEM approach. *International Journal of Procurement Management*, *11*(6), 748–776. doi:10.1504/IJPM.2018.095657

Gutemberg, E. (1968). *Economía de la Empresa. Teoría y práctica de la gestión empresarial*. Bilbao: Ed. Deusto.

Hachicha, Z. S., & Mezghani, K. (2018). Understanding Intentions to Switch Toward Cloud Computing at Firms' Level: A Multiple Case Study in Tunisia. *Journal of Global Information Management*, *26*(1), 136–165. doi:10.4018/JGIM.2018010108

Hadeel, A., & Sandhu, K. (2021). PLS Model Performance for Factors Influencing Students Acceptance of Elearning Analytics Recommender. *International Journal of Virtual and Personal Learning Environments* (IJVPLE): New Perspectives in Online Education: Environments, Approaches, Authoring Tools.

Hadeel, A., & Sandhu, K. (2021). PLS Model Performance for Factors Influencing Students Acceptance of Elearning Analytics Recommender. *International Journal of Virtual and Personal Learning Environments* (IJVPLE): New Perspectives in Online Education: Environments, Approaches, Authoring Tools. USA

Hailu, A. (2012). Factors influencing cloud-computing technology adoption in developing countries. (Doctoral dissertation), Capella University.

Hales, C. (2005). Rooted in Supervision, Branching into Management: Continuity and Change in the Role of First-Line Manager, *Journal of Management Studies, 42*(3), pp. 471-506.

Hambley, L., O'Neill, T., & Kline, T. (2007). Virtual team leadership: The effects of leadership style and communication medium on team interaction styles and outcomes. *Organizational Behavior and Human Decision Processes*, *103*(1), 1–20. doi:10.1016/j.obhdp.2006.09.004

Hamel, G., & Prahalad, C. K. (1989). Strategic Intent. *Harvard Business Review*, (July): 1. PMID:10303477

Hänninen, M., Smedlund, A., & Mitronen, L. (2018). Digitalization in retailing: Multi-sided platforms as drivers of industry transformation. *Baltic Journal of Management*, *13*(2), 152–168. doi:10.1108/BJM-04-2017-0109

Hardy, C. A., & Williams, S. P. (2008). E-government policy and practice: A theoretical and empirical exploration of public e-procurement. *Government Information Quarterly*, *25*(2), 155–180. doi:10.1016/j.giq.2007.02.003

Harris, G. (2014). Key steps of scenario planning and analysis Retrieved from https://www.wecc.org/Administrative/QPG%20Introduction%20to%20Scenario%20Planning.pdf

Hart, M., Bnner, K., Levie, J., & Heery, L. (2018). *GEM UK 2017 report: enterprising immigrants boosting prosperity in the UK*. Retrieved from https://www.gemconsortium.org/about/news/66

Hart, D. M., & Acs, Z. J. (2011). High-Tech Immigrant Entrepreneurship in the United States. *Economic Development Quarterly*, *25*(2), 116–129. doi:10.1177/0891242410394336

Hatch, M. (1993). The dynamics of organizational culture. *Academy of Management Review*, *18*(4), 657–693. doi:10.5465/amr.1993.9402210154

Hazelkorn, E. (2018). Reshaping the world order of higher education: The role and impact of rankings on national and global systems. *Policy Reviews in Higher Education*, *2*(1), 4–31. doi:10.1080/23322969.2018.1424562

Health, Q. (2017). *Digital Health Strategic Vision for Queensland 2026*.

Heeks, R. B. (2003). E-Government for development causes of e-government success and failure: Factor Model. Retrieved from http://www.egov4dev.org/causegap.htm

Heilbrunn, S., & Kushnirovich, N. (2008). The impact of policy on immigrant entrepreneurship and businesses practice in Israel. *International Journal of Public Sector Management*, *21*(7), 693–703. doi:10.1108/09513550810904514

Heilig, L., Lalla-Ruiz, E., & Voß, S. (2017). Digital transformation in maritime ports: Analysis and a game theoretic framework. *NETNOMICS: Economic Research and Electronic Networking*, *18*(2-3), 227–254. doi:10.100711066-017-9122-x

Helfat, C. E., & Raubitschek, R. S. (2018). Dynamic and integrative capabilities for profiting from innovation in digital platform-based ecosystems. *Research Policy*, *47*(8), 1391–1399. doi:10.1016/j.respol.2018.01.019

Helfat, C., & Martin, J. (2015). Dynamic managerial capabilities: a perspective on the relationship between managers, creativity, and innovation in organizations. In J. Zhou (Ed.), *The Oxford handbook of creativity, innovation, and entrepreneurship*. Oxford, UK: Oxford University Press.

Henden, A. (2019). Brexit latest: What is Black Swan theory? What does it mean for Brexit? *Express (Online)*.

Henfridsson, O., & Lind, M. (2014). Information Systems Strategizing, Organizational Sub-Communities, and the Emergence of a Sustainability Strategy, *The Journal of Strategic Information Systems, 23*(1), pp. 11-28.

Herscovitch, L. (2004). *Resistance to organizational change: Toward a multidimensional conceptualization (Doctoral dissertation)*, The University of Western Ontario, Canada. ProQuest Dissertations & Theses Global database.

Hershey, P. (1985). *The Situational Leader* (4th ed.). Warner Books.

Hess, T., Benlian, A., Matt, C., & Wiesböck, F. (2016). Options for Formulating a Digital Transformation Strategy. *MIS Quarterly Executive*, *15*(2), 123–139.

He, X. H. (2019). Digital Entrepreneurship Solution to Rural Poverty: Theory, Practice, and Policy Implications. *Journal of Developmental Entrepreneurship*, *24*(1), 32. doi:10.1142/S1084946719500043

Hill, C., & Hult, T. (2018). *Global business today* (10th ed.). New York: McGraw-Hill Education.

HIMSS-Elsevier Digital Healthcare Award Asia Pacific. (2017). *HIMSS-Elsevier Digital Healthcare Award Asia Pacific 2017 Winners Include Four Hospitals Recognizedfor Innovative use ofTechnology to Improve Patient Care*. PR Newswire.

Hodgkinson, G. P., & Sadler-Smith, E. (2018). The Dynamics of Intuition and Analysis in Managerial and Organizational Decision Making. *Academy of Management Perspectives 32*(4), Published Online (). doi:10.5465/amp.2016.0140

Hoerl, R., & Gardner, M. (2010). Lean Six Sigma, creativity, and innovation. *International Journal of Lean Six Sigma*, *1*(1), 30–38. doi:10.1108/20401461011033149

Hofmann, S., & Ogonek, N. (2018). Different But Still The Same? How Public and Private Sector Organisations Deal with New Digital Competences. *The Electronic Journal of E-Government*, *16*(2), 127–135.

Hogart, R. M. (1987). *Judgment and choice*. New York: Wiley & Sons.

Hohenthal, J. (2006). Integrating qualitative and quantitative methods in research on international entrepreneurship. *Journal of International Entrepreneurship*, *4*(4), 175–190. doi:10.100710843-007-0010-6

Holland, J. H. (1995). *Hidden order*. Reading, MA: Addison-Wesley Publishing.

Holland, J. H. (2014). *Complexity: a very short introduction*. Oxford, UK: Oxford University Press.

Horvat, A., & Filipovic, J. (2018). Service quality and maturity of health care organizations through the lens of complexity leadership theory. *Journal of Evaluation in Clinical Practice*, *24*(1), 301–301. doi:10.1111/jep.12789 PMID:28714561

HospitalityBiz. (2018). The Digital hotel has Arrived. Retrieved from https://www.hospitalitynet.org/news/4090589.html (Retrieved on 1 December 2019).

Hossain, M., & Heidemann Lassen, A. (2017). Q&A. How Do Digital Platforms for Ideas, Technologies, and Knowledge Transfer Act as Enablers for Digital Transformation? *Technology Innovation Management Review, 7*(9), 55–60.

Hulten, C. R., & Ramey, V. (2018). *Education, Skills, and Technical Change: Implications for Future U. S. GDP Growth*. University of Chicago Press. doi:10.7208/chicago/9780226567945.001.0001

Hunnius, S., & Schuppan, T. (2013). Competency Requirements for Transformational E-Government. In *Proceedings 2013 46th Hawaii International Conference on System Sciences*, 1664-1673. 10.1109/HICSS.2013.139

Ibarra, D., Ganzarain, J., & Igartua, J. I. (2018). Business model innovation through Industry 4.0: A review. *Procedia Manufacturing*, *22*, 4–10. doi:10.1016/j.promfg.2018.03.002

IHG. (2018). IHG Annual Report. Available at https://www.ihg.com/hotels/gb/en/ (Retrieved on 12 March 2019).

Ilhan, E., Sautner, Z., & Vilkov, G. (2019). Carbon Tail Risk. *Available at SSRN 3204420.*

Ilhan, N., & Rahim, M. (2017). Benefits of e-procurement systems implementation: Experience of an Australian municipal council. In Digital Governance and E-Government Principles Applied to Public Procurement (pp. 249–266). Hershey, PA: IGI Global. doi:10.4018/978-1-5225-2203-4.ch012

Illanes, P., Lund, S., Mourshed, M., Rutherford, S., & Tyreman, M. (2018). Retraining and reskilling workers in the age of automation. McKinsey & Company. Retrieved from https://www.mckinsey.com/~/media/McKinsey/Featured%20Insights/Future%20of%20Organizations/Retraining%20and%20reskilling%20workers%20in%20the%20age%20of%20automation/Retraining-and-reskilling-workers-in-the-age-of-automation.ashx

Institute of Hotel management Bhubaneswar, Ministry of Tourism, Government of India. (2017). Available at http://www.ihmbbs.org/upload/2)Classifi%20of%20Hotels.pdf (Retrieved on 5 August 2019).

International Council of Nurses. (2005). *Summary- Global shortage of registered nurses.*

International Monetary Fund. (2019). Sustainable finance, Chapter 6. *World Economic Outlook*, (October): 81–92.

Ivancic, L., Vukšic, V., & Spremic, M. (2019). Mastering the Digital Transformation Process: Business Practices and Lessons Learned. *Technology Innovation Management Review*, *9*(2), 36–50. doi:10.22215/timreview/1217

Iyawa, G. E., Herselman, M., & Botha, A. (2016). Digital Health Innovation Ecosystems: From Systematic Literature Review to Conceptual Framework. *Procedia Computer Science*, *100*, 244–252. doi:10.1016/j.procs.2016.09.149

Jain, A. K. (2010). Data clustering: 50 years beyond K-means. *Pattern Recognition Letters*, *31*(8), 651–666. doi:10.1016/j.patrec.2009.09.011

Jeffrey, D., Barden, R. R. D., Buckley, P. J., & Hubbard, N. J. (2002). What Makes for a Successful Hotel? Insights on Hotel Management Following 15 Years of Hotel Occupancy Analysis in England. *Service Industries Journal*, *22*(2), 73–88. doi:10.1080/714005078

Jensen, R. (2007). The Digital provide: Information (Technology), market performance and welfare in the south Indian fisheries sector. *The Quarterly Journal of Economics*, *122*(3), 879–924. doi:10.1162/qjec.122.3.879

JISC. (2014). Developing digital literacies. Retrieved from https://www.jisc.ac.uk/guides/developing-digital-literacies

Johnson, M. E., & Whang, S. (2002). E-business and supply chain management: An overview and framework. *Production and Operations Management*, *11*(4), 413–423. doi:10.1111/j.1937-5956.2002.tb00469.x

Johnson, S. (2001). *Emergence*. New York, NY: Scribner.

Jones, R. H., & Hafner, C. A. (2012). *Understanding digital literacies*. London, UK: Routledge. doi:10.4324/9780203095317

Jones, T. (1995). Instrumental stakeholder theory: A synthesis of ethics and economics. *Academy of Management Review*, *20*(2), 404–437. doi:10.5465/amr.1995.9507312924

Joo, B., Lim, D., & Kim, S. (2016). Enhancing work engagement: The roles of psychological capital, authentic leadership, and work empowerment. *Leadership and Organization Development Journal*, *37*(8), 1117–1134. doi:10.1108/LODJ-01-2015-0005

Kaka, N., Madgavkar, A., Kshirsagar, A., Gupta, R., Manyika, J., Bahl, K & Gupta, S. (2019*). Digital India: Technology to transform a connected nation*, McKinsey Global Institute, March.

Kampmeijer, R., Pavlova, M., Tambor, M., Golinowska, S., & Groot, W. (2016). The use of e-health and m-health tools in health promotion and primary prevention among older adults: A systematic literature review. *BMC Health Services Research*, *16*(S5Suppl 5), 290. doi:10.118612913-016-1522-3 PMID:27608677

Kane, G. C. (2015). How digital transformation is making health care safer, faster and cheaper. *MIT Sloan Management Review*, *57*(1).

Kane, G., Palmer, D., Phillips, A., Kiron, D., & Buckley, N. (2016). Aligning the organization for Its digital future. *MIT Sloan Management Review*, *58*(1), n.p.

Kane, G. C., Palmer, D., Nguyen Phillips, A., Kiron, D., & Buckley, N. (2016). *Aligning the Organization for Its Digital Future*. MIT Sloan Management Review and Deloitte University Press.

Kansal, M., Joshi, M., & Batra, G. S. (2014). Determinants of corporate social responsibility disclosures: Evidence from India. *Advances in Accounting, incorporating. Advances in International Accounting*, *30*(1), 217–229. doi:10.1016/j.adiac.2014.03.009

Kantur, D. Z., & Say, A. I. (2012). Organizational resilience: A conceptual integrative framework, *Journal of Management & Organizations, 18*(6), November.

Karunasena, K., & Deng, H. (2012). Critical factors for evaluating the public value of e-government in Sri Lanka. *Government Information Quarterly*, *29*(1), 76–84. doi:10.1016/j.giq.2011.04.005

Kellerman, B. (2008). *Followership: How followers are creating change and changing leaders*. Boston, MA: Harvard Business School Press.

Kelley, R. (2008). Rethinking followership. In R. Riggio, I. Chaleff, & J. Lipman-Blumen (Eds.), *The art of followership: How great followers create great leaders and organizations* (pp. 5–16). San Francisco, CA: Jossey-Bass.

Kelly, S. (2016). Creating a culture of continuous improvement and sustainable management systems at Abbott Diagnostics Longford. *Global Business & Organizational Excellence*, *36*(1), 6–24. doi:10.1002/joe.21753

Kennedy, G., Judd, T., Dalgarno, B., & Waycott, J. (2010). Beyond ndatives and immigrants: Exploring types of net generation students. *Journal of Computer Assisted Learning*, *26*(5), 332–343. doi:10.1111/j.1365-2729.2010.00371.x

Kernick, D. (2006). Wanted – new methodologies for health service research. Is complexity theory the answer? *Family Practice*, *23*(3), 385–390. doi:10.1093/fampra/cml011 PMID:16597669

Keszey, T. (2017). Information systems in transition economies: Does ownership matter? *Information Systems Management*, *34*(1), 65–84. doi:10.1080/10580530.2017.1254456

Keynes, J. M. (1930). *Economic possibilities for our grandchildren. Essays in persuasion* (pp. 358–373). New York: W. W. Norton & Co.

Khan, S. (2016). Leadership in the digital age: A study on the effects of digitalization on top management leadership. *Masters Dissertation, Stockholm Business School.*

Khan, G. F., Moon, J., Rhee, C., & Rho, J. J. (2010). E-government Skills Identification and Development : Toward a Staged-Based User-Centric Approach for Developing Countries. *Asia Pacific Journal of Information Systems.*, *20*(1), 1–31.

Khanna, T., & Palepu, K. (2000). Is Group Affiliation Profitable in Emerging Markets? An analysis of Diversified Indian Business Groups. *The Journal of Finance*, *55*(2), 867–891. doi:10.1111/0022-1082.00229

Khanna, T., & Palepu, K. G. (2005). *The Evolution of Concentrated Ownership in India: Broad Patterns and a History of the Indian Software Industry. A History of Corporate Governance Around the World: Family Business Groups to Professional Managers* (R. K. Morck, Ed.). University of Chicago Press.

Kirsch, L. J. (1996). The Management of Complex Tasks in Organizations: Controlling the Systems Development Process, *Organization Science, 7*(1), pp. 1-21.

Kitay, J., & Wright, C. (2003). Expertise and Organizational Boundaries: The Varying Roles of Australian Management Consultants, *Asia Pacific Business Review, 9*(3), pp. 21-40.

Kitzmuller, M., & Shimshack, J. (2012). Economic perspectives on corporate social responsibility. *Journal of Economic Literature*, *50*(1), 51–84. doi:10.1257/jel.50.1.51

Klein, H. K., & Myers, M. D. (1999). A Set of Principles for Conducting and Evaluating Interpretive Field Studies in Information Systems, *MIS Quarterly, 23*(1), pp. 67-93.

Kluge, E.-H. W. (2008). Ethical aspects of future health care: Globalization of markets and differentiation of societies-ethical challenges. *Studies in Health Technology and Informatics*, *134*, 77–87. PMID:18376035

Kokles, M., Romanová, A., & Hamranová, A. (2015). Information systems in the post-transition period in enterprises in Slovakia. *Journal of Global Information Technology Management*, *18*(2), 110–126. doi:10.1080/1097198X.2015.1052686

Koller, D., Ng, A., Do, C., & Chen, Z. (2013). Retention and Intention in Massive Open Online Courses: In Depth. Retrieved from https://er.educause.edu/articles/2013/6/retention-and-intention-in-massive-open-online-courses-in-depth

Korhonen, H., Still, K., Seppänen, M., Kumpulainen, M., Suominen, A., & Valkokari, K. (2017). The Core Interaction of Platforms: How Startups Connect Users and Producers. *Technology Innovation Management Review*, *7*(9), 17–29. doi:10.22215/timreview/1103

Kosmol, T., Reimann, F., & Kaufmann, L. (2019). You'll never walk alone: Why we need a supply chain practice view on digital procurement. *Journal of Purchasing and Supply Management*, *25*(4). doi:10.1016/j.pursup.2019.100553

Kotter, J. (1995). Leading change: Why transformation efforts fail. (cover story). *Harvard Business Review*, *73*(2), 59–67.

Kotter, J. (2012). *Leading change. Harvard Business Review*. Library of Congress Cataloging in Publication Data.

Kristanti, M. V., & Darma, G. S. (2019). The Role of Marketing & Competitive Intelligence. In Industrial Revolution 4.0. Jurnal Manajemen dan Bisnis, 16(1), Januari.

Kumar, V. (2013). *101 Design Methods* (1st ed.). Hoboken, NJ: John Wiley & Sons.

Kunisch, S., Keil, T., Boppel, M., & Lechner, C. (2019). Strategic initiative portfolios: How to manage strategic challenges better than one at a time. *Business Horizons*, *62*(4), 529–537. doi:10.1016/j.bushor.2019.03.007

Kushnirovich, N., & Heilbrunn, S. (2008). Financial Funding of Immigrant Businesses. *Journal of Developmental Entrepreneurship*, *13*(2), 167–184. doi:10.1142/S1084946708000910

Kuyucu, A. D. H. (2010). *The Playground of Cloud Computing in Turkey*. Elsevier. Retrieved August 7, 2019 from www.elsevier.com/locate/procedia

Kuza, C. C. (2017). Burnout among critical care nurses and the role of ICU telemedicine. (Doctoral thesis), University of Pittsburgh. ProQuest Dissertations & Theses Global database.

Laforet, S. (2013). Organizational Innovation Outcomes in SMEs: Effects of Age, Size, and Sector. *Journal of World Business, 48*(4), 490–502. doi:10.1016/j.jwb.2012.09.005

Lannan, K. (2015). *Interpretative phenomenological analysis: Exploring the formative experiences of community college students enrolled in remedial courses. (Doctoral dissertation)*, Colorado State University. Retrieved from https://dspace.library.colostate.edu/handle/10217/170325

Lan, T., & Zhu, S. J. (2014). Chinese apparel value chains in Europe: Low-end fast fashion, regionalization, and transnational entrepreneurship in Prato, Italy. *Eurasian Geography and Economics, 55*(2), 156–174. doi:10.1080/15387216.2014.948471

Larson, D., & Chang, V. (2016). A review and future direction of agile, business intelligence, analytics and data science. *International Journal of Information Management, 36*(5), 700–710. doi:10.1016/j.ijinfomgt.2016.04.013

Latterman, C., Fetscherin, M., Alon, I., & Li, S. (2009). CSR communication intensity in Chinese and Indian multinational companies. *Corporate Governance, 17*(4), 426–442. doi:10.1111/j.1467-8683.2009.00758.x

Lawler, E. E. (2014). Sustainable Effectiveness and Organization Development. *OD Practitioner, 46*(4), 65–67.

Le Moigne, J. L., & Bénard, A. (1976). *Los sistemas de decisión en las organizaciones.* Madrid, Spain: Tecniban, D. L.

Lech, P. (2019). Enterprise system implementations in transition and developed economies: Differences in project contracting and governance. *Information Technology for Development, 25*(2), 357–380. doi:10.1080/02681102.2018.1564726

Leek, S., Turnbull, P. W., & Naude, P. (2003). How is information technology affecting business relationships? Results from a UK survey. *Industrial Marketing Management, 32*(2), 119–126. doi:10.1016/S0019-8501(02)00226-2

Leimeister, J., Österle, H., & Alter, S. (2014). Digital services for consumers. *Electronic Markets, 24*(4), 255–258. doi:10.100712525-014-0174-6

Lewin, A. Y. (1999). Application of Complexity Theory to Organization Science, *Organization Science* (10:3), pp. 215-215.

Lewis, C. (2019). Does anyone in Australia think Newstart is adequate? *Welfare.* Retrieved from https://www.crikey.com.au/2019/04/03/newstart-federal-budget/

Li, P., Liu, S. W., Zhang, S. B., & Li, C. H. (2017). A Study on Factors Affecting Students' Entrepreneurial Intentions in HIT. In M. A. Tao (Ed.), *2017 24th Annual International Conference on Management Science & Engineering* (pp. 358-362). New York: IEEE.

Lichtenstein, B. B., Uhl-Bien, M., Marion, R., Seers, A., James, D., & Schreiber, C. (2006). *Complexity leadership theory: An interactive perspective on leading in complex adaptive systems, Institute for the Study of Coherence & Emergence.* Lawrence Erlbaum Associates.

Liebowitz, J. (2006). *Strategic Intelligence: Business intelligence, competitive intelligence, and knowledge management.* Boca Raton, FL: Auerbach Publications. doi:10.1201/9781420013900

Liere-Netheler, K., Packmohr, S., & Vogelsang, K. (2018). Drivers of Digital Transformation in Manufacturing. In *Proceedings of the 51st Hawaii International Conference on System Sciences* (pp. 3926–3935).

Lin, C.-L., Wang, Y.-N., & Tsai, H.-M. (2013). Innovative Thinking in Nursing Practice. *Hu Li Za Zhi, 60*(2), 97–102. PMID:23588694

LinkedIn. (2018). *2018 workplace learning report: The rise and responsibility of talent development in the new labor market*. Carpinteria, CA: LinkedIn Learning.

Lins, K. V., Servaes, H., & Tamayo, A. (2017). Social capital, trust and firm performance: The value of corporate social responsibility during the financial crisis. *The Journal of Finance*, *72*(4), 1785–1823. doi:10.1111/jofi.12505

Linski, C. M. III. (2013). Process consultation and law enforcement: A practitioner's case study report. *Organization Development Journal*, *31*(4), 63–73.

Litaker, D., Tomolo, A., Liberatore, V., Stange, K. C., & Aron, D. (2006). Using complexity theory to build interventions that improve health care delivery in primary care. *J Gen Intern Med.* 21(S2:S30).

Liu, D., Chen, S., & Chou, T. (2011). Resource fit in digital transformation : Lessons learned from the CBC Bank global e-banking project. *Management Decision*, *49*(10), 1728–1742. doi:10.1108/00251741111183852

Liu, D., Liao, H., & Loi, R. (2012). The dark side of leadership: A three-level investigation of the cascading effect of abusive supervision on employee creativity. *Academy of Management Journal*, *55*(5), 1187–1212. doi:10.5465/amj.2010.0400

Liu, J. Y., Nandhakumar, J., & Zachariadis, M. (2018). When guanxi meets structural holes: Exploring the guanxi networks of Chinese entrepreneurs on digital platforms. *The Journal of Strategic Information Systems*, *27*(4), 311–334. doi:10.1016/j.jsis.2018.10.003

Li, W. J., Du, W. Y., & Yin, J. M. (2017). Digital entrepreneurship ecosystem as a new form of organizing: The case of Zhongguancun. *Frontiers of Business Research in China*, *11*(1), 21. doi:10.118611782-017-0004-8

Lloyd, J., & Ferguson, S. L. (2017). Innovative information technology solutions: Addressing current and emerging nurse shortages and staffing challenges worldwide. *Nursing Economics*, *35*(4), 211–212.

Lloyd-Walker, B., & Walker, D. (2011). Authentic leadership for 21st century project delivery. *International Journal of Project Management*, *29*(4), 383–395. doi:10.1016/j.ijproman.2011.02.004

Lokuge, S., Sedera, D., Grover, V., & Dongming, X. (2019). Organizational Readiness for Digital Innovation: Development and Empirical Calibration of a Construct, *Information & Management, 56*(3), pp. 445-461.

Lokuge, S., Sedera, D., & Perera, M. (2018). The Clash of the Leaders: The Intermix of Leadership Styles for Resource Bundling, *Pacific Asia Conference on Information Systems (PACIS 2018)*, Yokohama, Japan: AIS.

Long, K. M., McDermott, F., & Meadows, G. N. (2018). Being pragmatic about healthcare complexity: Our experiences applying complexity theory and pragmatism to health services research. *BMC Medicine*, *16*(1), 94. doi:10.118612916-018-1087-6 PMID:29921277

López-Robles, J. R., Otegi-Olaso, J. R., Porto-Gómez, I., & Cobo, M. J. (2019). 30 years of intelligence models in management and business: A bibliometric review. *International Journal of Information Management*, *48*, 22–38. doi:10.1016/j.ijinfomgt.2019.01.013

Lucas, H. C. J. Jr, Agarwal, R., Clemons, E. K., El Sawy, O. A., & Weber, B. (2013). Impactful Research on Transformational Information Technology: An Opportunity to Inform New Audiences. *Management Information Systems Quarterly*, *37*(2), 371–382. doi:10.25300/MISQ/2013/37.2.03

Lusch, R. F. & Nambisan, S. (2015). Service Innovation: A Service-Dominant Logic Perspective, *MIS Quarterly* (39:1), pp. 155-175.

Lusch, R. F., & Nambisan, S. (2015). Service Innovation: A Service-Dominant-Logic perspective. *Management Information Systems Quarterly*, *39*(1), 155–175. doi:10.25300/MISQ/2015/39.1.07

Luthans, F. (2002). The need for and meaning of positive organizational behavior. *Journal of Organizational Behavior, 23*(6), 695–706. doi:10.1002/job.165

Luthans, F., & Avolio, B. (2003). Authentic leadership development. In K. Cameron, J. Dutton, & R. Quinn (Eds.), *Positive organisational scholarship: Foundations of a new discipline* (pp. 241–258). San Francisco, CA: Berrett Koehler Publishers.

Luu, T. (2019). *Linking authentic leadership to salespeople's service performance: The roles of job crafting and human resource flexibility. Industrial Marketing Management.* Advanced Online Publication.

Macdonald, I., Burke, C., & Stewart, K. (2018). *Systems Leadership. Creating Positive Organisations* (2nd ed.). London, UK: Routledge. doi:10.4324/9781315178486

Mackey, A., Mackey, T. B., & Barney, J. B. (2007). Corporate social responsibility and firm performance: Investor preferences and corporate strategies. *Academy of Management Review, 32*(3), 817–835. doi:10.5465/amr.2007.25275676

MacManus, S. A. (2002). Understanding the incremental nature of e-procurement implementation at the state and local levels. *Journal of Public Procurement, 2*(1), 5–28. doi:10.1108/JOPP-02-01-2002-B001

Mahajan, A., Islam, S. D., Schwartz, M. R., & Cannesson, M. (2017). A hospital is not just a factory, but a complex adaptive system—Implications for perioperative care. *Anesthesia and Analgesia, 125*(1), 333–341. doi:10.1213/ANE.0000000000002144 PMID:28614127

Makosa, P., & Mwangangi, P. (2018). Factors influencing e-procurement implementation public sector in South Sudan. *International Journal of Supply Chain and Logistics, 2*(2), 1–23.

Makrigiannakis, G., & Soteriades, M. (2015). Management Accounting in the Hotel Business. *International Journal of Hospitality & Tourism Administration, 8*(4), 47–76. doi:10.1300/J149v08n04_03

Manchiraju, H., & Rajgopal, S. (2017). Does corporate social responsibility (CSR) create shareholder value? Evidence from the Indian Companies Act 2013. *Journal of Accounting Research, 55*(5), 1257–1300. doi:10.1111/1475-679X.12174

Manson, S. M. (2001). Simplifying complexity: A review of complexity theory. *Geoforum, 32*(3), 405–414. doi:10.1016/S0016-7185(00)00035-X

Manuele, F. A. (2019). Systems/macro thinking a primer. *Professional Safety, 64*(1), 37–42.

Margolis, J. D., Effenbein, H. A., & Walsh, J. P. (2009). *Does it pay to be good and does it matter? A meta-analysis of the relationship between corporate social and financial performance.* Working Paper, Cambridge, MA: Harvard University.

Marion, R. (2008). Complexity theory for organizations and organizational leadership, *Complexity Leadership*, Part 1, 1-15.

Market Entry. (2018). Tourism and Hotel Industry in Taiwan: Analysis of Growth, Trends and Progress (2015-2020). Available at https://www.mordorintelligence.com/industry-reports/market-entry-tourism-and-hotel-industry-in-taiwan (Retrieved on 3 August 2019).

Marks, E. A., & Lozano, B. (2010). *Executive's Guide to Cloud Computing. SAD.* John Wiley & Sons.

Marschang, S. (2014). *Health inequalities and eHealth: 2014 eHealth Forum, Report of the e-Health stakeholder Group.* Retrieved from Athens: http://www.ehealth2014.org/wpcontent/

Marshak, R. J., & Grant, D. (2008). Organizational discourse and new organization development practices. *British Journal of Management, 19*(s1), S7–S19. doi:10.1111/j.1467-8551.2008.00567.x

Marston, S., Li, Z., Bandyopadhyay, S., Zhang, J., & Ghalsasi, A. (2010). *Cloud computing — The business perspective.* Elsevier. Retrieved September 7, 2019 from www.elsevier.com/locate/dss

Martin, A. M., & Roberts, K. R. (2015). Digital Native ≠ Digital Literacy. *Principal, 94*(3), 18–21.

Martin, A., Jones, E., & Callan, V. (2005). The role of psychological climate in facilitating employee adjustment during organizational change. *European Journal of Work and Organizational Psychology, 14*(3), 263–289. doi:10.1080/13594320500141228

Martin, A., Jones, E., & Callan, V. (2006). Status differences in employee adjustment during organizational change. *Journal of Managerial Psychology, 21*(2), 145–162. doi:10.1108/02683940610650758

Martin, C. (2018). Complex adaptive systems approaches in health care - A slow but real emergence? Journal of Evaluation in Clinical Practice. *Journal of Evaluation in Clinical Practice, 24*(1), 266–268. doi:10.1111/jep.12878 PMID:29589876

Mary Norine, W., & Rumsfeld, J. S. (2017). Leading the digital transformation of healthcare. *Journal of the American College of Cardiology, 70*(21), 2719–2722. doi:10.1016/j.jacc.2017.10.020 PMID:29169479

Masset, P., & Weisskopf, J. P. (2019). Top 10 Trends Reshaping. *Hospitals (Rockville), 2019*. Available at https://hospitalityinsights.ehl.edu/2019-top-hospitality-trends Retrieved 25 August 2019

McCabe, C., & Timmins, F. (2016). Embracing healthcare technology - What is the way forward for nurse education? *Nurse Education in Practice, 21*, 104–106. doi:10.1016/j.nepr.2016.10.007 PMID:27810650

McConalogue, E., Davis, P., & Connolly, R. (2019). Health Technology Assessment: The Role of Total Cost of Ownership. *Business Systems Research, 10*(1), 180–187. doi:10.2478/bsrj-2019-0013

McCullough, J. S., Casey, M., Moscovice, I., & Prasad, S. (2010). The effect of health information technology on quality in U.S. hospitals. *Health Affairs, 29*(4), 647–654. doi:10.1377/hlthaff.2010.0155 PMID:20368594

McDaniels, T. (2019). *Four Decades of Transformation in Decision Analytic Practice for Societal Risk Management. Risk Analysis - An International Journal*. Society for Risk Analysis, Wiley Online Library; doi:10.1111/risa.13332

McDermott, F. (2014). Complexity theory, trans-disciplinary working and reflective practice. In Applying complexity theory. Bristol: Policy Press; 2014. pp. 181–98.

McGonagle, J. J., & Vella, C. M. (1996). *A New Archetype for Competitive Intelligence*. West Port, CT: Green Wood Publishing Group.

McKinsey Global Institute. (2018). *Skill shift. Automation and the future of the workforce*, (accessed 12 June 2019) https://www.mckinsey.com/featured-insights/future-of-work/skill-shift-automation-and-the-future-of-the-workforce

McKinsey. (2016). *Indian's economy: Why the time for growth is now*, September, 1-11.

McLachlin, R. D. (1999). Factors for Consulting Engagement Success, *Management Decision, 37*(5), pp. 394-404.

McLaren, T. S., Head, M. M., Yuan, Y., & Chan, Y. E. (2011). A Multilevel Model for Measuring Fit between a Firm's Competitive Strategies and Information Systems Capabilities, *MIS Quarterly, 35*(4), pp. 909-929.

McWilliams, A., & Siegel, D. (2000). Research notes and communications- Corporate social responsibility and financial performance: Correlation or misspecification. *Strategic Management Journal, 21*(5), 603–609. doi:10.1002/(SICI)1097-0266(200005)21:5<603::AID-SMJ101>3.0.CO;2-3

Mehta, A. (2019, May 16). Innovative thinking needed to defang existing health issues: Ajoy Mehta. *eHealth*.

Melnikova, J., Grunwald, N., Ahrens, A., & Zascerinska, J. (2019). *Immigrants and Asylum Seekers' Digital Entrepreneurship Competence: Evaluation of the Theoretical Framework.* Paper presented at the 9th Balkan Region Conference on Engineering and Business Education and 12th International Conference on Engineering and Business Education Sibiu, Romania.

MENA Report. (2017). New Zealand: Techweek17 to encourage global thinking. *MENA Report.*

Menguzzato, M., & Renau, J. J. (1995). *La Dirección Estratégica de la Empresa.* Barcelona, Spain: Ed. Ariel Economía.

Meskó, B., Radó, N., & Győrffy, Z. (2019). Opinion leader empowered patients about the era of digital health: A qualitative study. *BMJ Open, 9*(3). doi:10.1136/bmjopen-2018-025267 PMID:30898816

Meyer, J., Land, R., & Bailie, C. (2010). *Threshold Concepts and Transformational Learning.* Rotterdam, The Netherlands: Sense Publishers. doi:10.1163/9789460912078

Micheli, P., & Manzoni, J. F. (2010). Strategic Performance Measurement: Benefits, Limitations and Paradoxes. *Long Range Planning, 43*(4), 465–476. doi:10.1016/j.lrp.2009.12.004

Milgrom, P., & Roberts, J. (1995). Complementarities and Fit Strategy, Structure, and Organizational Change in Manufacturing, *Journal of Accounting and Economics, 19*(2-3), pp. 179-208.

Milić, B., Grubić-Nešić, L., Kuzmanović, B., & Delić, M. (2017). The influence of authentic leadership on the learning organization at the organizational level: The mediating role of employees' affective commitment. *Journal for East European Management Studies, 22*(1), 9–38. doi:10.5771/0949-6181-2017-1-9

Min, H., & Galle, W. P. (2003). E-purchasing: Profiles of adopters and nonadopters. *Industrial Marketing Management, 32*(3), 227–233. doi:10.1016/S0019-8501(02)00266-3

Min, H., & Min, H. (1997). Benchmarking the quality of hotel services: Managerial perspectives. *International Journal of Quality & Reliability Management, 14*(6), 582–597. doi:10.1108/02656719710186209

Mintzberg, H. (2009). *Managing.* Oakland, CA: Berrett - Koehler Publication.

Mintzberg, H. (1994). *The Fall and Rise of Strategic Planning. Harvard Business Review.* January-February Issue.

Mishra, S., & Suar, D. (2010). Does corporate social responsibility influence firm performance of Indian companies? *Journal of Business Ethics, 95*(4), 571–601. doi:10.100710551-010-0441-1

Mjongwana, A., & Kamala, P. N. (2018). Non-financial performance measurement by small and medium sized enterprises operating in the hotel industry in the city of Cape Town. *African Journal of Hospitality, Tourism, and Leisure, 7*(1), 1–26.

Mohammed, F., Ibrahim, O., & Ithnin, N. (2016). Factors influencing cloud computing adoption for e-government implementation in developing countries: Instrument development. *Journal of Systems and Information Technology, 18*(3), 297–327. doi:10.1108/JSIT-01-2016-0001

Moon, M. J. (2005). E-procurement management in state governments: Diffusion of e-procurement practices and its determinants. *Journal of Public Procurement, 5*(1), 54–72. doi:10.1108/JOPP-05-01-2005-B003

Moore, G. M. (2001). Corporate social performance: An investigation in the UK supermarket industry. *Journal of Business Ethics, 34*(3/4), 299–315. doi:10.1023/A:1012537016969

Mordor Intelligence. (2018). Industry Report. Available at https://www.mordorintelligence.com/industry-reports (Retrieved on 8 March 2019).

Morgan, R. M., & Hunt, S. D. (1994). The commitment-trust theory of relationship marketing. *Journal of Marketing*, *58*(3), 20–38. doi:10.1177/002224299405800302

Morse, S. (2018). 82% of clinicians, IT leaders blame burnout on interruptions from tech tools. *Healthcare Finance*. Retrieved from https://www.healthcarefinancenews.com/news/82-clinicians-it-leaders-blame-burnout-interruptions-tech-tools

Mufudza, T. (2018). Dynamic Strategy in a Turbulent Business Environment. In IntechOpen Ed., Strategic Management - A Dynamic View, (DOI:) doi:10.5772/intechopen.81250

Mukherjee, A., & Bird, R. (2016). Analysis of mandatory CSR expenditure in India: A survey. *International Journal of Corporate Governance*, *7*(1), 32–59. doi:10.1504/IJCG.2016.077982

Mukherjee, A., Bird, R., & Duppati, G. (2018). Mandatory corporate social responsibility: The Indian experience. *Journal of Contemporary Accounting & Economics*, *14*(3), 254–265. doi:10.1016/j.jcae.2018.06.002

Mulgan, G. (2006). The process of social innovation. *Innovations: Technology, Governance, Globalization*, *1*(2), 145–162. doi:10.1162/itgg.2006.1.2.145

Mullins, L. J. (2005). *Management and Organizational Behaviour*. London, UK: Pitman Publishing.

Murawski, M., & Bick, M. (2017). Digital competences of the workforce – a research topic? *Business Process Management Journal*, *23*(3), 721–734. doi:10.1108/BPMJ-06-2016-0126

Muriuki, J. I., Guyo, W., Odhiambo, R., & Kinoti, J. (2019). Effect of electronic procurement technical support staff on procurement performance in energy sector state corporations in Kenya. *International Journal of Supply Chain Management, 4*(1), 20-38.

Murthy, V., & Abeysekera, I. (2008). Corporate social reporting practices of top Indian software firms. *The Australian Accounting Business & Finance Journal*, *2*(1), 36–59. doi:10.14453/aabfj.v2i1.4

Murugesan, S. (2011). Cloud computing gives emerging markets a lift. *IT Professional*, *13*(6), 60–62. doi:10.1109/MITP.2011.94

Myers, M. D., & Newman, M. (2007). The Qualitative Interview in IS Research: Examining the Craft, *Information & Organization, 17*(1), pp. 2-26.

Nachira, F. (2002). *Towards a Network of Digital Business Ecosystems Fostering The Local Development European Commission.*

Nahavandi, A. (2015). *The art and science of leadership* (7th ed.). New York: Pearson Education.

Nambisan, S. (2013). Information Technology and Product/Service Innovation: A Brief Assessment and Some Suggestions for Future Research, *Journal of the Association for Information Systems*, *14*(4), pp. 215-226.

Nambisan, S. (2017). Digital Entrepreneurship: Toward a Digital Technology Perspective of Entrepreneurship, *Entrepreneurship Theory and Practice, 41*(6), pp. 1029-1055.

Nambisan, S. (2017). Digital Entrepreneurship: Toward a Digital Technology Perspective of Entrepreneurship. *Entrepreneurship Theory and Practice*, *41*(6), 1029–1055. doi:10.1111/etap.12254

Nanaimo News Bulletin. (2017). *Climate change talk encourages global thinking*. Nanaimo News Bulletin.

Nancarrow, S. A. (2015). Six principles to enhance health workforce flexibility. *Human Resources for Health*, *13*(1), 9. doi:10.1186/1478-4491-13-9 PMID:26264184

Naseebullah, M., PDD, D., & Hassan, M. (2010). The role of factors in business-to-business e-procurement implementation.

Nataraj, S. (2014). The need for an introductory computer literacy course at the university level. [IJBMER]. *International Journal of Business Management and Economic Research*, *5*(4), 71–73.

Ndou, V. D. (2004). E - Government for Developing Countries : Opportunities and Challenges. *The Electronic Journal on Information Systems in Developing Countries*, *18*(1), 1–24. doi:10.1002/j.1681-4835.2004.tb00117.x

Nedelkoska, L., & Quintini, G. (2018). Automation, skills use and training. *OECD Social, Employment and Migration Working Papers No. 202*. Paris, France: OECD Publishing.

Neef, D. (2001). E-Procurement: From strategy to implementation: FT press.

Neuman, W. L. (2003). *Social Research Methods: Qualitative and Quantitative Approaches*. Massachusetts: Allyn & Bacon.

Neumeier, M. (2014). *The 46 rules of genius: an innovator's guide to creativity*. San Francisco, CA: New Riders.

Ngoasong, M. Z. (2018). Digital entrepreneurship in a resource-scarce context: A focus on entrepreneurial digital competencies. *Journal of Small Business and Enterprise Development*, *25*(3), 483–500. doi:10.1108/JSBED-01-2017-0014

Nicotra, M., Romano, M., Giudice, D. M., & Schillaci, E. C. (2018). The causal ralation between entrepreneurial ecosystem and productive entrepreneurship: A measurement framework. *The Journal of Technology Transfer*, *43*(3), 33. doi:10.100710961-017-9628-2

Noe, R. A., Hollenbeck, J. R., Gerhart, B., & Wright, P. M. (2017). *Human resource management: Gaining a competitive advantage*. New York: McGraw-Hill Education.

Nwankpa, J. K., & Roumani, Y. (2016). IT Capability and Digital Transformation: A Firm Performance Perspective. In *Proceedings of the Thirty Seventh International Conference on Information Systems*, 1-16. Academic Press.

Nylén, D., & Holmström, J. (2015). Digital Innovation Strategy: A Framework for Diagnosing and Improving Digital Product and Service Innovation, *Business Horizons, 58*(1), pp. 57-67.

Nylén, D., & Holmström, J. (2015). Digital innovation strategy: A framework for diagnosing and improving digital product and service innovation. *Business Horizons*, *58*(1), 57–67. doi:10.1016/j.bushor.2014.09.001

Odor, H. O. (2019). A literature review on organisational learning and learning organisations. *International Journal of Information Business and Management*, *11*(3), 281–295.

OECD. (2003). The e-government imperative. OECD e-government studies. Retrieved from www1.worlbank.org/publicsector/egov/E-governmentImperative.pdf.

OECD. (2014). Recommendation of the Council on Digital Government Strategies. Retrieved from http://www.oecd.org/gov/digital-government/Recommendation-digital-government-strategies.pdf

OECD. (2015). *Students, Computers, and Learning: Making the Connection, PISA*. OECD Publishing; doi:10.1787/9789264239555-

OECD. (2018). *Digital Government Review of Brazil: Towards the Digital Transformation of the Public Sector, OECD Digital Government Studies*. Paris, France: OECD Publishing; doi:10.1787/9789264307636-

OECD. (2018). Putting faces to the jobs at risk of automation. Policy brief on the future of work. Retrieved from https://www.oecd.org/employment/Automation-policy-brief-2018.pdf

OECD. (2019). Strengthening Digital Government. Retrieved from https://www.oecd.org/going-digital/strengthening-digital-government.pdf

Ojo, A., Janowski, T., Estevez, E., & Khan, I. K. (2007). Human Capacity Development for e-Government. *UNU-IIST-Report* (362). Retrieved from https://s3.amazonaws.com/academia.edu.documents/30890268/report362.pdf?

Olden, P. C. (2003). Hospital and community health: Going from stakeholder management to stakeholder collaboration. *Journal of Health and Human Services Administration*, *26*(1/2), 35–57. PMID:15330379

Oliech, C. O., & Mwangangi, P. (2019). Effect of strategic procurement management on performance of level five hospitals in Kenya. *International Journal of Supply Chain Management*, *4*(1), 39–62.

Oliveira, R. C. (2010). *The determinants of immigrant entrepreneurship and employment creation in Portugal: Chapter 4 of Migrant Entrepreneurship in OECD Countries.*

Ollo-López, A., & Aramendía-Muneta, M. E. (2012). ICT impact on competitiveness, innovation and environment. *Telematics and Informatics*, *29*(2), 204–210. doi:10.1016/j.tele.2011.08.002

Ommaya, A. K., Cipriano, P. F., Hoyt, D. B., Horvath, K. A., Tang, P., Paz, H. L., . . . Sinsky, C. A. (2018). Care-centered clinical documentation in the digital environment: Solutions to alleviate burnout. *NAM Perspectives. Discussion Paper, National Academy of Medicine, Washington, DC.* doi:10.31478/201801c

ORACLE. Oracle Data Sheet, Oracle Hospitality OPERA Cloud Services, Available at http://www.oracle.com/us/industries/opera-cloud-services-2870466.pdf (Retrieved on 20 January 2019).

Organisation for Economic Co-operation and Development. (2001). Understanding the digital divide. Paris: Directorate for Science, Technology and Industry, OECD.

Organizational Learning. (2014). Single and double loop learning. *Organizational Learning,* Retrieved from https://organizationallearning9.wordpress.com/single-and-double-loop-learning/

Orlitzky, M., Schmidt, F. L., & Rynes, S. L. (2003). Corporate social and financial performance: A meta-analysis. *Organization Studies*, *24*(3), 403–441. doi:10.1177/0170840603024003910

Özaralli, N. (2003). Effects of transformational leadership on empowerment and team effectiveness. *Leadership and Organization Development Journal*, *24*(6), 335–344. doi:10.1108/01437730310494301

Palacios, J. M. (2018). The Role of Strategic Intelligence in the Post-Everything Age. *The International Journal of Intelligence, Security, and Public Affairs, 20*(3), 181-203.

Panayiotou, N. A., Gayialis, S. P., & Tatsiopoulos, I. P. (2004). An e-procurement system for governmental purchasing. *International Journal of Production Economics*, *90*(1), 79–102. doi:10.1016/S0925-5273(03)00103-8

Panda, D., Sahu, G., & Gupta, P. (2010). *Promoting transparency and efficiency in public procurement: E-procurement initiatives by government of India.* Paper presented at the 7th International Conference on E-government (ICEG).

Pani, M. R., Agrahari, A., De, S., & Sahoo, G. (2011). Literature review and research issues in e-procurement. *Management and Labour Studies*, *36*(3), 225–246. doi:10.1177/0258042X1103600302

Parasuraman, A., Zeithaml, V. A., & Berry, L. L. (1985). A Conceptual Model of Service Quality and Its Implications for Future Research. *Journal of Marketing*, *49*(4), 41–50. Retrieved from http://www.jstor.org/stable/1251430%0Ahttp://www.jstor.org/%0Ahttp://www.jstor.org/action/showPublisher?publisherCode=ama. doi:10.1177/002224298504900403

Pardo, T. A. (2000). Realizing the Promise of Digital Government: It's More than Building a Web Site. *Center for Technology in Government*, 12. Retrieved from http://www.cisp.org/imp/october_2000/10_00pardo.htm

Parrado, S. (2005). Skills for Electronic Service Delivery in Public Agencies. In M. *Khosrow-Pour (Ed.), Practicing E-Government: A Global Perspective* (pp. 310–328). IGI Global; doi:10.4018/978-1-59140-637-2.ch014

Paul, K. (2019). Microsoft Japan tested a four-day work week and productivity jumped by 40%. *The Guardian.* Retrieved from https://www.theguardian.com/technology/2019/nov/04/microsoft-japan-four-day-work-week-productivity

Pavlatos, O. (2011). The impact of strategic management accounting and cost structure on ABC systems in hotels. *The Journal of Hospitality Financial Management, 19*(2), 37–55. doi:10.1080/10913211.2011.10653912

Pearcy, D. H., & Giunipero, L. C. (2008). Using e-procurement applications to achieve integration: What role does firm size play? *Supply Chain Management, 13*(1), 26–34. doi:10.1108/13598540810850292

Pellikka, J., & Ali-Vehmas, T. (2016). Managing Innovation Ecosystems to Create and Capture Value in ICT Industries. *Technology Innovation Management Review, 6*(10), 17–24. doi:10.22215/timreview/1024

Penrose, E. T. (1959). *The Theory of the Growth Of the Firm.* New York: Sharpe.

Pew Research Centre. (2016, May 11). America's shrinking middle class: A closer look at changes within metropolitan areas. *Social & Demographic Trends.* Retrieved from https://www.pewsocialtrends.org/2016/05/11/americas-shrinking-middle-class-a-close-look-at-changes-within-metropolitan-areas/

Piketty, T. (2015). *The Economics of inequality.* Cambridge, MA: President and Fellows of Harvard College. doi:10.4159/9780674915565

Pillutla, S., Yao, D.-Q., & Li, X. (2014). Data mining the adoption intention of e-procurement system for Chinese companies. *International Journal of Management and Decision Making, 13*(2), 207–219. doi:10.1504/IJMDM.2014.060872

Piotrowicz, W. (2015). Information technology and systems in the Visegrád group of countries (Czech Republic, Hungary, Poland, and Slovakia): A literature review. *Journal of Global Information Technology Management, 18*(2), 77–93. doi:10.1080/1097198X.2015.1052684

Pirttimäki, V. (2007). Business Intelligence as a Managerial Tool in Large Finnish Companies. Tampere University of Technology, Publication 646, Tampere.

Platform, P. (2019). Digital procurement and procurement transformation. Retrieved from https://www.procuredesk.com/digital-procurement/

Plsek, P. E., & Greenhalgh, T. (2001). The challenge of complexity in health care. *BMJ (Clinical Research Ed.), 323*(7313), 625–628. doi:10.1136/bmj.323.7313.625 PMID:11557716

Plum, A., Tanniru, M., & Khuntia, J. (2019). An Innovation Platform for Diffusing Public Health Practices across a Global Network, *HICSS-2019*, Hawaii, January.

Podsakoff, P., MacKenzie, S., Lee, J., & Podsakoff, N. (2003). Common method biases in behavioral research: A critical review of the literature and recommended remedies. *The Journal of Applied Psychology, 88*(5), 879–903. doi:10.1037/0021-9010.88.5.879 PMID:14516251

Poldrugovac, K., Tekavcic, M., & Jankovic, S. (2016). Efficiency in the hotel industry: An empirical examination of the most influential factors. *Economic Research-Ekonomska Istraživanja, 29*(1), 583–597. doi:10.1080/1331677X.2016.1177464

Pomffyova, M. (2018). Management Information Systems, *Infotech Open,* October. https://www.intechopen.com/books/management-of-information-systems

Pontefract, S. K., & Wilson, K. (2019). Using electronic patient records: Defining learning outcomes for undergraduate education. *BMC Medical Education, 19*(1), 30. doi:10.118612909-019-1466-5 PMID:30670000

Porter, M. E. (2008). *The Five Competitive Forces That Shape Strategy. Harvard Business Review.* January Issue.

Portes, A., & Sensenbrenner, J. (1993). Embeddedness and Immigration: Notes on the Social Determinants of Economic Action. *American Journal of Sociology, 98*(6), 1320.

Portes, A., & Yiu, J. (2013). Entrepreneurship, transnationalism, and development. *Migration Studies, 1*(1), 75–95. doi:10.1093/migration/mns036

Poulfelt, F., & Payne, A. (1994). Management Consultants: Client and Consultant Perspective, *Scandinavian Journal of Management, 10*(4), pp. 421-436.

Powell, W., Oberg, A., Korff, V., Oelberger, C., & Kloos, K. (2017). Institutional analysis in a digital era: Mechanisms and methods to understand emerging fields. In G. Krucken, C. Mazza, R. Meyer, & P. Walgenbach (Eds.), *New themes in institutional analysis: Topics and issues from European research.* Cheltenham, UK: Edward Elgar. doi:10.4337/9781784716875.00016

Prahlad, C. K. (2005). *The fortune at the bottom of the pyramid: eradicating poverty through profits.* New Jersey: Wharton School.

Prasad, G. C. (2019). Who gains the most from corporate charity, LiveMint, August 23, 2019. https://www.livemint.com/companies/news/who-gains-the-most-from-corporate-charity-1566536241870.html. (Accessed on November 22, 2019).

Prasad, M., Mishra, T., & Kalro, A. D. (2017). Environmental disclosure by Indian companies: An empirical study. *Environment, Development, and Sustainability, 19*(5), 1999–2022. doi:10.100710668-016-9840-5

Pricewaterhouse Cooper. (2017). *Workforce of the future. The competing forces shaping 2030.* (accessed 12 June 2019) https://www.pwc.com/gx/en/services/people-organisation/workforce-of-the-future/workforce-of-the-future-the-competing-forces-shaping-2030-pwc.pdf

Prieto-Carrón, M., Lund-Thomsen, P., Chan, A., Muro, A., & Bhushan, C. (2006). Critical Perspectives on CSR and development: What we know, what we don't know and what we need to know. *International Affairs, 82*(5), 977–987. doi:10.1111/j.1468-2346.2006.00581.x

Pruthi, S., Basu, A., & Wright, M. (2018). Ethnic ties, motivations, and home country entry strategy of transnational entrepreneurs. *Journal of International Entrepreneurship, 16*(2), 210–243. doi:10.100710843-017-0223-2

Pule, S., & Bakkabulindi, A. (2017). E-procurement implementation: Analysing the applicability of the EPOS method on the efficiency of public sector organisations-a comparative study. *International Journal of Procurement Management, 10*(2), 135–149. doi:10.1504/IJPM.2017.082783

Puranam, P., Singh, H., & Zollo, M. (2006). Organizing for Innovation: Managing the Coordination-Autonomy Dilemma in Technology Acquisitions, *Academy of Management Journal, 49*(2), pp. 263-280.

Puschmann, T., & Alt, R. (2005). Successful use of e-procurement in supply chains. *Supply Chain Management, 10*(2), 122–133. doi:10.1108/13598540510589197

Qian, C., Cao, Q., & Takeuchi, R. (2013). Top Management Team Functional Diversity and Organizational Innovation in China: The Moderating Effects of Environment, *Strategic Management Journal, 34*(1), pp. 110-120.

Qiu, R., & Gupta, V. (2015). A conceptual study of immigrant born globals and their multinationalization process. *Journal of Global Entrepreneurship Research, 5*(1), 1–18. doi:10.118640497-015-0018-7

Rahim, M., & Bantwal, A. (2012). Perceived benefits from a local government public procurement initiative: A diffusion of innovation perspective. In Inter-organizational information systems and business management: Theories for researchers (pp. 251–271). Hershey, PA: IGI Global. doi:10.4018/978-1-60960-768-5.ch016

Rahim, M., & Kurnia, S. (2014). *Understanding E-Procurement System Benefits using organisational Adoption Motivation Lens: a Case Study.* Paper presented at the PACIS.

Rahim, M., Chen, X., & Kurnia, S. (2008). Understanding benefits of electronic procurement systems adoption: experience of an Australian City Council.

Rahim, M., Ilhan, N., & Chen, X. (2011). *Factors affecting benefits of e-procurement systems: Experience of an Australian City Council.* Paper presented at the 11th International Conference on Electronic Business (ICEB 2011).

Rahim, M. (2008). Identifying factors affecting acceptance of e-procurement systems: An initial qualitative study at an Australian City Council. *Communications of the IBIMA*, *3*(1), 7–17.

Rahim, M., & As-Saber, S. N. (2011). E-procurement adoption process: An arduous journey for an Australian city council. *International Journal of Electronic Finance*, *5*(2), 150–168. doi:10.1504/IJEF.2011.040458

Rajan, R. (2010). How inequality fueled the crisis, Project Syndicate, July 9.

Ramey, C. (2019). The Boardroom Sage who was into good governance before it was cool, *Wall Street, Business Edition*, Nov. 15.

Ramirez, R., & Selsky, J. W. (2016). Strategic planning in turbulent environments: A social ecology approach to scenarios. *Long Range Planning*, *49*(1), 90–102. doi:10.1016/j.lrp.2014.09.002

Rantz, M. J., Scubic, M., Miller, S. J., & Krampe, J. (2008). *Using technology to enhance aging in place. Smart Homes and Health Telematics. ICOST 2008.* Berlin, Germany: Springer.

Rego, P., Lopes, M., & Simpson, A. (2017). The authentic-Machiavellian leadership grid: A typology of leadership styles. *The Journal of Leadership Studies*, *11*(2), 48–51. doi:10.1002/jls.21524

Reifsteck, M. W., & Reese, M. L. (2016). The digital health explosion. A comprehensive strategy is required to remain relevant and responsive. *Healthcare Executive*, *31*(3), 66–67. PMID:27319110

Renwick, D. (2003). Line Manager Involvement in Hrm: An inside View, *Employee Relations, 25*(3), pp. 262-280.

Reynolds, D. P., Bygrave, D. W., Autio, E., Cox, W. L., & Hay, M. (2002). *Global Entrepreneurship Monitor: 2002 Executive Report.*

Ribeiro, N., Duarte, A., Filipe, R., & Torres de Oliveira, R. (2019). How authentic leadership promotes individual creativity: The mediating role of affective commitment. *Journal of Leadership & Organizational Studies*; Advanced Online Publication. doi:10.1177/1548051819842796

Richards, G., Yeoh, W., Loong, A. Y., & Popovic, A. (2019). Business Intelligence Effectiveness and Corporate Performance Management: An Empirical Analysis. *Journal of Computer Information Systems, 59*(2), 188-196.

Rich, E., & Miah, A. (2014). Understanding digital health as public pedagogy: A critical framework. *Societies (Basel, Switzerland)*, *4*(2), 296–315. doi:10.3390oc4020296

Richter, C., Kraus, S., Brem, A., Durst, S., & Giselbrecht, C. (2017). Digital entrepreneurship: Innovative business models for the sharing economy. *Creativity and Innovation Management*, *26*(3), 300–310. doi:10.1111/caim.12227

Ridgeway, C. (2001). Gender, status, and leadership. *The Journal of Social Issues*, *57*(4), 637–655. doi:10.1111/0022-4537.00233

Ritter, T., & Lettl, C. (2018). The wider implications of business-model research. *Long Range Planning*, *51*(1), 1–8. doi:10.1016/j.lrp.2017.07.005

Robbins, S. (2009). Comportamiento Organizacional. Teoría y práctica 13th Ed. Naucalpan de Juárez: Ed. Pearson Educación México.

Robbins, S. P., & Judge, T. A. (2017). *Organizational behavior*. New York: Pearson Education.

Rockart, J. (1979). Chief Executives Define Their Own Information Needs. In *Harvard Business Review* (pp. 81–92). March/April.

Rockart, J. (1982). The Changing Role of the Information Systems Executive: A Critical Success Factors Perspective. *Sloan Management Review*, *23*(1), 3–13.

Rockart, J., & Van Bullen, C. (1986). A Primer on Critical Success Factors. In J. Rockart, & C. Van Bullen (Eds.), *The Rise of Management Computing*. Homewood: Irwin.

Rogers, E. (1995). Diffusion of Innovations, 4th ed. The Free Press New York.

Rogers, E. M. (1995). Diffusion of Innovations (4th ed.). New York: The Free Press.

Rogers, E. M. (2003). *Diffusion of innovations* (5th ed.). New York: Free Press.

Romme, G. L. A., & Witteloostuijn, A. (1999). Circular organizing and triple loop learning. *Journal of Organizational Change Management*, *12*(5), 439–453. doi:10.1108/09534819910289110

Ronchi, S., Brun, A., Golini, R., & Fan, X. (2010). What is the value of an IT e-procurement system? *Journal of Purchasing and Supply Management*, *16*(2), 131–140. doi:10.1016/j.pursup.2010.03.013

Rosenberg, J., & Mateos, A. (2011). *The Cloud at Your Service. SAD.* Manning Publications.

Ross, J., Sebastian, I. M., Beath, C., Scantlebury, S., Mockner, M., & Fonstand, N. … BCG. (2016). Designing Digital Organizations. *CISR Working Paper No. 406*. MIT Sloan School of Management. Retrieved from http://cisr.mit.edu/blog/documents/2016/03/10/mit_cisrwp406_designingdigitalorganzations_rosssebastianbeathscantleburymockerfonstadkaganmoloneykrusellbcg.pdf/

Ross, K. (2016). Corporate Stewardship: Achieving Sustainable Effectiveness. *People & Strategy*, *39*(1), 59–60.

Roszkowska, D. (2017). External Knowledge Sourcing and Innovation Processes in Modern Economic Environment. *International Journal of Management and Economics*, *53*(2), 39–56. doi:10.1515/ijme-2017-0011

Roth, M., & Keller, B. (2019). Modernization through digital transformation. *TD: Talent Development*, *73*(1), 32.

Ruggieri, R., Savastano, M., Scalingi, A., Bala, D., & D'Ascenzo, F. (2018). The impact of Digital Platforms on Business Models: An empirical investigation on innovative start-ups. *Management & Marketing*, *13*(4), 1210–1225. doi:10.2478/mmcks-2018-0032

Saadatmand, F., Lindgren, R., & Schultze, U. (2019). Configurations of platform organizations: Implications for complementor engagement. *Research Policy*, *48*(8), 103770. doi:10.1016/j.respol.2019.03.015

Sacconi, L., & Antoni, G. D. (2011). *Social capital, corporate responsibility, economic behavior and performance*. New York, NY: Palgrave MacMillan. doi:10.1057/9780230306189

Sahut, J.-M., Luca, I., & Frédéric, T. (2019). The age of digital entrepreneurship. *Small Business Economics*, 1–11. doi:10.100711187-019-00260-8

Salanova, M., Agut, S., & Peiró, J. M. (2005). Linking Organizational Resources and Work Engagement to Employee Performance and Customer Loyalty: The Mediation of Service Climate, *Journal of Applied Psychology, 90*(6), pp. 1217-1227.

Sali, M. (2011). *Exploring e-government skills and competencies required in public sectors: A literature study*. 1-29. Retrieved from https://pdfs.semanticscholar.org/fe54/2fb75dd5234007e39de3ef25484482421005.pdf

Sambasivan, M., Wemyss, G. P., & Che Rose, R. (2010). User acceptance of a G2B system: A case of electronic procurement system in Malaysia. *Internet Research*, *20*(2), 169–187. doi:10.1108/10662241011032236

Sandhu, K. (2020). Digital Systems Innovation for Health Data Analytics. Handbook of Research on Optimizing Healthcare Management Techniques. Editor: Nilmini Wickramasinghe (Swinburne University of Technology, Australia & Epworth HealthCare, Australia). doi:10.4018/978-1-7998-1371-2.ch019

Sandhu, K. (2020). Digital Systems Innovation for Health Data Analytics. Handbook of Research on Optimizing Healthcare Management Techniques. Hershey, PA: IGI Global.

Sandhu. (2020). Digital Systems Innovation for Health Data Analytics. In N. Wickramasinghe (Ed.), *Handbook of Research on Optimizing Healthcare Management Techniques*. (Swinburne University of Technology, Australia & Epworth HealthCare, Australia). IGI Global.

Sandhu, K. (2008). A Qualitative Approach to E-Services System Development. In T. Sobh (Ed.), *Advances in Computer and Information Sciences and Engineering* (pp. 236–241). New York: Springer; doi:10.1007/978-1-4020-8741-7_42

Sargent, K. (2011). Scenario planning. *Contract (New York, NY)*, *52*(5), 60.

Sarkar, J., & Sarkar, S. (2015). Corporate social responsibility in India- An effort to bridge the welfare gap. *Review of Market Integration*, *7*(1), 1–36. doi:10.1177/0974929215593876

Sarker, S., & Lee, A. S. (2000, January). *Using a positivist case research methodology to test three competing theories-in-use of business process redesign*. Paper presented at the International Conference on Information Systems (ICIS), Brisbane, Australia.

Scharmer, O., & Kaeufer, K. (2013). *Leading from the Emerging Future: From Ego-System to Eco-System Economies (Croatian E)*. Planetopija.

Scheer, V. (2016). Process Improvement for Implementing and Sustaining Change in Healthcare. (10125146 M.S.), The College of St. Scholastica, Ann Arbor. ProQuest Dissertations & Theses Global database.

Schein, E. H. (2004). Organizational culture and leadership (3rd ed.). San Francisco, CA: The Jossey-Bass Business & Management Series.

Schein, E. (1985). *Organizational culture and leadership*. San Francisco, CA: Jossey-Bass.

Schein, E. H. (1990). Process Consultation: Vol. I. *Its Role in Organizational Development*. Reading, MA: Addison-Wesley.

Schein, E. H. (2010). *Organizational Culture and Leadership*. John Wiley & Sons.

Schmidt, R., Möhring, M., Härting, R.-C., Reichstein, C., Neumaier, P., & Jozinović, P. (2015). Industry 4.0-potentials for creating smart products: empirical research results. In *Proceedings of the International Conference on Business Information Systems*. 10.1007/978-3-319-19027-3_2

Schoenherr, T., & Tummala, V. R. (2007). Electronic procurement: A structured literature review and directions for future research. *International Journal of Procurement Management*, *1*(1-2), 8–37. doi:10.1504/IJPM.2007.015353

Schulz, S., & Schuppan, T. (2012). Development of a European Framework for e-Government Competences, 14. Jörn von Lucke et al. (Hrsg.): Auf dem Weg zu einer offenen, smarten und vernetzten Verwaltungskultur Lecture Notes in Informatics (LNI), Gesellschaft für Informatik, Bonn. 47-60.

Schuppan, T. (2008). *Skill Requirements for e-Government*. EGPA Conference, September 3-5 in Rotterdam (The Netherlands). 1-18. Retrieved from https://www.scss.tcd.ie/disciplines/information_systems/egpa/docs/2008/Schuppan%20 EGPA%2008.pdf

Schuppan, T. (2010). E-Government Competencies : Looking beyond Technology. In C. Shea, & G. Garson (Eds.), *Public Administration and Public Policy* (Vol. 155)., doi:10.1201/EBK1439807569-c22

Schwandt, D. R. (2008). Individual and Collective Coevolution. In M. Uhl-Bien, & R. Marion (Eds.), *Complexity Leadership Part 1: Conceptual Foundations* (pp. 101–127). Charlotte, NC: Information Age.

Schwarzmüller, T., Brosi, P., Duman, D., & Welpe, I. M. (2018). How does the digital transformation affect organizations? Key themes of change in work design and leadership. *Management Review*, *29*(2), 114–138. doi:10.5771/0935-9915-2018-2-114

Sebastian, I. M., Ross, J. W., Beath, C., Mocker, M., Moloney, K. G., & Fonstad, N. O. (2017). How big old companies navigate digital transformation. *MIS Quarterly Executive*, *16*(3), 197–213. Retrieved from http://misqe.org/ojs2/index.php/misqe/article/viewFile/783/468

Sedera, D., Lokuge, S., Grover, V., Sarker, S., & Sarker, S. (2016). Innovating with Enterprise Systems and Digital Platforms: A Contingent Resource-Based Theory View, *Information & Management, 53*(3), pp. 366–379.

Seitovirta, L. C. (2011). *The Role of Strategic Intelligence in Corporate Decision Making. Archive number 12600*. Espoo: Aalto University.

Select Committee on Digital Skills. (2015). *Make or Break: The UK's Digital Future*. London, UK: Authority of the House of Lords.

Seligman, M., & Csikszentmihalyi, M. (2014). Positive psychology: An introduction. In M. Csikszentmihalyi (Ed.), *Flow and the foundations of positive psychology* (pp. 279–298). Dordrecht, The Netherlands: Springer.

Servaes, H., & Tamayo, A. (2013). The impact of corporate social responsibility on firm value: The role of customer awareness. *Management Science*, *59*(5), 1045–1061. doi:10.1287/mnsc.1120.1630

Shafran, R. F. (2017). Leading organizational change: A phenomenological study of the nonlinear strategies used by female leaders in global companies. (Doctoral thesis), Pepperdine University, Ann Arbor. ProQuest Dissertations & Theses Global database.

Sharma, S. K., Al-Badi, A. H., Govindaluri, S. M., & Al-Kharusi, M. H. (2016). Predicting motivators of cloud computing adoption: A developing country perspective. *Computers in Human Behavior*, *62*, 61–69. doi:10.1016/j.chb.2016.03.073

Shaw, V., & Braa, J. (2014). Approaches to participatory design in Africa in the age of cloud computing. In *Proceedings of the 13th Participatory Design Conference: Short Papers, Industry Cases, Workshop Descriptions, Doctoral Consortium papers, and Keynote abstracts* – Volume 2, (pp. 45-48). ACM 10.1145/2662155.2662180

Sheel, A. (2017). Hotel Industry Performance in 2016–2017 and the JHFM Index. *The Journal of Hospitality Financial Management*, *25*(2), 75–76. doi:10.1080/10913211.2017.1406743

Shepherd, J. (2004). What is the digital era? In G. Doukidis, N. Mylonopoulos, & N. Pouloudi (Eds.), *Social and economic transformation in the digital era* (pp. 1–18). IGI Global. doi:10.4018/978-1-59140-158-2.ch001

Shieh, H.-S. (2012). Does Cost Efficiency Lead to Better Financial Performance? A Study on Taiwan International Tourist Hotels. *The Journal of Hospitality Financial Management*, *20*(1), 17–30. doi:10.1080/10913211.2012.10721889

Shipton, H., Armstrong, C., West, M. & Dawson, J. (2008). The Impact of Leadership and Quality Climate on Hospital Performance, *International Journal for Quality in Health Care* (20:6), pp. 439-445.

Shirey, M., White-Williams, C., & Hites, L. (2019). Integration of Authentic Leadership Lens for Building High Performing Interprofessional Collaborative Practice Teams. *Nursing Administration Quarterly*, *43*(2), 101–112. doi:10.1097/NAQ.0000000000000339 PMID:30839447

Shkolnyk, I., Bukhtiarova, A., & Horobets, L. (2019). *The efficiency of electronic public procurement system in Ukraine*.

Shuck, B., Twyford, D., Reio, T. G. Jr, & Shuck, A. (2014). Human resource development practices and employee engagement: Examining the connection with employee turnover intentions. *Human Resource Development Quarterly*, *25*(2), 239–270. doi:10.1002/hrdq.21190

Siebel, T. M. (2017). Why digital transformation is now on the CEO's shoulders. *The McKinsey Quarterly*, *4*(3), 1–7.

Singh, D., & Ahuja, J. (1983). Corporate social reporting in India. *The International Journal of Accounting*, *18*, 151–169.

Sin, L. Y., Alan, C., Yau, O. H., Chow, R. P., Lee, J. S., & Lau, L. B. (2005). Relationship marketing orientation: Scale development and cross-cultural validation. *Journal of Business Research*, *58*(2), 185–194. doi:10.1016/S0148-2963(02)00493-9

Smart, A. (2010). E-procurement and its impact on supply management–evidence from industrial case studies. *International Journal of Logistics: Research and Applications*, *13*(6), 423–440. doi:10.1080/13675567.2010.486760

Smith, J. A. (2011). Evaluating the contribution of interpretative phenomenological analysis. *Health Psychology Review*, *5*(1), 9–27. doi:10.1080/17437199.2010.510659

Smollan, R. K., & Sayers, J. G. (2009). Organizational culture, change, and emotions: A qualitative study. *Journal of Change Management*, *9*(4), 435–457. doi:10.1080/14697010903360632

Soana, M. G. (2011). The relationship between corporate social performance and corporate financial performance in the banking sector. *Journal of Business Ethics*, *104*(1), 133–148. doi:10.100710551-011-0894-x

Sobowale, A., & Ortigara, A. R. C. (2017). From MDGs to SDGs: The need for Global Thinking and Local Actions in the Nigerian Water Sector. *Arid Zone Journal of Engineering, Technology, and Environment*, *13*(5), 620–629.

Soh, C., Chua, C. E. H., & Singh, H. (2011). Managing Diverse Stakeholders in Enterprise Systems Projects: A Control Portfolio Approach, *Journal of Information Technology*, *26*(1), pp. 16-31.

Soliman, K. S., Janz, B. D., Puschmann, T., & Alt, R. (2005). Successful use of e-procurement in supply chains. *Supply Chain Management*.

Somech, A., & Drach-Zahavy, A. (2013). Translating team creativity to innovation implementation: The role of team composition and climate for innovation. *Journal of Management*, *39*(3), 684–708. doi:10.1177/0149206310394187

Song, A. K. (2019). The Digital Entrepreneurial Ecosystem-a critique and reconfiguration. *Small Business Economics*, *53*(3), 569–590. doi:10.100711187-019-00232-y

Spigel, B., & Harrison, R. (2017). Toward a process theory of entrepreneurial ecosystem. *Strategic Entrepreneurship Journal*, *12*, 17.

Spigel, B., & Harrison, R. (2018). Toward a process theory of entrepreneurial ecosystems. *Strategic Entrepreneurship Journal*, *12*(1), 151–168. doi:10.1002ej.1268

Spremic, M. (2017). Governing digital technology – how mature IT governance can help in digital transformation? *International Journal of Economics and Management Systems*, *2*, 214–223. Retrieved from https://www.iaras.org/iaras/home/caijems/governing-digital-technology-how-mature-it-governance-can-help-in-digital-transformation

Springer, P. J., Clark, C. M., Faan, A., Strohfus, P., & Belcheir, M. (2012). Using transformational change to improve organizational culture and climate in a school of nursing. *The Journal of Nursing Education*, *51*(2), 81–88. doi:10.3928/01484834-20111230-02 PMID:22201273

Srai, J. S., & Lorentz, H. (2019). Developing design principles for the digitalisation of purchasing and supply management. *Journal of Purchasing and Supply Management*, *25*(1), 78–98. doi:10.1016/j.pursup.2018.07.001

Srivastava, A., Bartol, K., & Locke, E. (2006). Empowering leadership in management teams: Effects on knowledge sharing, efficacy, and performance. *Academy of Management Journal*, *49*(6), 1239–1251. doi:10.5465/amj.2006.23478718

Stacey, R. D. (1995). The Science of Complexity: An Alternative Perspective for Strategic Change Processes, *Strategic Management Journal* (16:6), pp. 477-495.

Stam, E., & Spigel, B. (2016). Entrepreneurial Ecosystems.

Stam, E. (2015). Entrepreneurial Ecosystems and RegionalPolicy: A Sympathetic Critique. *European Planning Studies*, *23*(9), 10. doi:10.1080/09654313.2015.1061484

Standing, G. (2011). *The Precariat: The New Dangerous Class*. London, UK: Bloomsbury Academic.

Stary, C. (2014). Non-disruptive knowledge & business processing in knowledge life cycles – aligning value network analysis to process management. *Journal of Knowledge Management*, *18*(4), 651–686. doi:10.1108/JKM-10-2013-0377

Steiner, G. (1979). *Strategic Planning: What Every Manager Must Know*. New York: Free Press.

Steininger, D. M. (2019). Linking information systems and entrepreneurship: A review and agenda for IT-associated and digital entrepreneurship research. *Information Systems Journal*, *29*(2), 363–407. doi:10.1111/isj.12206

Stensaker, I., & Meyer, C. (2011). *Change experience and employee reactions: Developing capabilities for change* (Vol. 41).

Subramaniam, R. K., Samuel, S. D., & Mathenthiran, S. (2015). Liquidity implications of corporate social responsibility disclosures: Malaysian evidence. *Journal of International Accounting Research*, *15*(1), 133–153. doi:10.2308/jiar-51204

Sultan, N. A. (2011). *Reaching for the "cloud": How SMEs can manage*. Elsevier. Retrieved June 27, 2019 from www.elsevier.com/locate/ijinfomgt

Suša Vugec, D., Pejic-Bach, M., & Spremić, M. (2019). IT Strategic Grid: A Longitudinal Multiple Case Study. In Ç. Doğru (Ed.), *Handbook of Research on Contemporary Approaches in Management and Organizational Strategy* (pp. 105–128). Hershey, PA: IGI Global; doi:10.4018/978-1-5225-6301-3.ch006

Sussan, F., & Acs, Z. J. (2017). The digital entrepreneurial ecosystem. *Small Business Economics*, *49*(1), 55–73. doi:10.100711187-017-9867-5

Svahn, F., Mathiassen, L., & Lindgren, R. (2017). Embracing Digital Innovation in Incumbent Firms: How Volvo Cars Managed Competing Concerns, *MIS Quarterly, 41*(1), pp. 239-253.

Swanson, E. B. (2012). The Managers Guide to IT Innovation Waves, *MIT Sloan Management Review, 53*(2), pp. 75-83.

Tait, I. E. (2013). Comparison of project management software tool use in healthcare and other industries. (Doctoral thesis), Capella University, Ann Arbor. ProQuest Central; ProQuest Dissertations & Theses Global database.

Tai, Y.-M., Ho, C.-F., & Wu, W.-H. (2010). The performance impact of implementing web-based e-procurement systems. *International Journal of Production Research*, *48*(18), 5397–5414. doi:10.1080/00207540903117915

Taleb, N. N. (April 2007). The black swan: The impact of the highly improbable. Retrieved from https://www.nytimes.com/2007/04/22/books/chapters/0422-1st-tale.html?pagewanted=all&_r=0

Tanniru. M., Khuntia, J., & Weiner, J. (2018). Hospital Leadership in Support of Digital Transformation, *Pacific Asia Journal of Association of Information Systems, (PAJAIS), 10*(3).

Tanniru, M. (2019). Engagement leading to empowerment - Digital Innovation Strategies for Patient Care Continuity [October.]. *J Hosp Management and Health Policy*, *3*, 28. doi:10.21037/jhmhp.2019.09.01

Tanniru, M., Jones, J., Kazziha, S., & Hornberger, M. (2019). Strategies to Address Care Transition - A Methodology and a Case Study. *Journal of Hospital Administration*, *8*(3), doi:10.5430/jha.v8n3p38

Tatsis, V., Mena, C., Van Wassenhove, L. N., & Whicker, L. (2006). E-procurement in the Greek food and drink industry: Drivers and impediments. *Journal of Purchasing and Supply Management*, *12*(2), 63–74. doi:10.1016/j.pursup.2006.04.003

Taylor, K., & Silver, L. (2019). Smartphone Ownership Is Growing Rapidly Around the World, but Not Always Equally. Retrieved from https://www.pewresearch.org/global/2019/02/05/smartphone-ownership-is-growing-rapidly-around-the-world-but-not-always-equally/

Themistocleous, M., Irani, Z., & O'Keefe, R. M. (2001). ERP and application integration: Exploratory survey. *Business Process Management Journal*, *7*(3), 195–204. doi:10.1108/14637150110392656

Thierauf, R. (2001). *Effective Business Intelligence Systems*. Westport, CT: Quorum Books.

Thimbleby, H. (2013). Technology and the future of healthcare. *Journal of Public Health Research*, *2*(3), 28. doi:10.4081/jphr.2013.e28 PMID:25170499

Thomas, D. (2014). Leading under pressure, coping practices of executives: A mixed methods study. (Doctoral thesis), Pepperdine University, Ann Arbor. ProQuest Central; ProQuest Dissertations & Theses Global database.

Tjikongo, R., & Uys, W. (2013). *The viability of cloud computing adoption in SME's in Namibia. In 2013 Ist-Africa conference & exhibition* (pp. 1–11). IEEE.

Torous, J., Chan, S., Luo, J., Boland, R., & Hilty, D. (2018). Clinical Informatics in Psychiatric Training: Preparing Today's Trainees for the Already Present Future. *Academic Psychiatry*, *42*(5), 694–697. doi:10.100740596-017-0811-4 PMID:29047074

Trkman, P., & McCormack, K. (2009). Estimating the benefits and risks of implementing e-procurement. *IEEE Transactions on Engineering Management*, *57*(2), 338–349. doi:10.1109/TEM.2009.2033046

Tsai, H., Song, H., & Wong, K. K. F. (2009). Tourism and Hotel Competitiveness Research. *Journal of Travel & Tourism Marketing*, *26*(5–6), 522–546. doi:10.1080/10548400903163079

Tumbas, S., Berente, N., & Brocke, J. (2017). Three Types of Chief Digital Officers and Reasons that Organizations Adopt the Role, MIS Q Executive, Jan., 16(2).

Turban, E., Aronson, J. E., & Liang, J. P. (2005). *Decision Support Systems and Intelligent Systems* (7th ed.). New Jersey: Ed. Prentice Hall.

Tytler, R., Bridgstock, R., White, P., Mather, D., McCandless, T., & Grant-Iramu, M. (2019). Jobs of the future. Burwood, Victoria, Ford Motor Company of Australia & Deakin University.

Ugboro, I., & Obeng, K. (2000). Top management leadership, employee empowerment, job satisfaction, and customer satisfaction in TQM organizations: An empirical study. *Journal of Quality Management*, *5*(2), 247–272. doi:10.1016/S1084-8568(01)00023-2

Uhl-Bien, M., & Marion, R. (2008). *Complexity Leadership Part I: Conceptual Foundations*. Charlotte, NC: Information Age Publishing.

Uhl-Bien, M., Marion, R., & McKelvey, B. (2011). *Complexity Leadership Theory: Shifting Leadership from the Industrial Age to the Knowledge Era, Leadership, Gender, and Organization* (pp. 109–138). Springer.

Ungureanu, E., & Avramescu, T. C. (2008). Strategic Diagnosis and the Part it Plays in Decision Making. *Annals of the University of Petrosani. Economics*, *8*(2), 269–274.

United Nations E-Government Survey. (2008). From e-Government to Connected. Governance.

United Nations E-Government Survey. (2012). E-Government for the People.

United Nations E-government Survey. (2014). E-government for the future we want.

United Nations E-government Survey. (2016). E-government in support of sustainable development.

United Nations E-government Survey. (2018). Gearing e-government to support Transformation towards Sustainable and resilient societies.

United Nations. (2002). Benchmarking e-government: A Global perspective, assessing the progress of the UN member states. Retrieved from www.golconferenceca/presentations/eGovernment_UN.pdf.

United Nations. (2003). World Public Sector Report 2003: E-Government at the Crossroads, New York.

United Nations. (2004). Global e-government readiness report 2004: Towards access for Opportunity, New York. Retrieved from http://unpan1.un.org/intradoc/groups/public/documents/un/unpan019207.pdf

USA Institute of Medicine. (2001). *Crossing the Quality Chasm: A New Health System for the 21st Century*. Washington, DC: The National Academies Press.

Valle-Cruz, D. (2019). Public value of e-government services through emerging technologies. *International Journal of Public Sector Management*, *32*(5), 530–545. doi:10.1108/IJPSM-03-2018-0072

Van de Ven, A. H., & Poole, M. S. (1995). Explaining Development and Change in Organizations, *Academy of Management Review* (20:3), pp. 510-540.

van der Zwan, P., Thurik, R., Verheul, I., & Hessels, J. (2016). Factors influencing the entrepreneurial engagement of opportunity and necessity entrepreneurs. *Eurasian Bus Rev, 6.*

van Deursen, A., & van Dijk, J. (2008). Measuring digital skills. Performance tests of operational, formal, information and strategic internet skills among the Dutch population. In *Proceedings of the 58th Conference of the International Communication Association*, 1-25, Academic Press.

van Deursen, A. J. A. M., & van Dijk, J. A. G. M. (2009). Improving digital skills for the use of online public information and services. *Government Information Quarterly*, *26*(2), 333–340. doi:10.1016/j.giq.2008.11.002

Van Fleet, D. D., & Peterson, T. O. (2016). Improving healthcare practice behaviors. *International Journal of Health Care Quality Assurance*, *29*(2), 141–161. doi:10.1108/IJHCQA-07-2015-0089 PMID:26959895

van Laar, E., van Deursen, A., van Dijk, J., & de Haan, J. (2017). The relation between 21st-century skills and digital skills: A systematic literature review. *Computers in Human Behavior*, *72*, 577–588. doi:10.1016/j.chb.2017.03.010

Van Tubergen, F. (2005). Self-Employment of Immigrants: A Cross-National Study Study of 17 Western Societies. *Social Forces*, *84*(2), 23. doi:10.1353of.2006.0039

Van Vugt, M. (2006). Evolutionary origins of leadership and followership. *Personality and Social Psychology Review*, *10*(4), 354–371. doi:10.120715327957pspr1004_5 PMID:17201593

Vasal, V. (1995). A spatio-temporal analysis on social reporting by companies-Evidence from Indian public sector. *Journal of Accounting and Finance*, *2*, 256–269.

Vecchiato, R. (2015). Strategic planning and organizational flexibility in turbulent environments. *Foresight*, *17*(3), 257–273. doi:10.1108/FS-05-2014-0032

Venkatesh, V., & Davis, F. D. (2000). A theoretical extension of the technology acceptance model: Four longitudinal field studies. *Management Science*, *46*(2), 186–204. doi:10.1287/mnsc.46.2.186.11926

Ventres, W., & Frankel, R. (1996). Ethnography: A stepwise approach for primary care researchers. *Family Medicine*, *28*(1), 52–56. PMID:8720228

Vieru, D., Bourdeau, S., Bernier, A., & Yapo, S. (2015). Digital Competence: A Multi-dimensional Conceptualization and a Typology in an SME Context. In *Proceedings 2015 48th Hawaii International Conference on System Sciences*, 4681-4690. 10.1109/HICSS.2015.557

Visconti, R. M., Larocca, A., & Marconi, M. (2017). Big data-driven value chains and digital platforms: From value co-creation to monetization. *Big Data Analytics: Tools and Technology for Effective Planning*, 355–371. doi:10.1201/b21822

Vitello-Cicciu, J. (2019). Am I an Authentic Nursing Leader for Healthy Workplace Environments? *Nurse Leader*, *17*(3), 201–206. doi:10.1016/j.mnl.2019.03.011

Vladimir, R.-B., Jorge, G.-M., Martín, L.-N., Yaroslava, R.-B., Javier, C.-R., Diego, Q.-P., ... Verónica Cevallos, L.-W. (2018). A Hybrid System to Support the Psychological Intervention for Children With Autism: A Proposal Based on Knowledge Modeling and Intelligent Technologies. [IJESMA]. *International Journal of E-Services and Mobile Applications*, *10*(4), 74–100. doi:10.4018/IJESMA.2018100105

Voellm, D. J. (2017). Taiwan Hotel Market Update. HVS Asia-Pacific. Available at https://www.hvs.com/article/8094-taiwan-hotel-market-update (Retrieved on 3 August 2019).

Vom Brocke, J., Maaß, W., Buxmann, P., Maedche, A., Leimeister, J. M., & Pecht, G. (2018). Future work and enterprise systems. *Business & Information Systems Engineering*, *60*(4), 357–366. doi:10.100712599-018-0544-2

Vrabie, C. (2012). Computer competencies necessary for an effective e-governance, 12.

Waagstein, P. R. (2011). The mandatory corporate social responsibility in Indonesia: Problems and implications. *Journal of Business Ethics*, *98*(3), 455–466. doi:10.100710551-010-0587-x

Wade, V. A., Taylor, A. D., Kidd, M. R., & Carati, C. (2016). Transitioning a home telehealth project into a sustainable, large-scale service: A qualitative study. *BMC Health Services Research*, *16*(1), 1–10. doi:10.118612913-016-1436-0 PMID:27185041

Wade, W. (2012). *Scenario planning- a field guide to the future*. Hoboken, NJ: John Wiley & Sons.

Wahee, S. J., & Bhardwaj, B. (2014). Entrepreneurial Motivation and Intentions: The Antecedent of Cyber Entrepreneurship. In V. Ribiere, & L. Worasinchai (Eds.), *Proceedings of the 2nd International Conference on Innovation and Entrepreneurship* (pp. 289-296). Nr Reading: Acad Conferences Ltd.

Walsh, T., Levy, N., Bell, G., Elliott, A., Maclaurin, J., Mareels, I., & Wood, F. (2019). The effective and ethical development of artificial intelligence: An opportunity to improve our wellbeing. Melbourne, Australia: Australian Council of Learned Academics (ACOLA).

Walsham, G. (1993). *Interpreting Information Systems in Organizations*. Chichester, UK: Wiley & Sons.

Wang, L., & Alexander, C. A. (2013). Automated identification technology and information security in digital hospitals. *International Journal of Information and Network Security*, *2*(5), 348–359. doi:10.11591/ijins.v2i5.3914

Wang, M., & Zhang, S. (2005). Integrating EDI with an E-SCM system using EAI technology. *Information Systems Management*, *22*(3), 31–36. doi:10.1201/1078/45317.22.3.20050601/88743.5

Wang, Q. F., & Liu, C. Y. (2015). Transnational activities of immigrant-owned firms and their performances in the USA. *Small Business Economics*, *44*(2), 345–359. doi:10.100711187-014-9595-z

Wang, Y. S., Lin, S. J., Yeh, C. H., Li, C. R., & Li, H. T. (2016). What drives students' cyber entrepreneurial intention: The moderating role of disciplinary difference. *Thinking Skills and Creativity*, *22*, 22–35. doi:10.1016/j.tsc.2016.08.003

Wan, J. P., Jiang, Q. W., & Xie, L. Q. (2017). *Research on Risk Factors of Entrepreneurship in Internet Industry*. Calgary, Canada: University Calgary Press.

Warner, K. S. R., & Wäger, M. (2019). Building dynamic capabilities for digital transformation: An ongoing process of strategic renewal. *Long Range Planning*, *52*(3), 326–349. doi:10.1016/j.lrp.2018.12.001

Watkins, M. J., Mohr, B. J., & Kelly, R. (2011). Appreciative Inquiry (AI) (2nd ed.). San Francisco, CA: John Wiley & Sons.

We are social. (2019). Digital 2019, Q2 Global Digital Statshot. Retrieved from https://wearesocial.com/blog/2019/04/the-state-of-digital-in-april-2019-all-the-numbers-you-need-to-know

Weill, P. (2011). What is your Digital Business Model? *MIT CISR Research Briefing, XI*(IX). Retrieved from http://cisr.mit.edu/blog/documents/2011/09/15/2011_0901_digitalbusinessmodels_weill.pdf/

Weiner, L. J. (2016). Improving physician engagement. *HealthLeaders, 19,* 8-8, 10.

Weinstein, R. S., Lopez, A. M., Krupinski, E. A., Beinar, S. J., Holcomb, M., McNeely, R. A., ... Barker, G. (2008). Integrating telemedicine and telehealth: Putting it all together. *Studies in Health Technology and Informatics*, *131*, 23–38. PMID:18305320

Weller, C. (2017, February 14). Elon Musk doubles down on universal basic income: 'It's going to be necessary'. *Money & Markets*. Retrieved from https://www.businessinsider.com.au/elon-musk-universal-basic-income-2017-2?r=US&IR=T

Weston, S. (2017). The real challenge for digital transformation is not your technology. *CIO (Framingham, MA)*.

Whitehead, N. P., Scherer, W. T., & Smith, M. C. (2015). Systems thinking about systems thinking a proposal for a common language. *IEEE Systems Journal*, *9*(4), 1117–1128. doi:10.1109/JSYST.2014.2332494

Wiedmer, T. (2015). Generations do differ: Best practices in leading Traditionalists, Boomers, and Generations X, Y, and Z. *Delta Kappa Gamma Bulletin*, *82*(1), 51–58.

Wieland, H., Vargo, S., Akaka, M. S., & Barbeau, B. (2018). A Unifying Perspective for the Technological, Business Model, and Market Aspects of Innovation, *Handbook on Service Dominant Logic*. Sage. *Lusch and Vargo, Ch.*, *29*, 508–521.

Williams, D. R., Costa, M. V., Odunlami, A. O., & Mohammad, S. A. (2008). Moving Upstream: How Interventions that Address the Social Determinants of Health can Improve Health and Reduce Disparities, *J Public Health Management Practice*, November 14 (Suppl), S8-17.

Williams, K. (2019). Chinese formula demand won't see Kiwi babies go without. *Stuff.co.nz*. Retrieved from https://www.stuff.co.nz/business/101470144/chinese-formula-demand-wont-see-kiwi-babies-go-without

Williams, K., Chatterjee, S., & Rossi, M. (2008). Design of emerging digital services: A taxonomy. *European Journal of Information Systems, 17*(5), 505–517. doi:10.1057/ejis.2008.38

Williams, M. I. (2010). *A Quick Start Guide to Cloud Computing*. United Kingdom: Kogan Page Limited.

Wimalasena, N. N., & Gunatilake, S. (2018). The readiness of construction contractors and consultants to adopt e-tendering: The case of Sri Lanka. *Construction Innovation, 18*(3), 350–370. doi:10.1108/CI-03-2017-0025

Winby, S., & Mohrman, S. A. (2018). Digital Sociotechnical System Design. *The Journal of Applied Behavioral Science, 54*(4), 399–423. doi:10.1177/0021886318781581

Wittmer, D. P. (2019). *Ethical Decision-Making. Chapter in Terry Cooper, Handbook of Administrative Ethics* (2nd ed.). New York: Routledge.

Wong, C., & Laschinger, H. (2013). Authentic leadership, performance, and job satisfaction: The mediating role of empowerment. *Journal of Advanced Nursing, 69*(4), 947–959. doi:10.1111/j.1365-2648.2012.06089.x PMID:22764828

Wootton, R., Kvedar, J., & Dimmick, S. L. (2006). *Home Telehealth: Connecting care within community*. London, UK: Royal Society of Medicine Press.

World Economic Forum. (2016). *The future of jobs. Employment, skills, and workforce strategy for the fourth industrial revolution*, (accessed 12 June 2019) https://www.weforum.org/reports/the-future-of-jobs

World Economic Forum. (2018). The Digital Enterprise. Moving from experimentation to transformation. Geneva.

World Health Organisation. (2016). *Global diffusion of eHealth: Making universal health coverage achievable. Report of the third global survey on eHealth*. Retrieved from Geneva, Switzerland: http://apps.who.int/iris/bitstream/10665/252529/1/9789241511780- eng.pdf?ua=1

World Health Organisation. (2019). *Global Strategy on Digital Health 2020-2024*.

World Health Organization. (2011). *mHealth: New horizons for health through mobile technologies; Global Observatory for eHealth series*.

World Travel & Tourism Council. (2018). Travel and Tourism Economic Impact 2018: World. Available at https://www.wttc.org/ (Retrieved on 15 March 2019).

Wright, P., & Ferris, S. (1997). Agency Conflict and corporate strategy: The effect of disinvestment on corporate value. *Strategic Management Journal, 18*(1), 77–83. doi:10.1002/(SICI)1097-0266(199701)18:1<77::AID-SMJ810>3.0.CO;2-R

Xie, X. F., Xie, X. M., & Martinez-Climent, C. (2019). Identifying the factors determining the entrepreneurial ecosystem of internet cultural industries in emerging economies. *The International Entrepreneurship and Management Journal, 15*(2), 503–522. doi:10.100711365-019-00562-z

Xi, Y., Liu, P., & Zhang, X. (2019). Living in the Present of the Future: In Memory of James G. March (1928–2018). *Management and Organization Review, 15*(4), 895–899.

Yablonsky, S. (2018). A Multidimensional Framework for Digital Platform Innovation and Management: From Business to Technological Platforms. *Systems Research and Behavioral Science, 35*(4), 485–501. doi:10.1002res.2544

Yang, Q., Miao, C., & Shen, Z. (2015). Digital services innovation for aging-in-place. In *Proceedings of 22nd IEEE International Conference on Industrial Engineering and Engineering Management* (IEEM). Singapore. DOI: 10.1109/IEEM.2015.7385711

Yeboah-Boateng, E. O., & Essandoh, K. A. (2014). Factors influencing the adoption of cloud computing by small and medium enterprises in developing economies. *International Journal of Emerging Science and Engineering*, *2*(4), 13–20.

Yin, R. K. (1994). *Case study research: Design and Methods* (2nd ed.). London, UK: Sage.

Yin, R. K. (1994). *Case Study Research: Design and Methods*. Newbury Park, CA: Sage.

Yin, R. K. (2010). *Qualitative Research from Start to Finish*. New York: Guilford Press.

Yogesan, K., Bos, L., Brett, P., & Gibbons, M. C. (2009). Handbook of Digital Homecare. Berlin, Gernmany: Springer-Verlag.

Yoo, Y., Boland, R. J. Jr., Lyytinen, K., & Majchrzak, A. (2012). Organizing for Innovation in the Digitized World, *Organization Science, 23*(5), pp. 1398-1408.

Yoo, Y., Boland, R. J. Jr, Lyytinen, K., & Majchrzak, A. (2012). Organizing for Innovation in the Digitized World. *Organization Science*, *23*(5), 1398–1408. doi:10.1287/orsc.1120.0771

You, T. L., & Zhou, M. (2019). Simultaneous Embeddedness in Immigrant Entrepreneurship: Global Forces Behind Chinese-Owned Nail Salons in New York City. *The American Behavioral Scientist*, *63*(2), 166–185. doi:10.1177/0002764218793684

Yue, H. (2017). *National Report on E-Commerce Development in China.*

Yu, L., & Huimin, G. (2007). An Analysis of Hotel Financial Management in China. *The Journal of Hospitality Financial Management*, *15*(2), 39–48. doi:10.1080/10913211.2007.10653841

Yu, X. Y., Roy, S. K., Quazi, A., Nguyen, B., & Han, Y. Q. (2017). Internet entrepreneurship and "the sharing of information" in an Internet-of-Things context The role of interactivity, stickiness, e-satisfaction and word-of-mouth in online SMEs' websites. *Internet Research*, *27*(1), 74–96. doi:10.1108/IntR-02-2015-0060

Zaharia-Rădulescu, A. M., & Radu, I. (2017). Cloud computing and public administration: approaches in several European countries. In *Proceedings of the International Conference on Business Excellence* - Volume 11, No. 1, pp. 739-749. De Gruyter Open. 10.1515/picbe-2017-0078

Zaheer, H., Breyer, Y., & Dumay, J. (2019). Digital entrepreneurship: An interdisciplinary structured literature review and research agenda. *Technological Forecasting and Social Change*, 148.

Ženko, Z., Mulej, M., & Potocan, V. (2017). Knowledge-cum-values Management belongs to the Way out from Global Crisis. *Business Systems Research*, *8*(1), 113–123. doi:10.1515/bsrj-2017-0009

Zennouche, M., Zhang, J., & Wang, B. W. (2013). Factors influencing innovation at individual, group and organisational levels: A content analysis. *International Journal of Information Systems and Change Management*, *7*(1), 23–42. doi:10.1504/IJISCM.2014.065052

Zhang, Q., Cheng, L., & Boutaba, R. (2010). *Cloud computing: state-of-the-art and research challenges*. Brazil: The Brazilian Computer Society.

Zhang, X., Fu, P., & Xi, Y. (2018). Promoting Management Education in China Through Developing Practice-Based Management Theories: An Interview with Practitioner-Scholar Youmin Xi. *Journal of Management Inquiry*, *27*(2), 212–223. doi:10.1177/1056492617708056

Zhukova, M. A. (2018). Development of the hotel industry based on the use of information technologies. Управление, (4), 62. doi:10.26425/2309-3633-2018-4-62-66

Ziadlou, D. (2019, September). Digital health leadership in healthcare organizations[YouTube]. Switzerland International Society for Telemedicine and e-Health (ISfTeH).

Zittrain, J. (2006). The generative Internet. *Harvard Law Review*, *119*(7), 1974–2040. Retrieved from http://portal.acm.org/citation.cfm?doid=1435417.1435426

Zoroja, J., & Pejic Bach, M. (2016). Impact of information and communication technology to the competitiveness of European countries-cluster analysis approach. *Journal of Theoretical and Applied Electronic Commerce Research*, *11*(1), i–x. doi:10.4067/S0718-18762016000100001

Zwass, V. (2006). The web-internet compound as the infrastructure of digital government. *Business Process Management Journal*, *12*(1), 7–12. doi:10.1108/14637150610643715

About the Contributors

Kamaljeet Sandhu is an active scholar, experienced & passionate about research in Digital Innovation & Strategy, Business Data Analytics, Digital Health, AI, IoT, Cryptocurrency, Blockchain, Cloud Computing, IT Startups, FinTech, Cybersecurity, Accounting, Corporate Governance & CSR, Supply Chain, ERP, SMEs, & Entrepreneurship. Editor of 4 Books & published internationally over 180 articles in peer-reviewed (refereed) journals, conferences, & book chapters. He earned his Ph.D. from Deakin University Melbourne campus Australia. Having held multiple Leadership positions including Professor at the University of Northern British Columbia Canada, Senior Research Fellow at the University of South Australia, and other Senior Academic positions at: University of Wollongong, Charles Darwin University, Deakin University, & at the University of South Pacific, Fiji. As a PhD Principal Supervisor (Advisor), have supervised & mentored several PhD Research students to successful completion & students have successfully gained academic & research appointments at leading international universities & other organizations. PhD students' research are funded by international government & industry scholarships. New scientific discoveries from supervising doctoral research have been implemented by government, private institutions and universities in Australia, Canada, USA, UK, Europe, China, India, Saudi Arabia, Jordan, & UAE. And have multidisciplinary experience working across different Faculties such as Business, Accounting, Finance, Management, Economics, IT & IS, Software, & Computer Science.

* * *

Mirjana Pejić Bach is a Full Professor at the Department of Informatics at the Faculty of Economics & Business. She graduated at the Faculty of Economics & Business – Zagreb, where she also received her Ph.D. degree in Business, submitting a thesis on "System Dynamics Applications in Business Modelling" in 2003. She is the recipient of the Emerald Literati Network Awards for Excellence 2013 for the paper Influence of strategic approach to BPM on financial and non-financial performance published in the Baltic Journal of Management. Mirjana was also educated at MIT Sloan School of Management in the field of System Dynamics Modelling, and at OliviaGroup in the field of data mining. She participates in number of EU FP7 projects and is an Expert for Horizon 2020. Mirjana Pejić Bach is editor in chief of two journals, and serves as the Associate editor and member of Editorial Board in several other journals, in the field of Information and Management, as well as conferences.

Fatma Bouaziz is assistant professor in Management at the Faculty of Economics and Management of Sfax - Tunisia. She got her Ph.D. in management in 2011 and her HDR in 2013 from the same faculty. Dr. Bouaziz is a member of the research laboratory "Perspectives and Research in Innovation,

Strategy, and Business Management (LR18ES24). Her doctoral thesis deals with the risk management of e-government projects in the case of the Tunisian government. Her research interests include the use of ICTs by government, their organizational impacts and abilities to improve public services delivery, management of human resources and organizational resilience in the context of transition.

Kerryn Butler-Henderson is the Associate Professor of Digital Innovation in Health and Health Pedagogy at the University of Tasmania, and holds PhD and Master of Business Administration qualifications. She is an internationally recognised for her research expertise in digital health workforce and capabilities. She has held two Australian Learning and Teaching grants examining digital innovation in learning and competencies, and provides leadership in digital capabilities in higher education. Kerryn has been a leading academic in digital health in Australia for more than a decade and has served on significant health informatics boards including: Health Information Management Association of Australia (HIMAA), Australasian College of Health Informatics (ACHI), Australian Health Informatics Education Council (AHIEC) and the Certified Health Informatician Australasia (CHIA).

Jo Coldwell-Neilson has built a strong research and teaching profile engaging students in and with technology. She has been involved in a variety of projects that are involved with implementing educational technologies in her teaching and more widely in the teaching and learning activities at Faculty and University levels. Jo has held the role of Associate Head (Teaching and Learning) in the School of IT for 7 years, during which time she initiated and supported a variety of projects including early identification of at-risk students, implementation of course wide portfolios to allow students to evidence their achievements of course learning outcomes. During her tenure as Associate Head, Jo introduced a number of initiatives into the curriculum the most recent being compulsory internships for all undergraduate IT students. Her current research reflects her teaching and learning interests and include investigations into gender issues in IT, digital technology uptake in schools and higher education, and preparing students for careers in a digital environment. In 2016 Jo was awarded an Australian Learning and Teaching Fellowship (previously OLT National Teaching Fellowship) entitled "Unlocking the code to digital literacy". Jo is currently Deputy Chair of Academic Board and Associate Dean Teaching and Learning in the Faculty of Science, Engineering and Built Environment.

Trudi Cooper leads the youth work professional education program and the social programme innovation research and evaluation (SPIRE) group at Edith Cowan University, in Australia. Her research includes digital literacy, higher education policy and curriculum research, teaching and learning, and youth work research. She is an Australian Learning and Teaching Fellow and has been recognised nationally for her university teaching.

Joseph Crawford is a leadership and behavior researcher who seeks to create empirical solutions for growing unethical behavior in leaders. He has provided conceptualizations for the authentic leader and follower, and published on best practice methodology for leadership measurement. Dr Crawford received his PhD in 2019 from the University of Tasmania, and received the Launceston Chamber of Commerce Young Professional of the Year in the same year for his contributions to research on leadership. He is the Co-Founder of award-winning sustainable grooming products company Fifth Estate Co. Joseph has served on numerous community and organizational boards, convened medium-large conferences, and is currently an Associate Editor of the Journal of University Teaching and Learning Practice and on the Editorial Board of the Journal of Applied Learning and Teaching.

Ajit Dayanandan is Professor of Finance, University of Alaska Anchorage.He is Ph.D in Financial Economics from University of Sydney, Sydney in Australia. Born in Kerala, he was gold medalist at University of Kerala in M.A. Economics. After a brief stint in teaching in various colleges in Kerala, he joined Reserve Bank of India where he became Director of Research (Department of Economic and Policy Research). Subsequently (in 2002), he went back to academics as a faculty in economics and finance at Thomson Rivers University and University of Northern British Columbia, British Columbia, Canada before moving to United States in 2014. He had published around 50 peer-reviewed articles in journals like Economic and Political Weekly, Global Finance Journal, Journal of Accounting & Information Management, Pacific Journal of Economics and Business, The International Journal of Accounting, Global Finance Journal, Journal of Accounting & Information Management etc. He is presently in the editorial board member of International Journal of Corporate Governance and is referee for a number of journals in the area of economics, finance etc. He has done consultancy for Asian Development Bank, World Bank, Securities and Exchange Board of India etc..

Carson Duan is a PhD candidate at the University of New England. His doctoral research project is about immigrant entrepreneurship process in relation to immigrants' host and home country entrepreneurial ecosystem. He has years of experiences as a digital entrepreneur engaged in document digitalization.

Sisira Edirippulige is a Senior Lecturer and the Program Director of Digital Health postgraduate programs delivered at the Centre for Online Health. His main professional interests include e-health/telemedicine, and education and training. In his role as senior academic at the centre, he has been involved in a number significant research projects and the development of a number of education and training programs relating to Telehealth/e-Health/m-Health. Over the past 10years, Dr Edirippulige has been involved in a number of research projects conducted within the Centre for Online Health, funded by national and International agencies. A number of PhD and Masters projects that Dr Edirippulige has supervised have also produced significant evidence in the field. Dr Edirippulige has published 1 monograph, 12 book chapters and 40 peer reviewed papers. He has presented in over 60 national and international conferences.

Nergiz Ilhan is working as a Lecturer (sessional) at Faculty of IT, Monash University. She has completed her Masters in Business Information Systems from Monash University. Her research interests include e-commerce, social media and big data. She has several research publications in the area of e-procurement.

Lucija Ivancic is a Teaching and Research Assistant in Business Computing, Business Process Management, and Data Management, and a PhD candidate in the Department of Informatics of the Faculty of Economics and Business at the University of Zagreb, Croatia. Her current research interests lie in business process modelling and management, IT management, data management, and intersection of these field in the process of digital transformation. She received two Dean's awards for previous papers on discrete-event simulation modelling and information systems auditing.

Bernice Kotey is Professor of Entrepreneurship at the University of New England. She has published on a range of management issues affecting small and medium entreprises (SMEs) and the impact of the macro-environment on their operations. She is editor of a book on how SMEs operate in the digital economy. In relation to context, her research examines SMEs in local, regional, developing and developed economies. Bernice is a Certified Practicing Accountant (Australia) and completed the Education Management Program in Harvard in 2010.

Sudershan Kuntluru is working as an Associate Professor in Finance and accounting area at Indian Institute of Management Kozhikode, Kerala, India. Prior to this, he was a post-doctoral Research Fellow at Indian School of Business, Hyderabad, India. Previously, he worked as a faculty at Narsee Monjee Institute of Management Studies, Mumbai. He was trained in teaching and research at Indian Institute of Management Ahmedabad. His primary areas of teaching and research include financial accounting, financial reporting and analysis, strategic finance and corporate governance. He was presented his research papers in both national and international conferences held in India and abroad. He has published in national and international journals. He had conducted several corporate training programs on IFRS and Finance for Non-Finance topics to various companies' executives in India and Dubai. He was a member of Accounting Standard Board of Institute of Chartered Accountants of India (2012-2013). He had participated in Global Colloquium on Participant-Centered Learning program organized by Harvard Business School, USA held at Boston, USA and Shanghai, China. He obtains Masters in Commerce and Ph.D. degree from Department of Commerce, Osmania University, Hyderabad, India.

Sachithra Lokuge is a Lecturer at the RMIT University in Melbourne, Australia. She received her PhD in Business Information Systems from the Queensland University of Technology in Brisbane, Australia. Her work has been published in Information & Management, Communications of the AIS, Information Technology & People, Journal of Information Technology Theory and Application (JITTA), The International Conference on Information Systems, Academy of Management conference and The International Conference on Information Systems. Her research interests are digital innovation, entrepreneurship, technology use and enterprise systems.

Manuel Antonio Fernández-Villacañas Marín was born in Quintanar de la Orden, Toledo, Spain in 1961. After finishing his military studies for officer at the Spanish Air Force Academy in San Javier, Murcia, Spain (1984) and was graduated as Lieutenant, he received a degree in statistics and operational research from the University of Granada, Andalucía, Spain (1987); a Bachelor degree in economics and business sciences from the University of Murcia, Murcia, Spain (1990); a Masters degree in Security and Defence form the University Complutense of Madrid, Madrid, Spain (1996); a Masters Degree in Business Studies from the IE University, Madrid, Spain (2011); an Executive Masters degree in Public Management from the IE University, Madrid, Spain (2011); a Ph.D. in Economic and Business Sciences from the University of Granada, Andalucía, Spain (1995) and a Ph.D. in Administration and Logistics of Security and Defence Systems from the Rey Juan Carlos University, Madrid, Spain (2012).

Shailesh Palekar holds a PhD in Information Systems from the Queensland University of Technology, Brisbane. His research focuses on Social Media and its impact on individuals and organizations.

Nicholas Patterson is a Senior IT Lecturer at Deakin University with publications in Cyber Security, Blockchain and EdTech. He also runs his own consulting company, that brings advanced research into industry to increase capability and outcomes. His expertise is underpinned by a PhD which earned him the Alfred Deakin medal and a range of advanced micro-credentials in Critical Thinking, Problem Solving, Communication, Teamwork, Data Analytics, Innovation, Digital Literacy and Digital Learning. To add to this Nick has received multiple Vice-Chancellors teaching awards for his innovative and evidence-based teaching work in 2016, 2017, 2018 and 2019.

MD Rahim completed PhD in Information Systems from the University of Melbourne. He is currently working as a Senior Lecturer at the Faculty of Information Technology, Monash University, Australia. His research interests include e-business, adoption of IT in organisations, and social media. He has published widely in many reputed journals and conference proceedings.

Darshana Sedera is a Professor at the Swinburne University of Technology in Melbourne, Australia. He has published over 120 publications in major refereed journals and conferences. His publications have appeared in The Journal of the AIS, The Journal of the Strategic Information Systems, Information & Management, Information Technology & People and the Communications of the AIS. His research interests include digital business transformation, innovation, entrepreneurship, outsourcing and technology use. For more details, please visit darshanasedera.com.

Buddhika Senanayake is a Telehealth Research Assistant, The University of Queensland, currently working as a Tele-health research assistant at Centre for Online Health, University of Queensland, Australia. Senanayake has a special interest in public health and telemedicine.

Mario Spremić is a Professor of ICT Governance and Digital Business in the Department of Informatics of the Faculty of Economics and Business at the University of Zagreb, Croatia. His current research interests are in digital transformation, digital technologies, ICT governance, cybersecurity, and IT auditing. Mario is teaching on these topics at the undergraduate and postgraduate levels, including some international engagements including as a guest lecturer at Imperial College London. He is a member of editorial boards and a reviewer of international scientific journals.

Mohan Tanniru is the Professor in the Division of Public Health Practice and Translational Research in the Mel and Enid Zuckerman College of Public Health, U of Arizona, Tucson/Phoenix and a senior investigator in the Global Health Initiative at Henry Ford Health System in Detroit, MI. He is also an emeritus professor of MIS of Oakland University. He was the former Dean of the School of Business and the Dept. Head of MIS at U of Arizona. He has published over 90 research articles and his work has appeared in journals such as ISR, MIS Quarterly, Decision Sciences, DSS, JMIS, IEEE Transactions in Eng. Management, Expert Systems and Applications, Information and Management, CACM, as well as Health Policy and Technology, J of Patient Satisfaction, J of Healthcare Management and J of Healthcare Administration. He worked with over 70 large and small corporations in support of research and teaching.

Kanitsorn Terdpaopong, Ph.D., is a lecturer in International Accounting Program at Rangsit University, Thailand. Her research interests are on small and medium-sized enterprises; succession plans for SMEs; business valuation; and management accounting practices. She has received grants from several public research organizations such as the Thai Research Fund; The Office of Higher Education Commission, Thailand; Australian Government, Department of Foreign Affairs and Trade, Australia; and Erasmus+. In 2018 she was awarded an Australian Endeavour Executive Fellowship from Australian Government, Department of Education and Training; and awarded a Taiwan Fellowship from Ministry of Foreign Affairs of the Republic of China (Taiwan).

Tomislav Videc has been working in IT and banking for almost 20 years. He worked with major technology companies such as Apple, IBM, EMC, Oracle, SAP, and Salesforce as well as with some leading banks from EU and GCC. He started as a programmer of banking solutions and through years shifted towards SW architecture, process design and business solutions creation. He was managing teams for development and presales, and he was the general manager of two companies. Tomislav's ongoing biggest contribution is the infusion of technology in business processes and strong impact on efficiency, performance, and innovations. His education background includes a Master's degree in Economics and MBA, a Master's degree in Computer science and a number of technology and business certificates.

Vesna Bosilj Vukšić is a Professor of Business Process Management and Business Computing in the Department of Informatics of the Faculty of Economics and Business, at the University of Zagreb, Croatia. Her current research interests are in modelling and the management of business processes, information systems development, and digital transformation. Vesna has a teaching experience in undergraduate and postgraduate courses. She participates actively in research within the framework of the Ministry of Science and Technology's scientific projects, and is a member of international scientific research projects. She is a member of editorial boards and a reviewer of international scientific journals.

Youmin Xi is Executive President of Xi'an Jiaotong-Liverpool University, Pro-Vice-Chancellor of University of Liverpool, Director of HeXie Management Research Centre, and also keeps his Professorship at the School of Management of Xi'an Jiaotong University. His research and teaching areas cover the development mechanism and strategic analysis of social and economic system, decision-making and decision support systems, organizational behavior and leadership, and education management. In 1987 Youmin Xi established HeXie theory and extended it into HeXie management theory gradually with his research team, an indigenous theory widely recognized in academic circle. As author or co-author, more than 300 academic papers and 20 books have been published. Currently, He serves as Director of the Business Administration Education Steering Committee in China, Rotating Chairman of Chinese Academy of Management and Vice President of the Chinese Association of Enterprise Studies, and deputy director of the national association of Sino-Foreign cooperative education and so on.

Dina Ziadlou is an expert specializing in digital healthcare transformation and has over 15 years of specialed expertise in the healthcare industry. She holds a Vice-Chair position at the Digital Transformation Leadership (DTL) working group in the International Society for Telemedicine and e-Health (ISfTeH) and part of the WHO Digital Health Technical Advisory Group (DHTAG) and Roster of Experts. She earned her Doctorate in Healthcare Management and Leadership from Colorado Technical University, USA, Ph.D. in Medical IT from TMU, MS in IT Management from TMU, and BS in Biomedical Engineering from SRU. She has been awarded for several successful projects in Telemedicine, HIT, and digital health projects and has widely been consulted internationally.

Jovana Zoroja, Ph.D., is an Assistant Professor at the Faculty of Economics and Business, University of Zagreb, Department of Informatics. She received Ph.D. in Information Systems at the Faculty of Economics and Business Zagreb with the dissertation thesis "Influence of the Information and Communication Technologies on the Competitiveness of the European Union Countries." She was also educated at the LSE in London in the field of Business Development and ICT Innovation. She participated in Erasmus-Preparatory visit programme in Rimini, Italy. Her main research interests are information and communication technology, e-learning, simulation games, and simulation modeling. She is actively engaged in a number of scientific projects (FP7-ICT, bilateral cooperation, national projects). Jovana Zoroja published several scientific papers in international and national journals and participated in many scientific international conferences.

Index

E

F

G

H

I

K

L

M

N

O

P

R

S

www.ingramcontent.com/pod-product-compliance
Lightning Source LLC
LaVergne TN
LVHW062135300926
849161LV00010B/2542